Quantitative Finance

An Object-Oriented Approach in C++

CHAPMAN & HALL/CRC
Financial Mathematics Series

Aims and scope:
The field of financial mathematics forms an ever-expanding slice of the financial sector. This series aims to capture new developments and summarize what is known over the whole spectrum of this field. It will include a broad range of textbooks, reference works and handbooks that are meant to appeal to both academics and practitioners. The inclusion of numerical code and concrete real-world examples is highly encouraged.

Series Editors

M.A.H. Dempster
Centre for Financial Research
Department of Pure
Mathematics and Statistics
University of Cambridge

Dilip B. Madan
Robert H. Smith School
of Business
University of Maryland

Rama Cont
Department of Mathematics
Imperial College

Published Titles

American-Style Derivatives; Valuation and Computation, *Jerome Detemple*

Analysis, Geometry, and Modeling in Finance: Advanced Methods in Option
Pricing, *Pierre Henry-Labordère*

Computational Methods in Finance, *Ali Hirsa*

Credit Risk: Models, Derivatives, and Management, *Niklas Wagner*

Engineering BGM, *Alan Brace*

Financial Modelling with Jump Processes, *Rama Cont and Peter Tankov*

Interest Rate Modeling: Theory and Practice, *Lixin Wu*

Introduction to Credit Risk Modeling, Second Edition, *Christian Bluhm,*
Ludger Overbeck, and Christoph Wagner

An Introduction to Exotic Option Pricing, *Peter Buchen*

Introduction to Risk Parity and Budgeting, *Thierry Roncalli*

Introduction to Stochastic Calculus Applied to Finance, Second Edition,
Damien Lamberton and Bernard Lapeyre

Monte Carlo Methods and Models in Finance and Insurance, *Ralf Korn, Elke Korn,*
and Gerald Kroisandt

Monte Carlo Simulation with Applications to Finance, *Hui Wang*

Nonlinear Option Pricing, *Julien Guyon and Pierre Henry-Labordère*

Numerical Methods for Finance, *John A. D. Appleby, David C. Edelman,*
and John J. H. Miller

Option Valuation: A First Course in Financial Mathematics, *Hugo D. Junghenn*

Portfolio Optimization and Performance Analysis, *Jean-Luc Prigent*

Quantitative Finance: An Object-Oriented Approach in C++, *Erik Schlögl*

Quantitative Fund Management, *M. A. H. Dempster, Georg Pflug, and Gautam Mitra*

Risk Analysis in Finance and Insurance, Second Edition, *Alexander Melnikov*

Robust Libor Modelling and Pricing of Derivative Products, *John Schoenmakers*

Stochastic Finance: A Numeraire Approach, *Jan Vecer*

Stochastic Financial Models, *Douglas Kennedy*

Stochastic Processes with Applications to Finance, Second Edition, *Masaaki Kijima*

Structured Credit Portfolio Analysis, Baskets & CDOs, *Christian Bluhm and Ludger Overbeck*

Understanding Risk: The Theory and Practice of Financial Risk Management, *David Murphy*

Unravelling the Credit Crunch, *David Murphy*

Proposals for the series should be submitted to one of the series editors above or directly to:
CRC Press, Taylor & Francis Group
3 Park Square, Milton Park
Abingdon, Oxfordshire OX14 4RN
UK

Chapman & Hall/CRC FINANCIAL MATHEMATICS SERIES

Quantitative Finance

An Object-Oriented Approach in C++

Erik Schlögl

CRC Press
Taylor & Francis Group
Boca Raton   London   New York

CRC Press is an imprint of the
Taylor & Francis Group, an **informa** business
A CHAPMAN & HALL BOOK

CRC Press
Taylor & Francis Group
6000 Broken Sound Parkway NW, Suite 300
Boca Raton, FL 33487-2742

For Jesu

Contents

Preface

In the forty years since the seminal article by Black and Scholes (1973), quantitative methods have become indispensable in the assessment, pricing and hedging of financial risk. This is most evident in the techniques used to price derivative financial instruments, but permeates all areas of finance. In fact, the option pricing paradigm itself is being increasingly applied in situations that go beyond the traditional calls and puts. In addition to more complex derivatives and structured financial products, which incorporate several sources of risk, option pricing techniques are employed in situations ranging from credit risk assessment to the valuation of real (e.g. plant) investment alternatives.

As quantitative finance has become more sophisticated, it has also become more computationally intensive. For most of the techniques to be practically useful, efficient computer implementation is required. The models, especially those incorporating several sources of risk, have also become more complex. Nevertheless, they often exhibit a surprising amount of modularity and commonality in the underlying method and approach. Ideally, one would want to capitalise on this when implementing the models.

C++ is the de facto industry standard in quantitative finance, probably for both of these reasons. Especially for models implemented "in–house" at major financial institutions, computationally intensive algorithms are typically coded in C++ and linked into a spreadsheet package serving as a front–end. The object–oriented and generic programming features of C++, when used properly, permit a high degree of code reusability across different models, and the possibility to encapsulate algorithms and data under a well–defined interface makes the maintenance of implemented models far easier.

Object–oriented Programming (OOP) is a loaded concept, and much has been written about its advantages and disadvantages, accompanied by scholarly efforts to delineate boundaries between this and other programming paradigms, such as Generic Programming. This book is not an ideological manifesto. If there is any dogma at all in the approach here, it might be the pursuit of "implement once," meaning that any particular functionality in a piece of software should be implemented in one place in the code only[1] — everything else is approached pragmatically, following the motto "horses for courses." While there are certainly many traditional "objects" in the code presented in this book, for example the C++ abstract base class `TermStructure`

---

[1] This often means that parts of the software have to be rewritten more than once during the development process.

in Chapter 2, C++ templates (a language element more of Generic Programming than OOP) are used extensively as well, especially where a more purist object–oriented solution would entail a performance penalty. Thus this book is about a practical approach to working solutions, and quite a few of those solutions are supplied in form of the C++ code accompanying this book. If there is a manifesto which influenced the code development, it is Gamma, Helm, Johnson and Vlissides (1995) — titled "Design Patterns: Elements of Reusable Object–Oriented Software," but whose design patterns in practice more often than not are implemented as templates.

The aim of this book is to provide a foundation in the key methods and models of quantitative finance, from the perspective of their implementation in C++. As such, it may be read in either of two ways.

On the one hand, it is a textbook, which introduces computational finance in a pragmatic manner, with a focus on practical implementation. Along with the fully functional C++ code, including additional C++ source files and further examples, the companion website[2] to this book includes a suite of practical exercises for each chapter, covering a range of levels of difficulty and problem complexity. While the code is presented to communicate concepts, not as a finished software product, it nevertheless compiles, runs, and deals with full, rather than "toy," problems.

The approach based on C++ classes and templates highlights the basic principles common to various methods and models, and the actual algorithmic implementation guides the student to a more thorough understanding. By moving beyond a purely theoretical treatment to the actual implementation of the models, using the de facto industry standard programming language, the student also greatly enhances her career opportunities in the field.

On the other hand, the book can also serve as a reference for those wishing to implement models in a trading or research environment. In this function, it provides recipes and extensible code building blocks for some of the most common methods in risk management and option pricing.

The book is divided into eight chapters. **Chapter 1** introduces the requisite elements of the C++ programming language. The aim is to keep the book as self–contained as possible, though readers without prior exposure to C++ may also wish to consult one of the many excellent textbooks on this topic.

Chapter 2 provides some basic building blocks, which will be used in the implementation of financial models in the subsequent chapters. The more generic (i.e. non-financial) building blocks are drawn to a large part from excellent open source projects, specifically the Boost libraries and for numerical arrays, Blitz++. Numerical integration, optimisation and root search are also discussed (and implemented) to the extent needed in the subsequent chapters. The remainder of this chapter deals with representing and manipulating term structures of interest rates. Term structure fitting, interpolation and the pricing of basic fixed income instruments is discussed at this opportunity.

---

[2] This is found at `http://www.schlogl.com/QF`

Almost all fundamental results of option pricing can be illustrated in lattice models and they are also effective numerical methods in their own right. This is the topic of **Chapter 3**. Two broad classes of lattice models are considered in a unified framework: binomial models for a single underlying and models for the term structure of interest rates.

Moving on to a continuous time framework, **Chapter 4** introduces the Black/Scholes model, which for a large range of applications still remains the dominant paradigm. The key assumption is that the underlying asset(s) exhibit deterministic proportional volatility. Whenever this assumption holds, Black/Scholes–type derivative pricing formulae typically follow. These pricing formulae are specific to each option payoff, but have strong commonalities, which can be exploited to price a large class of payoffs in a single ("quintessential") formula. The abstraction from the specific functional form of the deterministic proportional volatility is an important object–oriented feature.

As demonstrated in Chapter 4, the arbitrage–free prices of derivative financial instruments can be expressed as solutions to partial differential equations. In cases where these cannot be solved analytically, finite difference schemes provide one set of alternative numerical methods. **Chapter 5** presents a selection of such methods under a common interface.

Chapter 6 pays heed to the fact that many option contracts are now being liquidly traded in the market, and as such have become bearers of information, to which models need to be calibrated. In the extreme, given (market–quoted) option prices for a continuum of strikes, one can extract an implied risk–neutral distribution for the future price of the underlying asset. In practice, the construction of such implied distributions is linked to the arbitrage–free interpolation of implied volatilities. The fact that implied volatilities vary depending on the moneyness of the option invalidates the Black/Scholes assumption of deterministic proportional volatility, so stochastic volatility is considered as an alternative.

Monte Carlo simulation methods are presented in **Chapter 7**. In many cases, MC simulation is the easiest numerical method to implement, and for high–dimensional problems it is often the most efficient. This efficiency can be substantially improved by various variance reduction techniques. We will also consider cases where valuation by simulation is not straightforward, i.e. for options with the possibility of early exercise. The chapter closes with a discussion of quasi–random methods as an alternative to pseudo–random Monte Carlo.

As a capstone, **Chapter 8** covers the implementation of a tractable multifactor continuous–time model of the term structure of interest rates, the Gauss/Markov case of Heath, Jarrow and Morton (1992). It draws extensively on the building blocks introduced in previous chapters, e.g. the term structure of interest rates in Chapter 2, the deterministic proportional volatility analytical solutions from Chapter 4, and the Monte Carlo simulation engine of Chapter 7.

In the **appendix**, the reader will find useful supplementary information. Appendix A shows how to employ Microsoft Excel as a front end for the models implemented in C++, relieving the implementer of the tedious task of programming a graphical user interface. Generating HTML documentation from appropriately commented C++ source code using a freely available software tool is illustrated in Appendix B.

Quantitative Finance has become a very broad field, and the models and methods covered in this book are only an introductory subset. Beyond this subset there are key models and methods which are undoubtedly of very high practical importance as well. To mention just a few, this includes more sophisticated finite difference methods, local volatility models, upper bound Monte Carlo estimators for early exercise premia, and the LIBOR Market Model. Nevertheless, as in particular Chapter 8 demonstrates, the approach based on C++ classes and templates taken in this book allows reuse of what was introduced in a simpler context, for the solution of similar problems in a more complex context.

DISCLAIMER

THE SOFTWARE IN THIS BOOK AND ON THE COMPANION WEBSITE IS PROVIDED "AS IS" AND ANY EXPRESS OR IMPLIED WARRANTIES, INCLUDING, BUT NOT LIMITED TO, THE IMPLIED WARRANTIES OF MERCHANTABILITY AND FITNESS FOR A PARTICULAR PURPOSE ARE DISCLAIMED. IN NO EVENT SHALL THE AUTHOR OF THIS BOOK BE LIABLE FOR ANY DIRECT, INDIRECT, INCIDENTAL, SPECIAL, EXEMPLARY, OR CONSEQUENTIAL DAMAGES (INCLUDING, BUT NOT LIMITED TO, PROCUREMENT OF SUBSTITUTE GOODS OR SERVICES; LOSS OF USE, DATA, OR PROFITS; OR BUSINESS INTERRUPTION) HOWEVER CAUSED AND ON ANY THEORY OF LIABILITY, WHETHER IN CONTRACT, STRICT LIABILITY, OR TORT (INCLUDING NEGLIGENCE OR OTHERWISE) ARISING IN ANY WAY OUT OF THE USE OF THIS SOFTWARE, EVEN IF ADVISED OF THE POSSIBILITY OF SUCH DAMAGE.

Acknowledgements

Firstly, I would like to thank Sunil Nair and his team at CRC Press/Taylor & Francis Group for encouraging me to embark on this project, which has turned out to be a far longer journey than I had naively anticipated. Numerous colleagues and students were so kind as to provide constructive feedback on earlier versions of the manuscript over the years. In particular, I am grateful to Kristoffer Glover, Mark Capogreco, Boda Kang, Henry Leung, Niko Kunelas and Karl Gellert for their helpful comments. Lastly, and most importantly, I thank my wife Jesu, without whose unwavering support and understanding this book would never have seen the light of day.

Erik Schlögl, Sydney, July 2013

## CHAPTER 1

A brief review of the C++ programming language

The purpose of this chapter is to introduce to the reader the elements of the C++ programming language used throughout the book. In order to keep the book as self-contained as possible, the key features of the language are briefly discussed. However, readers without prior exposure to C++ may wish to supplement this chapter by consulting one of the many excellent textbooks, such as Shtern (2000), or the book on C++ by its inventor, Bjarne Stroustrup.[1] Readers already familiar with C++ can safely skip this chapter. All the C++ source code discussed in this and in subsequent chapters is available for download from the website for this book, at `http://www.schlogl.com/QF`.

1.1 Getting started

The canonical way to introduce a programming language is to present a "Hello World!" program, which can then be modified to try out various elements of the language. Consider Listing 1.1. The C++ statement, which displays the "Hello World!" message on the screen, is

$$\texttt{cout << "Hello World!" << endl;} \tag{1.1}$$

This is done by pushing `"Hello World!"` to the standard output stream `cout`, followed by an end–of–line marker `endl`. A C++ program statement ends with a semicolon.

The body of our C++ program is contained in the function

$$\texttt{int main (int argc, char *argv[])}$$

Essentially, this is the function called by the operating system when the program is invoked.

Comments can be included in the source code in two ways: Anything to the right of a double forward slash (`//`) is ignored by the compiler, and sections of the source file delimited by `/*` and `*/` (i.e. multiline comments) are also ignored.[2]

The functionality, which allows us to write `"Hello World!"` to the standard

---

[1] See e.g. Stroustrup (1997).

[2] The particular format of the multiline comment at the top of the listing allows the use of Doxygen, a tool for the automatic generation of program documentation from C++ source code. This tool is discussed in Appendix B.

```
/** \file  Hello.cpp
    \brief "Hello world!" program.
    */


#include <cstdlib>
#include <iostream>


using namespace std;


int main(int argc, char *argv[])
{
    // Display "Hello World!" message
    cout << "Hello World!" << endl;
    return EXIT_SUCCESS;
}
```

Listing 1.1: A "Hello World!" program

output stream in statement (1.1), is defined and implemented in the C++ Standard Library. In order to use this functionality, we have to include the header file `iostream`. The preprocessor directive

```
#include <iostream>
```

effectively causes the compiler to read the contents of the file `iostream` as if it were inserted into the source code at this location. All the required definitions (of `cout`, `endl` and the stream operator `<<`) are found in this *header file*.[3] Any functionality used in a particular C++ source code file must either be defined within that file, or its definition included via a header file. Different parts of the Standard Library require the use of different header files, and producing a custom C++ library, as we will proceed to do in the course of this book, requires creating C++ header files *declaring* functionality, as well as C++ source code files *implementing* it.

Note that the program we have implemented here is what in Microsoft Windows parlance is called a *console application*, i.e. a program without a graphical user interface (GUI), which is invoked from a command window and writes its output to the same. We will use such console applications as the test bed for our C++ library. For user–friendly applications, Microsoft Excel will serve as the front–end for the library code. Interfacing between C++ and Excel is discussed in Appendix A.

C++ applications are typically written using an Integrated Development Environment (IDE). On the website, the reader will find the project files

---

[3] In fact, `cout` and `endl` are defined within the namespace `std` of the Standard Library. The statement `using namespace std;` makes variables and functions defined in this namespace easily accessible in the current context. See Section 1.6.

```
/** \file   Declare.cpp
    \brief Example program for variable declarations.
           Copyright 2005 by Erik Schloegl
    */


#include <cstdlib>
#include <iostream>


using namespace std;

int main(int argc, char *argv[])
{
    int i = -3;
    cout << "An integer: " << i << endl;
    double x = 3.5;
    cout << "A double precision floating point number: ";
    cout << x << endl;
    bool b = true;
    cout << "A Boolean value: " << b << endl;
    char c = '$';
    cout << "A character: " << c << endl;
    return EXIT_SUCCESS;
}
```

Listing 1.2: Examples of variable declarations

required to compile and link the C++ programs discussed in this book under the Eclipse CDT and Microsoft Visual Studio IDEs.

1.2 Procedural programming in C++

While C++ is a language, which fully supports object–oriented programming (and we will be making extensive use of this in subsequent chapters), algorithms still need to be implemented using iteration, conditional statements, and the like. These basic features of the language will be introduced in this section.

1.2.1 Built-in data types and the definition and declaration of variables

In C++, the *definition* of a variable associates a name with a type and allocates space in machine memory for the variable. In contrast, a *declaration* only associates a name with a type; the requisite space for the variable is assumed to have been allocated elsewhere. Each variable must be defined exactly once, while its declaration may (need to) be repeated. In most cases, declaration and definition are identical, and we will use the two roughly synonymously in the sequel.[4]

The following fundamental data types are available in C++:

- Integers, declared as `int`.

- Floating point numbers, declared as `float` or `double` (these days, most programs use `double`).

- Booleans (variables which can hold the value `true` or `false`), declared as `bool`.

- Characters, declared as `char`.

Integers can also be declared using the additional modifiers `unsigned` (for non-negative integer variables) and `long` or `short`. Variables declared as `long` or `short` may represent a greater or smaller range of possible integers, but the actual implementation is machine and/or compiler dependent (See Stroustrup (1997)).

Within a given scope, all variables used in a C++ program must be declared exactly once. Variables can be declared anywhere within a program, but the *scope* of a declaration is limited by the brackets { }. The outermost scope is given by a source file itself — variables declared outside any { } block are called *global*. Variables are assigned values using the operator =. Listing 1.2 gives examples of variable declarations.

C++ also allows the declaration of pointers and references to variables. A *pointer* is the machine memory address holding a variable of a particular type, and a *reference* refers to contents of a particular block of machine memory. The statement

```
double& x;
```

declares x to be a reference to a variable of type `double`. Thus the following statements

```
double y = 0.0;
double& x = y;
x = 1.0;
cout << y << endl;
```

will print 1, whereas omitting the ampersand & in the declaration of x would result in 0 being displayed.

The statement

```
double* p;
```

declares p to be a pointer variable to hold the address of a `double`. The ampersand operator can be used to determine the pointer to an existing variable. Conversely, a pointer is *de-referenced* (i.e. the contents of the memory

---

[4] The notable exception is when a declaration is prefixed with the keyword `extern` to declare a variable that is defined in a different source code file.

to which it points is accessed) using the operator *. Thus

```
double y = 0.0;
double* x = &y;
*x = 1.0;
cout << y << endl;
```

will again print 1.

Arrays of any data type can be created using square brackets. Thus

```
int integer_array[5];
```

creates an array of five integers. Array elements can be accessed using the operator [ ], e.g.

```
integer_array[2] = 213;
```

Note that the base index of such arrays is zero, i.e. the first element of the above array is accessed by `integer_array[0]`. The variable `integer_array` itself is a pointer to the first element in the array; thus

```
int integer_array[5];
int* p = integer_array;
p[0] = 105;
cout << integer_array[0] << endl;
```

will display 105. Note also that the size of the array must be a constant known at compile time. Arrays, the size of which is only known at run time, must be dynamically allocated using the **new** operator (see Section 1.2.5). However, the object–oriented features of C++ allow such dynamic memory allocation to be encapsulated in utility classes,[5] e.g. for vectors and matrices, obviating the need to worry about these issues when writing higher level programs.

Constants can be declared using the qualifier `const`, e.g.

```
const int one = 1;
```

They must be initialised (assigned a value) at the time of declaration and cannot be modified. Additionally, sets of constant integer values can be defined as so–called enumerations using **enum**, e.g.

```
enum currency {USD, EUR, JPY, AUD};
```

defines a new type `currency`, which may take the values `USD`, `EUR`, `JPY` or `AUD` (defined by default to be represented as integers 0, 1, 2, 3[6]). We can then declare a variable `domestic` to be of type `currency` and initialise it to the value `USD` with the statement

```
currency domestic = USD;
```

[5] See Section 1.3.1 for a first example of this.
[6] This can be modified, see Stroustrup (1997).

Operator	Description	Usage
+	addition	expr + expr
+	unary plus	+ expr
−	subtraction	expr − expr
−	unary minus	− expr
*	multiplication	expr * expr
/	division	expr / expr
%	modulo division (remainder)	expr % expr

Table 1.1 *Arithmetic operators in C++*

1.2.2 Statements, expressions and operators

Each operation in a program is expressed by a *statement*. A statement may be a declaration, as in the previous section, a function call (see Section 1.2.4), or, in the most common case, an assignment of the form

lvalue   =   expression;

An *lvalue* is a variable that can be assigned a value (i.e. not a constant), while expressions can be formed using the various C++ operators, e.g.

x + y

is an expression.

C++ provides a variety of operators to produce the expressions, which make up the program statements. The assignment operator =, the "address of" operator &, the de-referencing operator * and the subscripting operator [ ] were already introduced in the previous section. The purpose of this section is to review the commonly used operators on built-in data types. Note that it is important to distinguish between unary operators (those that take one argument) and binary operators (those that take two, one on each side). For example, the unary operator * (de-referencing) has a quite different meaning than the binary operator * (multiplication).

Table 1.1 summarises the arithmetic operators available in C++.[7]

We can override operator precedence using parentheses, i.e.

a + b*c

means

a + (b*c)

---

[7] Note that C++ does not provide an exponentiation operator. The operator ^ commonly used for this purpose is defined as bitwise exclusive OR in C++. Bit–level operators are not needed for our purposes, and thus not discussed here.

Operator	Description	Usage
$+ =$	add and assign	lvalue $+ =$ expr
$- =$	subtract and assign	lvalue $- =$ expr
$* =$	multiply and assign	lvalue $* =$ expr
$/ =$	divide and assign	lvalue $/ =$ expr
$\% =$	modulo division and assign	lvalue $\% =$ expr

Table 1.2 *Arithmetic assignment operators in C++*

(no parentheses required). Thus parentheses are needed for

```
(a + b)*c
```

The arithmetic operator symbols can also be combined with assignment operator symbol = to form operators such as

```
x += 2;      (add and assign)
```

This is equivalent to

```
x = x + 2;
```

See Table 1.2 for a list of operators of this type. Furthermore, there are special unary operators to increment or decrement a value by 1. The statements

```
x = x + 1;
x += 1;      (add and assign)
x++;      (post-increment)
++x;      (pre-increment)
```

all have the same effect on x, and similarly

```
y = y - 1;
y -= 1;      (subtract and assign)
y--;      (post-decrement)
--y;      (pre-decrement)
```

all have the same effect on y.

Expressions typically result in a value, while assignments result in a reference. Thus

```
a = b = 2;
```

which is interpreted as

$$a = (b = 2); \qquad (1.2)$$

first assigns the value 2 to b. (b = 2) results in a reference to b, thus in this

Operator	Description
==	equal
!=	not equal
<	less than
<=	less than or equal
>	greater than
>=	greater than or equal

Table 1.3 *Comparison operators in C++*

example the value of both a and b will be 2. This is what distinguishes the post-increment x++ from the pre-increment ++x, i.e.

$$x = 3;$$
$$y = ++x;$$

will set both x and y to 4.

The comparison operators available in C++ are summarised in Table 1.3. These operators return Boolean values, i.e. true or false. Note the difference between the assignment operator = and the equality comparison operator ==, i.e.

$$(x = 2)$$

will set x to the value 2 and the expression returns a reference to x (as in the example 1.2, above), which evaluates to true as x has value 2 and any non-zero value converts to a Boolean value of true. On the other hand,

$$(x == 2)$$

does not modify x and returns true only if x has the value 2. The operators required to form Boolean expressions are summarised in Table 1.4. Using Boolean expressions, we can also create conditional expressions of the form

```
expr1 ? expr2 : expr3
```

which returns the value of the expression expr2 if expr1 evaluates to true; otherwise it returns the value of expr3. Thus

```
mx = (x>y) ? x : y;
```

will set mx to the greater of x and y.

1.2.3 Control flow

Control flow statements, i.e. conditional execution and loops, are the key elements of a programming language, which allow us to express algorithms as computer code. Simple conditional execution is controlled using if/else blocks, i.e.

Operator	Description
&&	logical "and"
||	logical inclusive "or"
!	logical "not"

Table 1.4 *Logical operators in C++*

```
switch (c) {
  case 1: cout << "Sunday" << endl;
          break;
  case 2: cout << "Monday" << endl;
          break;
  case 3: cout << "Tuesday" << endl;
          break;
  case 4: cout << "Wednesday" << endl;
          break;
  case 5: cout << "Thursday" << endl;
          break;
  case 6: cout << "Friday" << endl;
          break;
  case 7: cout << "Saturday" << endl;
          break;
  default: cout << "Invalid input" << endl;
          break;                            }
```

Listing 1.3: Example usage of a `switch` statement

```
if (x>y) {
    cout << x << endl; }
else {
    cout << y << endl; }
```

will print the greater of x and y. Each block delimited by curly brackets { } may contain any number of statements, including nested `if`/`else` statements (or other control flow statements). The `else` block may be omitted. If only one statement follows `if` or `else`, the brackets may be omitted, i.e.

```
if (x>y) cout << x << endl;
else cout << y << endl;
```

Conditional execution with multiple cases can be implemented using a `switch` statement, as demonstrated in Listing 1.3. Execution jumps to the `case`, which matches the value of the variable on which the `switch` statement conditions. If no match is found, execution jumps to the `default` label. Note that if the `break;` statement at the end of each `case` is removed, execution would simply continue until a `break;` is encountered or the `switch` statement ends. Thus if c is 5 in Listing 1.3, the program will print "Thursday"; however, if the `break;` statements are removed, it will print

Thursday
Friday
Saturday
Invalid input

Loops may be implemented using either `for`, `while` or `do/while`. The syntax of a `for` loop is

```
for (initialisation; continuation condition; step expression)
{
program statement;
...
}
```

The initialisation expression is executed prior to entry into the loop. The program statements in the body of the loop (delimited by the brackets { }) are repeatedly executed as long as the continuation condition evaluates to `true`. The step expression is evaluated at the end of each iteration of the loop. Thus

$$\text{for (i=1; i<=10; i++) cout << i << endl;} \qquad (1.3)$$

prints all integers from 1 to 10. Note that if the body of the loop consists of only one statement, the brackets { } are not required. Initialisation, continuation condition and/or step expression may be omitted and a `break;` statement is an alternative way of terminating the loop. Thus

```
i = 1;     // replaces initialisation
for (;;) {
   cout << i << endl;
   i = i + 1;     // replaces step expression
   if (i>10) break; }     // replaces continuation condition
```

achieves the same outcome as (1.3), though in a somewhat less elegant manner.

The syntax for a `while` loop is

```
while (continuation condition) {
program statement;
...}
```

where the program statements in the body of the loop are repeatedly executed as long as the continuation condition evaluates to `true`. The equivalent statements to (1.3) using a `while` loop are

```
i = 1;
while (i<=10) cout << i++ << endl;
```

Alternatively, one can have the continuation condition evaluated at the end of the loop using `do/while`, i.e.

```
do {
program statement;
...} while (continuation condition)
```

Implementing the equivalent of (1.3),

```
int i = 1;
do {
    cout << i++ << endl; } while (i<10);
```

The difference between a while loop and a do/while loop is that the body of a do/while loop is executed at least once, because the continuation condition is evaluated at the end of the loop, whereas the body of a while loop will not be executed at all if the continuation condition initially evaluates to false.

1.2.4 Functions

Functions allow us to structure our programs by packaging the implementation of (often repeatedly required) tasks in subprograms with a well–defined interface. Here it is important to distinguish between function declaration and definition (implementation). A function must be declared before it can be used. The declarations of functions not defined, but used in a particular source file, are typically imported into that source file by an #include of the appropriate header file. For example, the statement

```
double my_function(double x, int i);
```

declares my_function() to take two arguments, the first of which is of type double, the second of type int. The type keyword double preceding the function name declares the return value of the function to be of type double. If the function does not return any value, this is indicated by the keyword void preceding the function name in the declaration.

A function is defined (i.e. implemented) in a block of program statements (delimited by curly brackets { }) following a declaration of the function[8], e.g.

```
int sum (int a, int b)
{
    return a + b;
}
```

The statement return expr; causes the function to exit and return the value of the expression expr. Note that return may be used anywhere inside the function, e.g.

```
int max(int a, int b)
{
    if (a>b) return a;
    else return b;
}
```

---

[8] As is the case with variables, functions are defined exactly once, whereas the declaration may be repeated.

As they are declared above, the functions `my_function()`, `sum()` and `max()` are passed their arguments *by value*. The program `funccall.cpp` on the website presents three alternative ways of passing function arguments: by value, by reference and by pointer. Passing an argument by value has the effect of creating a new variable with this value. This variable exists only inside the called function and any modifications to its value do not have any effect outside the scope of the function. In contrast, an argument passed by reference is a handle on an existing variable; thus any modifications made to the variable inside the called function persist after the function exits. Similarly, if an argument is passed by pointer, any modifications to the value of variable to which the pointer points will also persist.

Function return types may also be declared as references or pointers. This becomes particularly useful in combination with some object–oriented features of the language, so we defer examples of this until Section 1.3.1.

As a function body represents a { }–delimited scope, any variables declared inside a function are local to this function and exist only while the function is being executed. The value of such local variables can be made to persist between function calls using the keyword `static` in the variable declaration. The program `Static.cpp` on the website gives an example of this.

1.2.5 Dynamic memory allocation

As discussed previously, the subscripting operator [ ] allows us to declare arrays like

$$\text{double my\_array[4];} \qquad (1.4)$$

However, such declarations are only valid if the array size is a constant known at compile time. To create arrays of variable size known only at run time, memory must be allocated dynamically using the `new` operator. The following code creates an array of `n` double precision floating point numbers (where `n` is an integer):

```
double* my_array = new double[n];
```

We can then use `my_array` in the same way as if it were declared using (1.4), e.g. (provided that $n > 3$)

```
my_array[3] = 3.14;
```

or, equivalently, as a pointer,

```
*(my_array + 3) = 3.14;
```

Memory allocated using the `new` operator must be deallocated when it is no longer required — otherwise our program has what is called a memory leak, using more and more computer memory until no more is available. The deallocation operator is `delete`, i.e. to deallocate the memory allocated above,

```
delete[] my_array;
```

where the square brackets appended to `delete` indicate that an array is to be deallocated.

Memory can also be allocated for non-array types, e.g.

```
double* px = new double;
```

allocates memory for one variable of the type `double`. This is not very useful for the built-in fundamental types, but is often used to instantiate classes (see Section 1.3.1, below). Memory allocated in this way is correspondingly deallocated via

```
delete px;
```

Important warning: Mistakes in memory management are one of the most common causes of C++ program crashes (or seemingly inexplicable program behaviour). This includes

- Writing values to an array beyond its allocated length, e.g.

```
double* my_array = new double[4];
my_array[4] = 1.0;
```

- Writing values to or `delete`-ing an array, which has not been successfully allocated. Especially when allocating large arrays one should always check the pointer returned by the `new` operator. An unsuccessful memory allocation request returns a `NULL` (value zero) pointer. For example, the following is one safe way to proceed:

```
double* my_array;
if (NULL != (my_array = new double [n])) {
    ...    // perform some operations using my_array
    delete[] my_array; }
```

Memory management caused headaches for a generation of programmers in C, the non-object–oriented precursor of C++. Luckily, the object–oriented features of C++ allows us to encapsulate this in programmer–defined types, obviating the need to worry about memory management (too much) when writing higher–level programs.

1.3 Object–oriented features of C++

The C++ language provides support for the key concepts of object–oriented programming, including

- aggregation and encapsulation
- operator overloading
- inheritance and polymorphism

Each of these concepts will be discussed below. Furthermore, using templates to create meta-types and exceptions to handle errors will also be covered. The various language features will be illustrated by creating user–defined types for matrices. On the one hand, a matrix is a relatively simple and basic object, thus facilitating the exposition, but on the other hand it is not included in the Standard Template Library for C++.[9]

1.3.1 Classes

A *class* is a user–defined type. Thus defining

```
class Matrix {
/* ... */
};
```
(1.5)

we can declare variables of the type `Matrix`, e.g.

```
Matrix neo;
```
(1.6)

In addition to creating a user–defined type, classes allow us to associate data and operations on this data, i.e. a class has *data members* and *member functions*. Typically, the data members are used to represent the object in question. A matrix is an array of numbers and is characterised by the number of rows and columns. Thus, we might specify the data members of our `Matrix` class as

```
class Matrix {
    int r;   ///< number of rows
    int c;   ///< number of columns
    double* d; ///< array of doubles for the matrix contents
    /* ...*/ };
```

In order for this class to be useful, we must add some functionality in the form of member functions. Firstly, we must ensure that the internal representation of a variable of type `Matrix` is well defined whenever such a variable is created. This is done using a special member function called a *constructor*. A constructor is a member function with name identical to the name of the class, e.g.

```
class Matrix {
    /* ...*/
    Matrix(int nrows, int ncols, double ini = 0.0);
};
```
(1.7)

We define this function to initialise a matrix of `nrows` rows and `ncols` columns, with all entries in the matrix set to the value `ini`:

---

[9] As there is no generally accepted standard definition of numerical array types, different implementations abound, including freely available ones such as newmat (see `http://www.robertnz.net/`), the Template Numerical Toolkit (see `http://math.nist.gov/tnt/`) or the implementation proposed in the Boost C++ libraries (see `http://www.boost.org/`). Subsequent chapters will make use of the Blitz++ template–based implementation (see `http://blitz.sourceforge.net/`), which is discussed in Section 2.3.

```
Matrix::Matrix(int nrows, int ncols, double ini)
{
    int i;
    r = nrows;
    c = ncols;                                          (1.8)
    d = new double [nrows*ncols];
    double* p = d;
    for (i = 0; i < nrows*ncols; i++) *p++ = ini;
}
```

Thus the constructor takes care of allocating sufficient memory for a **nrows** by **ncols** matrix of double precision floating point numbers. Note that in the declaration (1.7) of the constructor we have defined a default value for the third argument of the constructor. If the constructor is called with only two arguments, **ini** will be set to 0.0 (which is arguably the most common case, where all elements of the matrix is initialised to zero). In code making use of this class we can now create an $n \times m$ matrix simply by

```
Matrix mat(n,m);                                        (1.9)
```

with no need to explicitly perform any memory allocation. Creating a variable of a type given by a class is also called *instantiating* the class. **mat** is an *instance* of a **Matrix**.

From Section 1.2.5, the reader may recall that any memory allocated using the **new** operator must also be deallocated by calling the **delete** operator when the allocated memory is no longer needed. In the case of our **Matrix** type, the allocated memory is no longer needed when the variable ceases to exist, for example when a function in which this variable has been declared exits. When a variable is destroyed, a special member function called a *destructor* is automatically invoked. A destructor is a member function with name identical to the name of the class, prefixed by a tilde ˜. It does not take any arguments. In the case of the class **Matrix**, we define

```
Matrix::~Matrix()
{
    delete[] d;                                         (1.10)
}
```

Now we are in a position to use variables of type **Matrix** without having to worry about memory management issues. However, in order to make effective use of a type representing a matrix, we need (at the very least) a subscripting operator. C++ allows us to define the behaviour of operators on user–defined types (this is called *operator overloading*). We choose to define the function call operator () rather than the subscripting operator [] for this purpose. This is because matrix indices consist of two integers (the row and column index) and the subscripting operator [] takes only one argument, whereas the operator () can be defined to take any number of comma–separated arguments. Thus, for the class **Matrix**, define subscripting as follows

```
class Matrix {
    /* ...*/
    inline double operator()(int i, int j) const;        (1.11)
    inline double& operator()(int i, int j);
};
```

The operator () is overloaded twice. The first version serves to read a value from a Matrix in row i and column j and is implemented as

```
inline double Matrix::operator()(int i, int j) const
{
    return d[i*c + j];                                   (1.12)
}
```

while the second version returns a reference to an element of the matrix:

```
inline double& Matrix::operator()(int i, intj)
{
    return d[i*c + j];                                   (1.13)
}
```

Though the two versions are superficially similar, the key difference is that (1.13) returns an *lvalue*, i.e. something which can be assigned a value (thus changing the contents of the Matrix), while (1.12) just returns a number, leaving the contents of the Matrix unchanged. As (1.12) does not modify an instance of the class, this member function is declared const. For class instances declared as const, one can only call const member functions.[10] Various legal usage of the overloaded operator () is illustrated in Listing 1.4. Finally, note that (1.12) and (1.13) are declared as inline. This means that the compiler does not create a function call whenever (1.12) or (1.13) is invoked; rather, the body of the operator definition is inserted into the code as it is being compiled. This makes the resulting program larger, but saves the execution time required for a function call — this is more efficient, especially for short operations which are expected to be called often, in particular in loops. The definition (implementation) of functions and operators declared as inline must be available whenever code using such functions/operators is compiled. For this reason, the definition of inline functions/operators is typically located in the appropriate header file.

Traditional (i.e. non-member) functions and operators can also be over-loaded. For example, we may wish to define the operator + when applied to two instances of Matrix as element–wise addition as in[11]

```
Matrix operator+(const Matrix& A,const Matrix& B);       (1.14)
```

and furthermore as element–wise addition when applied to a Matrix and a double, as in

---

[10] Strictly speaking (1.12) is not needed for code, which does not use const instances of Matrix. However, it is good programming practice to use const to express any assumption that an object is not modified by a segment of code. This way, this assumption is enforced by the compiler and any violations are caught at compile time, possibly preventing unintended behaviour at run time.

[11] For the implementation of these two functions, see the file Matrix.cpp on the website.

```
/** \file  MatrixIdx.cpp
    \brief Example program demonstrating usage of overloaded
           operator () for subscripting.
           Copyright 2005 by Erik Schloegl
    */


#include <iostream>
#include "Matrix.h"


using namespace std;


int main(int argc, char *argv[])
{
    int i,j;
    Matrix A(3,2);
    // assign values to matrix elements
    for (i=0;i<3;i++) {
      for (j=0;j<2;j++) A(i,j) = 0.1*i*j; }
    // access matrix elements
    double sum = 0.0;
    for (i=0;i<3;i++) {
      for (j=0;j<2;j++) sum += A(i,j); }
    cout << "The sum of the matrix elements is ";
    cout <<  sum << endl;


    return EXIT_SUCCESS;

}
```

Listing 1.4: Usage of the overloaded operator () for matrix subscripting

$$\text{Matrix operator+(const Matrix\& A,double x);} \qquad (1.15)$$

Functions and operators may be overloaded multiple times, as long as each version can be distinguished by the types of its arguments. Class member functions/operators can additionally be distinguished by being declared const, cf. (1.12) and (1.13), above. Note it is not legal to overload functions/operators by varying the return type alone, e.g.

```
double my_func(int i);
double& my_func(int i); /* illegal, the compiler cannot
                           distinguish this version of
                           my_func() from the one
                           above. */
```

Now that the constructor takes care of creating an appropriate internal representation of the class and member functions and operators allow us to manipulate the class without needing to refer to its internal representation, we have what is called *encapsulation*. For example, a user of the class Matrix accessing Matrix elements via the operator () does not need to know (or

```
class Matrix {
private:
   int      r;      ///< number of rows
   int      c;      ///< number of columns
   double* d;       ///< array of doubles for matrix contents
public:
   Matrix(int nrows,int ncols,double ini = 0.0);
   ~Matrix();
   inline double operator()(int i,int j) const;
   inline double& operator()(int i,int j);
   friend Matrix operator+(const Matrix& A,const Matrix& B);
   friend Matrix operator+(const Matrix& A,double x);
   /* ... */
};
```

Listing 1.5: Matrix class definition (excerpt)

care) whether the contents of the Matrix is represented internally row by row (as in the present implementation) or column by column. In fact, one should be able to change the internal representation without impacting on the users' code, as long as the functionality supplied by the member functions/operators remains the same. The member functions/operators present the *interface* of the class to the user; this is all that should be accessible. This encapsulation improves the maintainability of the code by localising the impact of changes in implementation and providing a library of well–defined (and testable) building blocks. Encapsulation is enforced in C++ via the keywords private and public (cf. Listing 1.5). Only those function, operator and data members under the label public can be accessed outside the class. Function, operator and data members under the label private can only be accessed by member functions of the class itself. For the overloaded operator +, which for efficiency reasons needs to access the internal representation of Matrix directly, we must make an explicit exception. This operator, which is not a member, is declared to be a friend in the class Matrix. Functions and operators declared friend have the same access privileges as class members. Arguably, this is justifiable in the case of overloading an operator such as this, as defining arithmetic operations on a type can be seen as an integral part of the functionality of that type. However, friend should be used sparingly in order to maintain encapsulation as much as possible.

Given a set of arithmetic operators defined element–wise[12] on instances of Matrix, we may wish to evaluate expressions like

$$\text{Matrix C(A + B);} \tag{1.16}$$

or

$$\text{C = A + B*C;} \tag{1.17}$$

---

[12] We choose to define the operator * as element–wise multiplication, rather than matrix multiplication, which is defined separately.

```
/// Copy constructor.
Matrix::Matrix(const Matrix& mat)
{
  // Initialise data members.
  r = mat.r;
  c = mat.c;
  d = new double[r*c];
  // Copy data.
  int i;
  for (i=0;i<r*c;i++) d[i] = mat.d[i];
}


/// Assignment operator.
Matrix::operator=(const Matrix& mat)
{
  // Make sure we're not assigning a matrix to itself.
  if (this!=&mat) {
    // Check size and allocate new memory if necessary.
    if (r*c<mat.r*mat.c) {
      delete[] d;
      d = new double[mat.r*mat.c]; }
    // Initialise data members.
    r = mat.r;
    c = mat.c;
    // Copy data.
    int i;
    for (i=0;i<r*c;i++) d[i] = mat.d[i]; }
  return *this;
}
```

Listing 1.6: Matrix copy constructor and assignment operator

where A and B are instances of **Matrix** of equal dimensions. The definition of the operators alone is insufficient for this to work. Additionally, a constructor needs to be defined, which takes an instance of a **Matrix** as an argument and creates a copy — the so-called *copy constructor*. Furthermore, since C already exists in (1.17), the assignment operator needs to be defined for the class **Matrix**. The corresponding declarations to be added to class **Matrix** are

```
class Matrix {
  /* ...*/
public:
  Matrix(const Matrix& mat);                    (1.18)
  Matrix& operator=(const Matrix& mat);
  /* ...*/
};
```

The copy constructor is invoked whenever an object is passed by value (such as for example the return value of operator + in (1.16)) or when an object is explicitly constructed from an instance of the same class (as for example `Matrix C` in (1.16)).

The implementation of the copy constructor and `Matrix` assignment operator is given in Listing 1.6. Note that the assignment operator is defined to return a reference to a `Matrix`. Specifically, the return statement reads

$$\text{return *this;} \tag{1.19}$$

`this` is a pointer to the current instance of the class, thus the expression (1.17) returns a reference to the instance `C` of the class `Matrix`. This is consistent with the behaviour of the assignment operator on built–in types as described in Section 1.2.1. The assignment operator also uses the pointer `this` to check for self–assignment (i.e. whether the objects on either side of the assignment operator = are identical), in which case no action needs to be taken.

Additional member functions may be added to broaden the available functionality of the class `Matrix`. For example, we may define a member function

$$\text{Matrix transpose() const;} \tag{1.20}$$

to create a transpose of a matrix. Member functions are called using the operator . (dot). Thus

$$\text{Matrix A(3,2);} \tag{1.21}$$

will create a `Matrix` with 3 rows and 2 columns. Subsequently,

$$\text{Matrix A\_T = A.transpose();} \tag{1.22}$$

will set `A_T` to the transpose of `A`. Member functions can also be accessed using the operator `->` on pointers to classes, e.g.

$$\begin{aligned} &\text{Matrix* pA= \&A;} \\ &\text{Matrix A\_T = pA->transpose();} \end{aligned} \tag{1.23}$$

1.3.2 Composition and inheritance

Classes are meant to represent conceptual units in the problem domain. Thus we may define classes to represent mathematical concepts such as matrices or probability distributions, or financial objects such as assets or a term structure of interest rates. Semantically, the most basic relationships between objects at different levels of abstraction can be expressed in terms of "has a" or "is a". For example, a matrix has a certain number of rows and an option has an underlying asset, while the identity matrix is a matrix and a down–and–out call option is an option. In C++, "has a" relationships are typically represented by composition, i.e. if object X "has an" object Y, Y is a data member of X. In the `Matrix` implementation of the previous section, the number of rows was a data member of the class `Matrix`. Any defined type, including user–defined classes, can be used as data members of a class.

If a class contains user–defined types as data members, care must be taken to ensure that instances of the class are constructed properly. If a default constructor (a constructor which does not take any arguments) has been defined for each such data member, this is not an issue: The default constructor for the data members is invoked automatically when the enclosing class is constructed. However, consider the following

```
class my_class {
  Matrix A;
  Matrix B;
};
```
(1.24)

where the default constructor of the class `Matrix` is not well–defined. Then any constructor of `my_class` must explicitly call a `Matrix` constructor for `A` and `B`. This is done using the following syntax:

```
my_class::my_class()
  : A(4,4), B(3,2)
{
  /* ...*/
}
```
(1.25)

This causes `A` to be initialised to a 4 by 4 `Matrix` and `B` to a 3 by 2 `Matrix` every time this constructor is used to instantiate `my_class`.

"is a" relationships are typically expressed in C++ via inheritance in a class hierarchy. A class can inherit from another class (called its *base class*) all members and member functions. A class may have multiple base classes from which it inherits, and more than one different class may be derived from any given base class. Derived classes may become base classes of other derived classes. The total of all inheritance relationships between a given set of classes make up the *class hierarchy*. Inheritance, in combination with polymorphism (to be discussed below) is the most important mechanism for creating a truly object–oriented library of classes, and thus will be used extensively throughout the book.

To begin with a simple example, consider a class to represent a square matrix. A square matrix is a matrix, and all the operations that were defined for the class `Matrix` are also valid for this special case, but additional operations not valid for a general matrix can also be defined. Defining a derived class `SquareMatrix` by

```
class SquareMatrix : public Matrix {
public:
  SquareMatrix(int nrows);
  double determinant() const;
};
```
(1.26)

provides a class which has all the functionality of a `Matrix`, and additionally allows the determinant to be calculated via the member function

```
SquareMatrix::determinant()
```
(1.27)

Thus the overloaded operator () used for accessing elements of the matrix is already defined, and if `A` and `B` are instances of `SquareMatrix`, the expression `A + B` will call the overloaded[13] (non-member)

---

[13] When a function or operator is called and there is none defined matching the exact

```
class Matrix {
protected:
  int      r;      ///< number of rows
  int      c;      ///< number of columns
  double* d;       ///< array of doubles for matrix contents
public:
  Matrix(int nrows,int ncols,double ini = 0.0);
  ~Matrix();
  inline double operator()(int i,int j) const;
  inline double& operator()(int i,int j);
  friend Matrix operator+(const Matrix& A,const Matrix& B);
  friend Matrix operator+(const Matrix& A,double x);
  /* ... */
};
```

Listing 1.7: Matrix class definition with protected data members

$$\text{operator+(const Matrix\& A, const Matrix\& B)} \qquad (1.28)$$

Note that in this case A + B will return a Matrix, not a SquareMatrix, so one may wish to additionally overload the operator + as

$$\begin{aligned}\text{SquareMatrix operator+(const SquareMatrix\& A,}\\ \text{const SquareMatrix\& B);}\end{aligned} \qquad (1.29)$$

The inheritance relationship between SquareMatrix and Matrix is defined using the syntax

$$\text{class SquareMatrix : public Matrix \{} \qquad (1.30)$$

where the keyword public indicates that all members of Matrix declared as public will also be public members of SquareMatrix.[14]

Given the way the class Matrix was defined in Listing 1.5, member functions of SquareMatrix cannot access the data members of Matrix — they are private. Thus SquareMatrix::determinant() would need to be implemented using the public Matrix::operator()() to access elements of the matrix. If this is not desired, i.e. if we want to grant derived classes access to certain members without making these members public, we can declare them as protected instead. Thus, replacing the previous definition of Matrix with the one given in Listing 1.7 will allow SquareMatrix::determinant() to access the data members of Matrix. However, this is not necessarily a good idea, as this means that any change to the internal representation of Matrix will potentially affect all derived classes.

Finally, note that in (1.26), a constructor for SquareMatrix is declared even

---

types of the arguments, the compiler searches for the corresponding function or operator taking the base classes as an argument. Depending on what sort of constructors or type conversions are defined, the compiler may also perform an implicit type conversion in order to find a matching function or operator. See Stroustrup (1997) for details.

[14] This is the most common case, though one could alternatively declare the inherited members to be private or protected.

though this class definition has no data members of its own.[15] Here, the sole purpose of the constructor is to call the constructor of the base class, using the syntax

```
SquareMatrix::SquareMatrix(int nrows)
   : Matrix(nrows, nrows) { }
```
(1.31)

1.3.3 Polymorphism

Suppose now that we want to add a class representing diagonal matrices to our library. A diagonal matrix is a square matrix, so one possibility would be to derive a class DiagonalMatrix from SquareMatrix. However, as the matrix class definitions currently stand, this would not be efficient, because Matrix allocates memory for all elements of a matrix, including those off the diagonal, which are known to be zero in the case of a diagonal matrix. Nor would it be safe, because a user could alter an off–diagonal element of a DiagonalMatrix to be non-zero using the overloaded operator (). Essentially, we want DiagonalMatrix to inherit the *interface* (i.e. the declaration of the public member functions) from its parent class, while being able to override the *implementation* of this functionality. Naively, one may do this by defining

```
class DiagonalMatrix : public SquareMatrix {
public:
   DiagonalMatrix(int nrows);
   double& operator()(int i, int j);
   double operator()(int i, int j) const;
};
```
(1.32)

The declaration of operator()(int i, int j) as a member of Diagonal-Matrix would allow us to implement this operator in a way specific to Diago-nalMatrix — it will "hide" the Matrix::operator()() defined in the base class. However, consider the following program statements:

```
DiagonalMatrix D(n);
Matrix& M = D;
M(1,2) = 1.3;
```
(1.33)

This is legal C++ code, since a DiagonalMatrix "is a" Matrix (i.e. Matrix is a base class of DiagonalMatrix). However, the last statement will call Matrix::operator()(), not DiagonalMatrix::operator()(), potentially leading to unintended consequences. This can be avoided by declaring any member function, which a derived class may override, as virtual. If the definition of Matrix reads

---

[15] Of course, derived classes may in general add data members to those inherited from the base class. In this case, these data members would typically be initialised in the constructor.

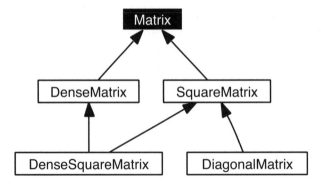

Figure 1.1 *Matrix class hierarchy*

```
class Matrix {
  /* ...*/
  virtual double& operator()(int i, int j);
  virtual double operator()(int i, int j) const;
  /* ...*/
};
```

(1.34)

and the keyword `virtual` is also added in front of the operator declarations in (1.32), then the last statement in (1.33) will call `DiagonalMatrix::operator()()`, as intended. This is *polymorphism*: the base class defines the interface, and the correct member function of a derived class instance is called even though we may be referring to the instance by its base class. The advantage of this is that we may now write code, which only relies on the public member functions of the base class being properly implemented — this code will then work without modification on all classes derived from the base class.

The problem that remains to be solved is that the internal representation of `Matrix` is rather wasteful if `DiagonalMatrix` fills it with mostly zeroes — i.e. one may want a different internal representation of a derived class as opposed to its base class(es). The way to approach this is to define `Matrix` as an *abstract base class*, as in Listing 1.8. The class definition of `Matrix` lists member functions, which represent operations on an arbitrary matrix. These functions are not implemented at this level of abstraction, they are *pure virtual functions*, as indicated by adding = 0 to the member function/operator declaration. Classes with pure virtual functions are *abstract*; they cannot be instantiated — only *concrete* derived classes, which implement all functions defined as pure virtual in the base class, can be instantiated. The actual implementation of the internal representation of a matrix is delayed to more specific derived classes. Consider the class hierarchy in Figure 1.1. `SquareMatrix` we now define as

```
class Matrix {
public:
  virtual ~Matrix();
  virtual double operator()(int i,int j) const = 0;
  virtual double& operator()(int i,int j) = 0;
  virtual Matrix transpose() = 0;
  /* ... */
};
```

Listing 1.8: `Matrix` as an abstract base class

```
class SquareMatrix : public Matrix {
public:
  double determinant() const;
};
```
(1.35)

It only supplies an additional member function `determinant()`, but does not implement the pure virtual functions declared in its parent class; thus `SquareMatrix` is also an abstract class. Note that `determinant()` does not need to be virtual, as it can be implemented by accessing the elements of the matrix via the operator `()`, which will be supplied by the concrete derived classes.[16] `DenseMatrix` is a concrete class with implementation basically identical to our original implementation of `Matrix`. Multiple inheritance can be used to create a concrete class `DenseSquareMatrix`, and `DiagonalMatrix` is a concrete class with a different internal representation. Note that since we now have different internal representations of matrices, the memory for which must be allocated in the constructor and deallocated in the destructor, the destructor must also be declared virtual in order to ensure that the correct destructor is called for derived classes.

1.4 Templates

Suppose now that we wish to implement a collection of matrix classes, where the contents of a matrix is of some other type than `double`, say of type `complex`, where `complex` is a class defined to represent complex numbers. If all the required operators have been overloaded to work correctly for arguments of the type `complex`, the implementation of a class hierarchy as in Figure 1.1 for complex matrices is the same as the one for matrices of `doubles`, the only difference being that the type `complex` replaces `double` in all declarations. However, implementing the complex matrices by copying and pasting from the implementation of our `double` matrices would be time–consuming, inelegant and prone to errors. Fortunately, C++ provides a language facility for *templates*, meta-classes and meta-functions which take one or more types as parameters. For example, implementing `Matrix` as a template instead of

---

[16] For implementation details of the matrix classes in the hierarchy discussed here, see `MatrixHierarchy.h` and `MatrixHierarchy.cpp` on the website.

```
template<class T> class Matrix {
public:
  virtual ~Matrix();
  virtual T operator()(int i,int j) const;
  virtual T& operator()(int i,int j);
  virtual Matrix<T> transpose();
  /* ... */
};
```

Listing 1.9: Matrix as a class template

a class, we replace Listing 1.8 by 1.9, and analogously for the other classes. Matrices of doubles and complex numbers are then instantiated by[17]

```
DenseMatrix<double> D(n,m);
DenseMatrix<complex> C(n,m);                          (1.36)
```

The usage of templates does mean that some care must be taken in the way the source code is organised.[18] One possibility is to collect the full definition of a class template in a header file, which is included whenever the template is used. This is the strategy used in many implementations of the *Standard Template Library* (STL)[19] and for the class templates created in this book.

Given a template definition such as in Listing 1.9, the code for a Matrix-<double> is generated by the compiler when a Matrix<double> is needed, by substituting double for every occurrence of the placeholder T in the template. A separate set of code for Matrix<complex> is generated in the same manner, if Matrix<complex> is used somewhere in the program. A template definition may allow for more than one placeholder, separated by commas, e.g.

```
template <class T,class U>
class Example2 { /* ...*/ };                           (1.37)
```

C++ permits specialisation of templates overriding the definitions in the "master" template for specific cases. Unlike classes derived from a base class, there is no requirement for a template specialisation to have the same interface as the "master" template (though it is often good programming practice to maintain the same interface). Defining

```
template <>
class Matrix<int> {
  /* ...*/                                             (1.38)
};
```

would supersede Listing 1.9 when the compiler generates code for Matrix<int>. Templates can also be specialised partially by specifying only a particular member function, i.e. if void func() is a member function of the DenseMatrix template, one could define a specific implementation of func() only applicable to a DenseMatrix of complex by

---

[17] The full source code is given on the website.
[18] See Stroustrup (1997) for a more detailed discussion.
[19] Section 2.1 presents the parts of the STL used in subsequent chapters of this book.

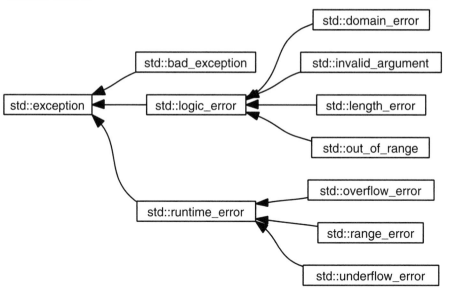

Figure 1.2 *Hierarchy of standard library exceptions*

```
template <>
void DenseMatrix<complex>::func() {
    /* ...*/
}
```
                                                                        (1.39)

An example where partial template specialisation is used in the code accompanying the book is the member function `max_stddev()` of the class template `MCGatherer` in Chapter 7.

1.5 Exceptions

In writing a library of classes and functions, it is desirable to have some facility for these to report errors to the client code in a robust manner. For example, in our `Matrix` classes we may want to implement range checking such that the indexing error in

```
DenseMatrix D(4,5);
int i = 4;
D(i,i) = 1.0;
```
                                                                        (1.40)

will be caught at runtime, instead of writing a value outside the bounds of allocated memory and possibly causing the program to crash (or at the very least exhibit unintended behaviour). Similarly, a function which opens a file may need to report an error if the file cannot be found. *Exceptions* provide a mechanism in C++ for functions to report errors to the client code.

Any concrete C++ class can be used as an exception, though typically exception classes are created explicitly to represent errors. Thus, for the example (1.40) we may wish to create a

```
class index_error {
   /* ...*/                                                    (1.41)
};
```

A function can report an error by "throwing" an exception, e.g.

```
double& DenseMatrix::operator()(int i, int j)
{
   if ((i>=rows)||(j>=cols)||(i<0)||(j<0))
      throw index_error();                                     (1.42)
   else return d[i*cols+j];
}
```

The client code is responsible for handling errors by "catching" exceptions. In this case, (1.40) becomes

```
try {
   DenseMatrix D(4,5);
   int i=4;
   D(i,i) = 1.00; }
catch (index_error& idx) {                                     (1.43)
   cout << "Index error" << endl;
   // ...some other statements to handle the error
}
```

The code in the catch { } block is executed if an **index_error** is thrown by any function called directly or indirectly within the **try** { } block. If no exception is thrown, the effect is that the statements in the **try** block are executed as if there were no enclosing **try** block and the code in the **catch** block is ignored.

Exceptions, being classes, may be part of a class hierarchy. One such hierarchy is provided by the Standard Library (see Figure 1.2). By structuring the catch statements appropriately, we can selectively handle different types of exceptions. Thus

```
try {
  // some statements
}
catch (domain_error& domerr) {
  // handle domain errors
}
catch (length_error& lenerr) {
  // handle length errors
}
catch (logic_error& logerr) {
  // handle all logic errors not caught previously        (1.44)
}
catch (exception& except) {
  // handle all exceptions that have
  // the class exception as their base
  // and have not been caught previously
}
catch (...) {
  // catch (...) catches all exceptions
  // (regardless of their class)
  // not caught previously
}
```

A catch statement will catch all exceptions of a particular class and any exceptions derived from this class. Therefore, if we had put

$$\text{catch(logic\_error\& logerr)} \tag{1.45}$$

as the first catch statement after the try block, the separate catch blocks for domain_error and length_error would never be invoked, i.e. an exception will be caught by the first matching catch statement. However, an exception may be re-thrown after it has been caught, e.g.

```
catch (logic_error& logerr) {
  // do some error handling              (1.46)
  throw; }
```

This is potentially useful when there are several levels of nested try blocks, e.g. a function is called from within a try block, and this function itself contains a try block.

1.6 Namespaces

In a large software project, especially one involving several developers or using third-party libraries, there is invariably a danger of name clashes — the same class or function names being defined differently (and possibly for a different purpose) in different parts of the project. One could try to avoid name clashes by prefixing names by some hopefully unique identifier, e.g.

```
#include <iostream>

int main(int argc, char *argv[])
{
    std::cout << "Hello World!" << std::endl;
    std::cout << "It's a fine day!" << std::endl;
    return EXIT_SUCCESS;
}
```

Listing 1.10: Accessing names in the namespace std

ErikSchloglMatrix, instead of **Matrix** but this becomes rather verbose and tedious. To facilitate the logical grouping of declarations under a unique common identifier, C++ provides the **namespace** mechanism. For example, the functions and classes provided by the C++ standard library are grouped in the **namespace std**.[20] We can create our own namespace by enclosing our declarations in a **namespace** block, e.g.

$$\begin{aligned}
&\texttt{namespace my\_namespace \{} \\
&\texttt{double func1(double x);} \\
&\texttt{class Matrix \{} \\
&\texttt{/* ...*/ \};} \\
&\texttt{class string \{} \\
&\texttt{/* ...*/ \};} \\
&\texttt{\}}
\end{aligned} \tag{1.47}$$

ensuring that our **string** class does not clash with the standard library **string** and our **Matrix** class not with some other implementation of **Matrix**. Outside a given namespace, names are accessed using the double colon :: syntax, e.g.

$$\begin{aligned}
&\texttt{std::string A;} \\
&\texttt{my\_namespace::string B;}
\end{aligned} \tag{1.48}$$

declares A to be a standard library **string**, while B is a **string** as implemented in **my_namespace**.

Inside the scope of a class or function of a particular namespace, all names not explicitly qualified are assumed to be from that namespace by default, i.e.

$$\begin{aligned}
&\texttt{double my\_namespace::func1(double x)} \\
&\texttt{\{} \\
&\texttt{ std::string A;} \\
&\texttt{ string B;} \\
&\texttt{ /* ...*/} \\
&\texttt{\}}
\end{aligned} \tag{1.49}$$

will declare A and B as in (1.48).

---

[20] See Section 2.1 in Chapter 2. Note that some compiler implementations do not comply exactly with the standard, so that certain functions, which according to the standard should be in the namespace **std**, are not.

```
#include <iostream>

int main(int argc, char *argv[])
{
    using std::cout;
    using std::endl;

    cout << "Hello World!" << endl;
    cout << "It's a fine day!" << endl;
    return EXIT_SUCCESS;
}
```

Listing 1.11: Creating local synonyms with *using–declarations*

If a name from a particular namespace is used frequently, it can be useful to shorten the notation by creating a local synonym with a *using–declaration*. The program in Listing 1.10 writes some output to the standard output stream `std::cout` and terminates each output line with `std::endl`. In Listing 1.11 the two using–declarations create local synonyms for `cout` and `endl` within the scope of the function `main()`, arguably making the program less tedious to code. A similar effect can be achieved with a *using–directive*, see e.g. Listing 1.1. This has the effect of importing *all* names in the namespace `std` into the present context, in this case the entire source file. As Stroustrup (1997) notes, the using–directive is mainly designed to be applied when compiling legacy code that was written before the advent of namespaces; in all other situations it is preferable to be explicit about which names should have local synonyms by applying using–declarations.

Similarly, using–directives can be used to import entire namespaces into a namespace, and using–declarations can be used to import individual names.[21] Thus

```
namespace my_namespace {
using namespace std;                                          (1.50)
/* ...*/ }
```

means that all names in the namespace `std` are also in the namespace `my_namespace`, while

```
namespace my_namespace {
using std::cout;                                              (1.51)
/* ...*/ }
```

imports only the name `cout`.

Lastly, note that namespaces (unlike classes) are *open*, which means names can be added to a namespace in several namespace declarations — this is in fact what happens for the namespace `std` in the standard library `#include` files.

---

[21] Stroustrup (1997) calls this namespace *composition* and *selection*, respectively.

CHAPTER 2

Basic building blocks

Learning the language elements is only the first step toward using C++ to implement quantitative models in finance. Many of the basic building blocks for an implementation are already available in well-tested code libraries, making it unnecessary to "reinvent the wheel," as the saying goes. This allows us to concentrate on the task of financial model implementation at a higher level, selectively using existing solutions where re-implementation would not add value in terms of the object–oriented approach to financial modelling adopted in this book. While there are certainly useful commercial C++ libraries produced by various vendors, we will focus on those that are freely available, either as part of the C++ Standard Library, or as the outcome of an open source development effort.

In subsequent chapters, the use of library templates, classes and functions will be covered in the context in which the need arises, i.e. for the purpose of solving a particular problem in quantitative finance. However, this chapter provides some basic building blocks, starting with especially relevant parts of the Standard Template Library. This is followed by a brief introduction to some of the Boost libraries, an open source project which provides libraries for very useful generic functionality not found in the Standard Library. Vectors, matrices and the basic operations of linear algebra are fundamental to mathematical modelling and are therefore presented next. Furthermore, the time value of money, as represented by the term structure of interest rates, is one of the fundamental concepts of finance. This basic building block is required in all subsequent chapters in order to discount future cash flows and its introduction here also serves as a first case study of object–oriented design. Term structure fitting, interpolation and the pricing of basic fixed income instruments is discussed at this opportunity.

2.1 The Standard Template Library (STL)

As in its predecessor C, the C++ "vocabulary" is quite parsimonious, and much of the functionality required for higher–level programming is provided by the Standard Library — literally standard in the sense that the contents of the library is subject to ISO standards. The C++ Standard Library includes the C Standard Library, and much of the additional, C++–specific functionality is provided in the form of C++ templates in the Standard Template Library (STL). The purpose of this section is to introduce some of the features of the Standard Library widely used in the code covered in the subsequent chapters.

2.1.1 Mathematical functions

A number of standard mathematical functions are declared in the header file `<cmath>`. Typically, these functions are overloaded to accept `float`, `double` or `long double` arguments, and return `float`, `double` or `long double` results, respectively. They include: `sqrt()`, returning the square root of its argument, which must be non-negative; `exp()`, returning e^x for an argument x; `log()`, returning the natural (base e) logarithm of its argument; `abs()`, returning the absolute value of its argument; and `pow()`, which takes two arguments (say `pow(x,y)`) and returns x^y. The latter function is further overloaded to provide versions in which the second argument is an `int`. In good library implementations, these versions execute substantially faster than the versions for non-integer powers, thus where possible one should ensure that y is declared as `int` when calling `pow(x,y)`.

2.1.2 Complex numbers

A class template representing complex numbers is declared in the header file `<complex>`. Template specialisations are provided for `complex<float>`, `complex<double>` and `complex<long double>`. The constructor takes two arguments; the first for the real part and the second for the imaginary part of the complex number. For an existing instance of `complex`, these parts can be accessed, respectively, via the member functions `real()` and `imag()`. An assignment constructor is provided and all arithmetic operators are appropriately defined, including the arithmetic assignment operators such as `+=`, as well as all comparison operators. A subset of the mathematical functions, which are well–defined for complex arguments, are also provided via the header file `<complex>`, including `sqrt()`, `exp()`, `log()`, `abs()` and `pow()`.

2.1.3 Strings

A common aggregate data type often found in other programming languages (such as BASIC) is a *string*, which is essentially an array of characters. C++ is fully backwards compatible with C, so it supports C–style strings such as

      char cstring1[] = "Hello world!";   (2.1)

which, for example, can be written to `cout` with

      std::cout << cstring1 << std::endl;   (2.2)

As in C, the Standard Library function `atoi()` converts a valid character representation of an integer into an `int`, e.g.

      char cstring2[] = "123";
      int i = std::atoi(cstring2);   (2.3)
      std::cout << i+1 << std::endl;

will print 124. Similarly, `atof()` converts a string to a floating point number.

In C++, the Standard Template Library provides the `string` class, defined in the header `<string>`. An STL `string` can be constructed from a C–style string, e.g.

```
std::string my_string(cstring1);
```
(2.4)

or

```
std::string my_string("Hello world!");
```
(2.5)

The **string** class provides convenient functionality through its member functions, for example **my_string.length()** will return the number of characters in **my_string**. The + operator is overloaded to implement concatenation, so after

```
std::string str1("Very");
std::string str2(" ");
std::string str3("good");
std::string str4 = str1 + str2 + str3;
```
(2.6)

str4 will contain "Very good". The comparison operator < is overloaded in the appropriate manner to allow sorting (see Section 2.1.5, below) of strings. For further **string** functionality, see the STL documentation or Lischner (2003).

2.1.4 Containers and iterators

Containers serve to store collections of objects and iterators serve to access the elements of a container. The C++ Standard Library implements a number of containers as class templates, allowing the containers to be created for any type, built–in or user–defined, as long as objects of that type can be copied and assigned freely,[1] with some containers stipulating additional requirements. The standard containers used in this book are **list**, **map** and **vector**.

Each container has a default constructor, which initialises the container to be empty. A copy constructor initialising a container with a copy of all objects in the original container is also defined. For further constructors[2] there is a distinction between *sequence* and *associative* containers. The former store objects in the order in which they were added to the container, while the latter maintain their contents in ascending order to speed up searching, where the order relation can be specified by the user through an additional argument in the constructor. For our purposes, the default and copy constructors are typically sufficient.

To access elements in a container, there are five categories of iterators: input, output, forward, bidirectional, and random access. Essentially, an iterator is an abstraction of a pointer — in fact, a random access iterator behaves as a C++ pointer. It can be dereferenced with the * operator, and incremented or decremented with the ++ and -- operators (to iterate to the next or previous elements in the container, respectively). It supports the [] to access any index in the sequence relative to the present position of the random access iterator, and the iterator can be moved forward or backward any number of steps by

---

[1] Note that pointers to objects will meet this requirement even in cases where the objects themselves do not.

[2] There are also constructors allowing the user to specify an allocator to replace the operators **new** and **delete** for managing memory in a container. However, this goes beyond the scope of the present discussion. For details, see e.g. Lischner (2003).

adding or subtracting integer values. Like pointers, iterators are inherently unsafe — there is no guarantee that they point to a valid element in a valid container. It is up to the user of the iterator to ensure that one doesn't iterate beyond the bounds of a container, or dereference an iterator that points to a container that no longer exists.

The other four categories of iterators are restrictions of the random access iterator. A bidirectional iterator supports only the dereferencing operator `*` and the increment and decrement operators `++` and `--`. A forward iterator restricts this further by not allowing decrementation `--`. Input and output iterators are forward iterators that respectively only permit reading from or writing to a sequence in one pass. The properties of an iterator supported by a particular container depend on the type of container.

The `list` template is defined in the header file `<list>`. This sequence container allows the user to rapidly insert or delete objects at any position, but random access is not supported. To illustrate some key functionality, consider the following example of a list of integers:[3] First declare a `list` of integers with

$$\texttt{std::list<int> my\_list;} \tag{2.7}$$

This list is initially empty. The list member function `push_back()` appends elements to the back of the list. If `i` is an `int`, the list can be populated with integers from 1 to 10 by

$$\texttt{for(i=1;i<=10;i++) my\_list.push\_back(i);} \tag{2.8}$$

To access elements in the list, declare a `list` iterator,

$$\texttt{std::list<int>::iterator my\_iter;} \tag{2.9}$$

Using this iterator, one could then proceed to print the elements of the list,

$$
\begin{aligned}
&\texttt{for(my\_iter=my\_list.begin();} \\
&\qquad \texttt{my\_iter!=my\_list.end();my\_iter++)} \\
&\qquad \texttt{std::cout << *my\_iter << ' ' << std::flush;}
\end{aligned}
\tag{2.10}
$$

where the member function `begin()` of `list` returns an iterator pointing to the first element of the list and `end()` returns an iterator that points one past the last item in the list. Note that the continuation condition in the `for` loop must be

$$\texttt{my\_iter!=my\_list.end();} \tag{2.11}$$

rather than

$$\texttt{my\_iter<my\_list.end();} \tag{2.12}$$

as the `list` iterator object does not support $<$ comparison.

The member function `insert()` serves to insert an element in a list, i.e.

$$\texttt{my\_list.insert(my\_iter,j);} \tag{2.13}$$

inserts `j` into the list after the element to which `my_iter` is currently pointing. Conversely,

---

[3] For a more detailed documentation of the full functionality, see e.g. Lischner (2003). An example program containing the code discussed here can be found in the file `ListExample.cpp` on the website for this book.

```
my_list.erase(my_iter);
```
                                                                  (2.14)

erases from the list the element to which `my_iter` is currently pointing.

The `vector` template is defined in the header file `<vector>`. This sequence container supports all operations supported by `list`, and also permits random access.[4] Thus one could create a vector of integers 1 to 10 by

```
std::vector<int> my_vector(10);
for(i=1;i<=10;i++) my_vector[i-1] = i;
```
                                                                  (2.15)

and print all the elements in a vector by

```
for(i=0;i<my_vector.size();i++)
    std::cout << my_vector[i] << ' ' << std::flush;
```
                                                                  (2.16)

where the member function `size()` returns the number of elements in the vector. Note that assigning a value to an element in a vector using the random access operator `[]` as in (2.15) requires that an element already exists at that position, which is why the vector was constructed explicitly to a size of 10 (as opposed to using the default constructor, which would have resulted in a vector of length zero). Also, if a vector is resized, either explicitly by a call to the member function `resize()` or implicitly because an `insert()` operation requires it, the contents of the entire vector becomes invalid, so `resize()` and `insert()` operations on vectors should typically be avoided.

The `map` template is defined in the header file `<map>`. This associative container stores pairs of *keys* and *values*, where each key is assumed to be unique. Elements in the map can be accessed via their keys using the operator `[]`. If a key is accessed, which does not exist in a map, an element with that key is created. Thus one can create an empty map, say where the keys are `strings` and the values are `doubles`, by

```
std::map<std::string,double> my_map;
```
                                                                  (2.17)

and proceed to fill the map with elements by assigning values to unique keys, as Listing 2.1 demonstrates.

A `map` organises its elements in ascending order, which by default is determined by comparing keys using the operator `<`.[5] Thus the `for` loop at the end of Listing 2.1 will iterate over the elements in alphabetical order of the keys. (This is how the operator `<` is overloaded for `std::string` in the Standard Template Library.) Dereferencing the map iterator returns a `std::pair` of a key and the corresponding value, accessed respectively by the member functions `first()` and `second()` of `pair`.

2.1.5 The `<algorithm>` header

This header file defines a variety of generic algorithm function templates, in particular those that operate on iterators, but also standard operations such as the minimum or maximum of two values, respectively declared as

---

[4] An example program can be found in the file **VectorExample.cpp** on the website.
[5] This behaviour can be modified (for details see e.g. Lischner (2003)), but this is beyond the scope of the present discussion.

```
#include <iostream>
#include <string>
#include <map>

int main(int argc,char *argv[])
{
  // Create an empty map
  std::map<std::string,double> my_map;
  // Add stocks and their prices to the map
  my_map["Mended Hill Properties"] = 25.2;
  my_map["Yoyodyne Industries"] = 105.1;
  my_map["Peak Petroleum Corporation"] = 803.5;
  my_map["Titanic Shipping Ltd."] = 3.15;
  // Elements of the map can be accessed through their keys
  std::cout << "The share price of Yoyodyne Industries is "
    << my_map["Yoyodyne Industries"] << std::endl;
  // Print elements in order
  std::map<std::string,double>::iterator my_iter;
  for (my_iter=my_map.begin();
         my_iter!=my_map.end();my_iter++)
    std::cout << "The share price of " << my_iter->first
      << " is " << my_iter->second << std::endl;
}
```

Listing 2.1: Using a `map` container

$$
\begin{aligned}
&\texttt{template <typename T>} \\
&\qquad \texttt{const T\& min(const T\& a,const T\& b);} \\
&\texttt{template <typename T>} \\
&\qquad \texttt{const T\& max(const T\& a,const T\& b);}
\end{aligned} \tag{2.18}
$$

Similarly, to swap the values of two variables of the same type, use the template

$$\texttt{template <typename T> void swap(T\& a,T\& b);} \tag{2.19}$$

As an example of an algorithm that operates on iterators, suppose one wishes to apply a function `f` to each element in a given range within a container, given by input iterators pointing to the first and one past the last element in a range. One can use the

$$
\begin{aligned}
&\texttt{template <typename InIter, typename Func> Func} \\
&\texttt{for\_each(InIter first,InIter last,Func f);}
\end{aligned} \tag{2.20}
$$

Note that the function object (functor) is passed into `for_each` and returned *by value*, so it must have a valid copy constructor. As an example, the class template in Listing 2.2 implements a functor, which when applied to each element in a range counts the number of occurrences of each value in that collection of elements. The typecast operator `std::map<T,int>&()` is overloaded to allow access to the `map` containing the result once the count is completed. Listing 2.3 shows how the functor is applied to a list of integers using the `for_each()` template function.[6]

---

[6] For the full, working source code see `for_eachExample.cpp` on the website.

```
template <typename T>
class number_of_occurences {
private:
  std::map<T,int> count_occurences;
public:
  inline void operator()(const T& x) {
    count_occurences[x]++; };
  inline operator std::map<T,int>&() {
    return count_occurences; };
};
```

Listing 2.2: Functor for counting occurrences of a value in a collection of elements

```
int main(int argc,char *argv[])
{
  int i;
  // Declare list.
  std::list<int> my_list;
  // Fill with integers 1 to 10.
  for (i=1;i<=10;i++) my_list.push_back(i);
  // append another 5 and another 3
  my_list.push_back(5);
  my_list.push_back(3);
  // instantiate counting functor
  number_of_occurences<int> f;
  // count occurences of each integer using the for_each
  // algorithm template
  number_of_occurences<int> g =
    for_each(my_list.begin(),my_list.end(),f);
  std::map<int,int>& occurences = g;
  // Print number of occurences of each integer
  std::map<int,int>::iterator my_iter;
  for (my_iter=occurences.begin();
       my_iter!=occurences.end();my_iter++)
    std::cout << "The integer " << my_iter->first
              << " occurs " << my_iter->second
              << " times in the list." << std::endl;
}
```

Listing 2.3: Counting occurrences of a value in a collection of elements using for_each()

Some of the code discussed in this and subsequent chapters also makes use of the sort function template, e.g.

```
template <typename RandIter>
void sort(RandIter first,RandIter last);
```
(2.21)

```
#include <iostream>

std::ostream& operator<<(std::ostream& os,const Matrix&
A)
{
  int i,j;
  for (i=0;i<A.rows();i++) {
    for (j=0;j<A.columns()-1;j++) os << A(i,j) << ',';
    os << A(i,j) << std::endl; }
  return os;
}
```

Listing 2.4: Overloaded operator for writing a `Matrix` to an `ostream`

where `RandIter` is required to be a random access iterator. Applying `sort()` to a range within a container delimited by [first, last) results in the elements in that range being sorted *in place* (i.e. the order of elements in the original container is modified). The template function sorts in ascending order using the `<` operator. This behaviour can be modified in

$$\text{template <typename RandIter,typename Compare> void} \atop \text{sort(RandIter first,RandIter last,Compare comp);}$$ (2.22)

where the functor `Compare comp(a,b)` should return `true` if `a` should precede `b` in the sorted range, and false otherwise.

2.1.6 Streams

Input and output in C++ is managed in the Standard Template Library by the `stream` abstraction, providing a functionality too extensive to fully present here. Instead, let us restrict ourselves to the features of C++ streams used in later chapters. We have already encountered the standard output stream object `cout`, declared in the header `<iostream>`. Objects written to `cout` using the overloaded operator `<<` appear in the console window, e.g.

        std::cout << "Hello world!" << std::endl;   (2.23)

where `std::endl` signals an end–of–line to `std::cout`. The operator `<<` is appropriately defined for all built–in data types (e.g. `char`, `int`, `double`, etc.) and also some STL classes such as `string`. For user–defined classes, a custom overload of the operator `<<` can be created.

Consider for example the class `Matrix` defined in Section 1.3.1. Listing 2.4 defines an operator `<<` which writes a `Matrix` to an `ostream` (like `cout`), one row per line and each value separated by a comma. This is the CSV (comma–separated value) format; if this output is directed to a file, the file can be loaded directly into a spreadsheet program such as Microsoft Excel. The easiest way to produce such a file is to write the output to `cout`, as in the program `MatrixOut.cpp` (Listing 2.5). If `MatrixOut.cpp` is compiled into a program `MatrixOut.exe` and run from a command line window as in Figure

```
#include <iostream>
#include "Matrix.h"

using namespace std;

int main(int argc, char *argv[])
{
    int i,j;
    Matrix A(3,4),B(3,4);
    // assign values to matrix elements
    for (i=0;i<3;i++) {
      for (j=0;j<4;j++) {
        A(i,j) = 0.1*i*j;
        B(i,j) = 1.0 - A(i,j); }}
    cout << "Matrix A:\n" << A;
    cout << "Matrix B:\n" << B;

    return EXIT_SUCCESS;

}
```

Listing 2.5: Writing matrices to cout

Figure 2.1 *Running* MatrixOut.exe *from a command line window*

2.1, the contents of the Matrix is displayed in the window. However if the output is redirected into a file, as in

$$\text{MatrixOut > Matrix.csv} \qquad (2.24)$$

then the contents of the Matrix is written to the file Matrix.csv rather than the display. This is typically sufficient for creating test output during the development phase of a project.[7]

---

[7] Once the code has been sufficiently tested, it can be integrated into a more user–friendly environment, e.g. as an XLL for use from Microsoft Excel — see Appendix A.

```
#include <fstream>
#include "Matrix.h"

using namespace std;

int main(int argc, char *argv[])
{
    int i,j;
    Matrix A(3,4),B(3,4);
    // assign values to matrix elements
    for (i=0;i<3;i++) {
      for (j=0;j<4;j++) {
        A(i,j) = 0.1*i*j;
        B(i,j) = 1.0 - A(i,j); }}
    // open output file
    std::ofstream fout("Matrix.csv");
    fout << "Matrix A:\n" << A;
    fout << "Matrix B:\n" << B;

    return EXIT_SUCCESS;
}
```

Listing 2.6: Writing matrices to a file

Alternatively, one can direct the output, directly to a file using an instance ofstream (declared in the header <fstream>) instead of cout, as in Listing 2.6. Note that the constructor of ofstream takes a file name as its argument.

To control the formatting of output through an ostream, one can either use the stream object's member functions, or instead pass *manipulator* objects to the stream. In Listing 2.7 a setprecision manipulator (declared in the header <iomanip>) is passed to cout prior to the Matrix B, setting the output precision of the stream to 12 significant digits.

Data can be read from input streams using the overloaded operator >>. The header <iostream> provides cin to read from the command prompt window and <fstream> declares an ifstream class. However, for all but the simplest inputs it is best to use library classes (not provided by the STL) which facilitate parsing input data into C++ data structures.[8]

2.2 The Boost Libraries

The Boost project (www.boost.org) provides a collection of high quality, expertly designed and peer–reviewed C++ libraries to supplement the Standard Library. Some of the Boost libraries have been included in the Library Tech-

---

[8] The library code accompanying this book (provided on the website) includes a template function CSV2Array() to parse a CSV file into a C++ matrix.

```
#include <iostream>
#include <iomanip>
#include <cmath>
#include "Matrix.h"

using namespace std;

int main(int argc, char *argv[])
{
    int i,j;
    Matrix A(3,4),B(3,4);
    // assign values to matrix elements
    for (i=0;i<3;i++) {
      for (j=0;j<4;j++) {
        A(i,j) = 0.1*i*j;
        B(i,j) = std::sqrt(1.0 - A(i,j)); }}
      cout << "Matrix B:\n" << std::setprecision(12) << B;

    return EXIT_SUCCESS;
}
```

Listing 2.7: Applying a manipulator object to a stream

nical Report of the C++ Standards Committee[9] as a first step toward becoming part of a future version of the Standard Library. In the chapters that follow, we will mainly make use of three features from the Boost libraries: functors (classes representing functions), smart pointers (wrapper classes for C++ pointers that facilitate memory management), and classes representing probability distributions. This part of Boost is primarily template–based, and thus all that is required is to #include the appropriate header files, with no need to link to library files.

2.2.1 Smart pointers

Smart pointers are conceived as "owning" the object to which they refer. Thus responsibility for deleting a dynamically allocated object can be delegated to the smart pointer, i.e. smart pointers keep track of references (via smart pointers) made to a particular object, and delete it when the last smart pointer referring it is destroyed. Consequently, smart pointers should *only* be used to point to dynamically allocated objects; and when dynamically creating objects using the operator **new**, it is good practice to *always* wrap the resulting "plain" pointer in a smart pointer.

Smart pointers which allow shared ownership of dynamically allocated objects are implemented in the **shared_ptr** (for single objects) and **shared_array** (for arrays of objects) class templates. They can be copied freely and thus

[9] See International Standards Organization (2005).

```
#include <iostream>
#include <boost/shared_ptr.hpp>

boost::shared_ptr<double> create_a_double(double d)
{
  boost::shared_ptr<double> result(new double(d));
  return result;
}

int main(int argc,char* argv[])
{
  boost::shared_ptr<double> creation =
    create_a_double(5.3);
  std::cout << *creation << std::endl;
  /* Note: No need to explicitly call "delete" - the
     shared_ptr object takes care of this. */
}
```

Listing 2.8: shared_ptr usage example

permit dynamically allocated objects to be passed safely between functions, including as return values. A toy example is given in Listing 2.8. For a proper usage example in the context of the library code developed and discussed in the chapters that follow, see for example the class MCPayoffList in Chapter 7, declared in MCEngine.hpp and implemented in MCEngine.cpp on the website for this book.

2.2.2 Functions

There are two primary reasons to use classes to represent functions (such classes are also called *functors*). One is that it allows functions to be easily passed as arguments of other functions, and the other is that in this way one can associate functions with data (such as function coefficients). The header file <boost/function.hpp> provides a template which facilitates and standardises the former: Suppose you wish to declare a functor f which wraps a function that takes two integer arguments and returns a double. The preferred syntax[10] is

$$boost::function<double(int\ x,\ int\ y)>\ f; \qquad (2.25)$$

The Boost functor is empty by default, it can be initialised with a global function or a class as demonstrated in Listing 2.9, where f can then be used as if it were a C++ function, or passed to another function.

---

[10] The "preferred syntax" works with most modern compilers, including Microsoft Visual C++ 7.1 and above. However, <boost/function.hpp> also provides a "portable syntax," which should work on all compilers. The portable syntax of the above example is boost::function2<double, int, int>.

```
#include <iostream>
#include <boost/function.hpp>

void calculate_and_print(boost::function<double (int x,int y)> g)
{
    std::cout << "Result: " << g(2,3) << std::endl;
}

double divide(int a,int b)
{
  return ((double)a)/b;
}

class multiply {
public:
    double operator()(int a,int b) { return a*b; };
};

int main(int argc,char* argv[])
{
  boost::function<double (int x,int y)> f;
  f = divide;
  calculate_and_print(f);
  f = multiply();
  calculate_and_print(f);
}
```

Listing 2.9: `boost::function` usage example

In order to associate a Boost functor with data, it must be "bound" to a particular *instance* of a class. This can be achieved with the help of `boost::bind` and `boost::mem_fn`, respectively defined in the header files `<boost/bind.hpp>` and `<boost/mem_fun.hpp>`. Consider the toy example in Listing 2.10. The function `boost::mem_fun` is applied to a pointer to a member function (in this case `operator()()`) to obtain a function object suitable for use with boost:bind.[11] This function object takes one more argument than the member function that it represents: Its first argument is a pointer to an instance of the class `multiply`, and its second argument is the original argument of the member `operator()()` — consequently, `boost::mem_fn` applied to a member function with two arguments would return a function object taking three arguments, and so on. Invoking `boost::bind` in Listing 2.10 then binds the first argument to a pointer to an instance of `multiply` (`&mult`) and the placeholder `_1` leaves the second argument open — it becomes the first and only argument of the functor `f`.

---

[11] The function object returned by `boost::mem_fn` is compatible with most situations in the STL and in Boost where such objects are required, for example in the `std::for_each` algorithm.

```
#include <iostream>
#include <boost/function.hpp>
#include <boost/bind.hpp>
#include <boost/mem_fn.hpp>

void calculate_and_print(boost::function<double (int x)> g)
{
  std::cout << "Result: " << g(3) << std::endl;
}

class multiply {
private:
  double coefficient;
public:
  multiply(double x) : coefficient(x) { };
  double operator()(int a) { return a*coefficient; };
};

int main(int argc,char* argv[])
{
  boost::function<double (int x)> f;
  multiply mult(3.5);
  f = boost::bind(boost::mem_fn(&multiply::operator()),&mult,_1);
  calculate_and_print(f);
}
```

Listing 2.10: boost::bind usage example

In general, the purpose of boost::bind is to bind any argument of a function to a specific value and to route input arguments into arbitrary positions. Consider a function

$$\text{int } g(\text{int } x, \text{ int } y, \text{ int } z) \tag{2.26}$$

Setting

$$\begin{aligned}&\texttt{boost::function<int (int x, int y)> f;}\\&\texttt{f = boost::bind(g,\_1,8,\_2);}\end{aligned} \tag{2.27}$$

means that the function call f(2,3) is equivalent to g(2,8,3). If instead one sets

$$\texttt{f = boost::bind(g,8,\_2,\_1);} \tag{2.28}$$

then f(2,3) is equivalent to g(8,3,2).

For usage examples of boost::function, boost::mem_fn and boost::bind in the context of the library code developed in the subsequent chapters, see Listing 3.2 (binomial lattice models) on page 101 and Listing 6.1 (calculation of Black/Scholes implied volatility) on page 186.

2.2.3 Probability distributions

A number of probability distributions play an important role in quantitative finance applications, most notably the normal (Gaussian) distribution, the binomial distribution and the noncentral chi–squared distribution. These and other univariate distributions are implemented in the Boost.Math library.[12] Operations on distributions are implemented as global, rather than member, functions. The first argument of these functions is an object representing the distribution, followed by any additional arguments needed to evaluate the operation.

For example, to evaluate the cumulative distribution function (CDF) of the standard normal distribution, first create an instance of the class representing the normal distribution,[13]

$$\texttt{boost::math::normal norm;} \tag{2.29}$$

which by default constructs a normal distribution with mean zero and unit standard deviation. For a `double x`,

$$\texttt{boost::math::cdf(norm,x)} \tag{2.30}$$

will return the value of the standard normal CDF at `x`. Similarly, for a double `p` in $[0, 1]$,

$$\texttt{boost::math::quantile(norm,p)} \tag{2.31}$$

will return the inverse of the CDF at `p`.

$$\texttt{boost::math::pdf(norm,x)} \tag{2.32}$$

returns the value of the probability density function (PDF) at `x`. Furthermore, the library implements accessors for various distributional properties, such as `mean`, `standard_deviation`, `skewness` and `kurtosis`, which are applied to the object representing a distribution, e.g.

$$\texttt{boost::math::mean(norm)} \tag{2.33}$$

For a full list of available functions, see the Boost.Math documentation.

2.3 Numerical arrays

A class hierarchy of matrices served as an illustrative example with which to introduce the object–oriented features of C++ in the previous chapter. Among other things, this example made extensive use of polymorphism. While this approach does result in a reasonably user–friendly library of matrix classes, the computer code it generates is relatively inefficient, which is a disadvantage in numerically intensive applications. The main sources of this inefficiency are the computational overhead involved in virtual function calls and frequent calls of the `new` and `delete` operators when matrix objects are created and destroyed.

---

[12] Multivariate distributions are not yet available — alternative ways to evaluate the cumulative distribution function of the multivariate normal distribution are discussed in Section 2.4.3.

[13] Declared in the header file `<boost/math/distributions/normal.hpp>`.

The first problem can be avoided by using a library of class templates instead of classes derived from an abstract base class: Template meta-programming can obviate the need for polymorphism.[14]

The second problem typically results from the excessive creation of temporary objects (by the compiler) when expressions involving matrix objects are evaluated. Suppose that A, B, C and D are matrices of identical size and element–wise arithmetic operations have been implemented using operator overloading as suggested in Section 1.3.1. Then the statement

```
D = A + B*C;                                                (2.34)
```

will result in the creation of two (or possibly more, depending on the compiler) temporary matrix objects: One as the result of B*C, and the other as the result of adding A to the result of B*C. Additionally, the code generated by statement (2.34) will involve multiple loops. Ideally, one would want a single loop over all matrix elements, performing the operation (2.34) on each element. This can be achieved (and temporary matrix objects avoided) using *expression templates*.[15]

Widespread dissatisfaction with the numerical arrays and basic linear algebra operations implemented via the `valarray<>` template in the Standard Template Library (STL) has led to an abundance of alternative implementations, many of them freely available. An open–source effort, which makes use of expression templates is *Blitz++*.[16] As Lischner (2003) puts it, "In some respects, it is what `valarray<>` should have been." Blitz++ is used in this book whenever numerical arrays are required. This library also supplies some of the random number generators used in Chapter 7.

2.3.1 Manipulating vectors, matrices and higher–dimensional arrays

Listing 2.11 gives a simple example of basic Blitz++ usage. As a template library, most of the implementation is in header files, which reside in the `blitz/` directory.[17] Blitz++ definitions are contained in the `blitz` namespace; it is often convenient to create local synonyms with `using` declarations. Arrays are declared using the `Array<T,N>` template, where T is the (numeric) type of the array elements and N is an integer determining the dimension of the array (i.e. 1 for a vector, 2 for a matrix, etc.). The size of the array along each

---

[14] See Section 2.5 for an illustration of this.

[15] Expression templates are an advanced use of the template features of C++, and their detailed discussion is beyond the scope of this book. For present purposes, it is sufficient to note that a library implementing numerical array operations using expression templates is desirable from the viewpoint of computational efficiency. For further reading on expression templates, see e.g. Veldhuizen (1995, 1998) and Veldhuizen and Gannon (1998).

[16] The Blitz++ home page is found at `http://blitz.sourceforge.net`.

[17] Note that the parent directory of `blitz/` must be in the compiler's include path. The Blitz++ library file contains some global data, and it must be linked to the final executable. See the Blitz++ documentation for installation instructions and compiler settings. Alternatively, Microsoft Visual Studio project files with the correct settings to compile the sample code are available on the website for this book.

```
#include <iostream>
#include <blitz/array.h>

int main(int argc, char *argv[])
{
  using std::cout;
  using std::endl;
  using blitz::Array;

  Array<double,2> A(3,3);
  Array<double,2> B(3,3);
  Array<double,2> C(3,3);
  Array<double,2> D(3,3);

  A = 1.0;
  B = 1.0, 0.0, 1.0,
      0.0, 1.0, 0.0,
      1.0, 0.0, 1.0;
  blitz::firstIndex i;
  blitz::secondIndex j;
  C = 1.0 + j + i*3.0;
  D = A + B*C;
  D(0,0) = 0.0;
  cout << A << endl << B << endl << C << endl << D << endl;

  return EXIT_SUCCESS;
}
```

Listing 2.11: Simple Blitz++ usage

dimension is specified in the constructor; thus A, B, C and D are 3×3 matrices in the current example. Matrices can be filled with values in various ways: A is assigned a scalar, which means that all elements of A are set equal to this value. B is initialised with a list of elements — the comma operator has been overloaded appropriately by Blitz++. Note that formatting the source code for this statement across three lines is for readability only. The initialisation of matrix C is effected using a Blitz++ feature called *index placeholders*. The statements

```
    blitz::firstIndex i;
    blitz::secondIndex j;                                              (2.35)
    C = 1.0 + j + i*3.0;
```

declare i and j to be index placeholders and the third statement is evaluated for all i and j along the first and second dimension of C, respectively, with the values resulting on the right–hand side being assigned to the corresponding matrix elements C(i,j). The matrix D is assigned the result of the statement (2.34) evaluated element–wise. Finally, matrix elements can be accessed directly via the overloaded operator(), i.e.

```
    D(0,0) = 0.0                                                       (2.36)
```

sets the element in the upper left–hand corner of D to zero.

Note that constructing an `Array` from another `Array` causes both `Arrays` to reference the same data, and any change to one will change the other.[18] Thus

```
Array<double,2> A(3,3);
A = 1.0;
Array<double,2> B = A;
B(1,2) = 2.0;
cout << A(1,2);
```
(2.37)

will print 2, not 1. To give B its own copy of the data, use the member function `copy()`:

```
Array<double,2> A(3,3);
A = 1.0;
Array<double,2> B = A.copy();
B(1,2) = 2.0;
cout << A(1,2);
```
(2.38)

will print 1.

To access subarrays and slices of `Arrays`, Blitz++ provides the `Range` class. Suppose that A is a 4×4 matrix, and we want to access the 2×2 submatrix in the lower right–hand corner. This can be achieved by

```
Array<double,2> B = A(Range(2,3),Range(2,3));
```
(2.39)

Again, B will be referencing part of the data of A, so

```
A = 1.0;
B(0,0) = 2.0;
cout << A(2,2);
```
(2.40)

will print 2, not 1. The special `Range` object `Range::all()` represents the full range of indices along a particular dimension. Thus

```
Array<double,2> B = A(Range::all(),Range(3,3));
```
(2.41)

will result in B referencing the last column of A. However, B is still a matrix, of dimensions 4×1. To create a slice of A representing the last column as a vector, use an integer (instead of a `Range`) as the column index:

```
Array<double,1> B = A(Range::all(),3);
```
(2.42)

Blitz++ overloads the ANSI C++ math functions to work element–wise on `Arrays`. Thus `C = sqrt(A)` sets the elements of the `Array` C equal to the square root of the corresponding elements of the matrix A. Similarly, `C = pow(A,B)` computes a^b for each element a in A and the corresponding element b in B (the dimension of A and B must coincide). User–defined functions can easily be declared to exhibit the same sort of behaviour using the `BZ_DECLARE_FUNCTION` macros provided by Blitz++. Thus the code in Listing 2.12 would allow us to use the function `add_one()` element–wise on `Arrays`:

---

[18] This feature is particularly tricky when passing `Arrays` to functions by value: Since this invokes the construction of an `Array` from an `Array`, operations modifying the `Array` inside the function will also affect the original `Array`.

```
using namespace blitz;
double add_one(double x)
{
  return x+1.0;
}
BZ_DECLARE_FUNCTION(add_one)
```

Listing 2.12: Defining a scalar function to operate element–wise on Arrays

C = add_one(A). Further examples where the return type differs from the argument type, or where the scalar function takes two arguments, are given in the sample program BlitzScalarFunction.cpp on the website.

Furthermore, a suite of functions is provided to carry out reductions on Arrays. *Complete reductions* on an Array expression result in a scalar; for example[19] sum(A) calculates the sum of all elements of the Array A. *Partial reductions* transform an Array of N dimensions to one with $N-1$ dimensions. For example, the following statements calculate the sum of the elements in each row of a matrix A:

```
blitz::secondIndex j;
Array<double,1> rowsum(A.extent(blitz::firstDim));      (2.43)
rowsum = sum(A,j);
```

Thus the partial reduction is performed by the overloaded function sum(), which takes as its second argument an index placeholder for the dimension along which the reduction is to be carried out. As of the current implementation,[20] partial reduction can only be carried out along the last dimension. If one instead wishes to reduce A by taking the sums of each column, one must first reorder the dimensions, i.e. (2.43) becomes

```
blitz::firstIndex i;
blitz::secondIndex j;
Array<double,1> colsum(D.extent(blitz::secondDim));     (2.44)
colsum = sum(D(j,i),j);
```

Note also that the variable to take the reduced Array must be correctly dimensioned, e.g. in (2.43), rowsum is a one–dimensional Array of size equal to the number of rows in A. This can be determined using the member function extent(), which returns the size of an Array along a chosen dimension, where blitz::firstDim represents the first dimension, blitz::secondDim the second, and so on.

2.3.2 Linear algebra

The Blitz++ features discussed in the previous section, in particular index placeholder objects, element–wise and reduction operations, make it very easy

---

[19] For a full list of supported functions for complete and partial Array reductions, see the Blitz++ documentation.
[20] Blitz++ version 0.9.

to code basic linear algebra operations in a way that will assist the compiler in producing efficient code. The dot product of two vectors x and y, for example, can easily be calculated like this:

```
dot_product = sum(x*y);
```
$$(2.45)$$

Matrix multiplication of the matrices A and B can be coded as
```
Array<double,2>
  C(B.extent(blitz::firstDim),
    A.extent(blitz::secondDim));
C = sum(B(i,k)*A(k,j),k);
```
$$(2.46)$$

where C holds the matrix product of B times A.

Many numerical algorithms in quantitative finance require higher–level linear algebra operations, such as matrix inversion or linear equation solvers. However, relatively little such code in C++ is freely available, whereas LAPACK provides highly efficient FORTRAN routines for exactly these purposes. LAPACK's FORTRAN routines have been automatically translated to C using f2c and made available on the Internet under the name of CLAPACK. Thus the easiest way to access efficient implementations of the high–level linear algebra operations for the algorithms discussed in subsequent chapters was to create user–friendly C++ interfaces based on Blitz++ arrays for the required CLAPACK routines.

The header file `InterfaceCLAPACK.hpp`[21] declares the interface routines. For example, consider the linear system of equations

$$AX = R \qquad (2.47)$$

where A is an $N \times N$ matrix and X and R are $N \times M$ matrices. To determine X given A and R, use the function
```
void SolveLinear(const Array<double,2>& A,
                 Array<double,2>& X,
                 const Array<double,2>& R)
```
$$(2.48)$$

Note that the target matrix X is passed into the function as a non-const reference, and needs to be properly dimensioned as an $N \times M$ matrix to hold the solution. For a usage example, see `BlitzLinearAlgebra.cpp` on the website.

Similarly, the functions
```
void SolveTridiagonal(const Array<double,2>& A,
                      Array<double,2>& X,
                      const Array<double,2>& R);
void SolveTridiagonalSparse(const Array<double,2>& A,
                            Array<double,2>& X,
                            const Array<double,2>& R);
```
determine X to solve the linear system $AX = R$, but under the assumption that A is a tridiagonal matrix. `SolveTridiagonalSparse()` differs from

---

[21] This file is part of the C++ code library associated with this book, and it is used widely in the code discussed in the chapters that follow. The header files for this C++ code library are found in the folder Include on the website.

`SolveTridiagonal()` in that in the former the $N \times N$ tridiagonal matrix A is represented as an $N \times 3$ array `A`, where the middle column is the diagonal and the first and third columns are the sub- and superdiagonals, respectively. `A(1,0)` $= A_{2,1}$ is the first element of the subdiagonal and `A(0,2)` $= A_{1,2}$ is the first element of the superdiagonal. Thus `A(N-1,2)` and `A(0,0)` are not used.

The functions

```
void Cholesky(const Array<double,2>& A,
              Array<double,2>& triangular,
              char LorU);
void PositiveSymmetricMatrixInverse(
              const Array<double,2>& A,
              Array<double,2>& inverseA);
double PositiveSymmetricMatrixDeterminant(
              const Array<double,2>& A);
```

$$(2.49)$$

operate on positive definite matrices A. `Cholesky()` returns the Cholesky decomposition of A in the output `Array triangular`, i.e. if `LorU = 'L'`, `triangular` contains a lower triangular matrix L such that

$$LL^\top = A$$

while if `LorU = 'U'`, `triangular` contains an upper triangular matrix U such that

$$U^\top U = A$$

`PositiveSymmetricMatrixInverse()` returns the matrix inverse of A in the output `Array inverseA`, and `PositiveSymmetricMatrixDeterminant()` returns the determinant $|A|$ of A.

For more general $M \times N$ matrices A, we have

```
void SingularValueDecomposition(const Array<double,2>& A,
                                Array<double,2>& U,
                                Array<double,1>& sigma,
                                Array<double,2>& V);
```

which computes an $M \times M$ orthogonal matrix U, and $M \times N$ matrix Σ, which is 0 except for its $\min(M, N)$ diagonal elements, and an $N \times N$ orthogonal matrix V, such that

$$U\Sigma V^\top = A$$

The diagonal elements Σ_{ii} of Σ are output as the `Array<double,1>& sigma`. These are the singular values of A; they are real and non-negative, and are in descending order in `sigma`. If one defines the diagonal matrix Σ^{-1} by

$$\Sigma^{-1}_{ii} = \begin{cases} \frac{1}{\Sigma_{ii}} & \text{if } \Sigma_{ii} > 0 \\ 0 & \text{otherwise} \end{cases} \tag{2.50}$$

Then

$$A^{-1} := V\Sigma^{-1}U^\top \tag{2.51}$$

gives the *Moore/Penrose inverse* of A. This generalised matrix inverse coincides with the classical matrix inverse when the latter exists, i.e. when A is square and has full rank. The function

```
void MoorePenroseInverse(const Array<double,2>& A,
                         Array<double,2>& inverseA,          (2.52)
                         double eps = 1e-6);
```

calculates the Moore/Penrose inverse, where the parameter `eps` means that all elements of Σ less than `eps` are considered to be numerically zero.

The eigenvalues and eigenvectors of a symmetric matrix A are calculated by the function

```
void SymmetricEigenvalueProblem(const Array<double,2>& A,
                                Array<double,1>& eigval,
                                Array<double,2>& eigvec,          (2.53)
                                double eps = 1e-12);
```

The output `Array<double,1>& eigval` contains the non-zero eigenvalues of A and the corresponding columns of `eigvec` contain the corresponding eigenvectors. The parameter `eps` sets the magnitude threshold below which an eigenvalue is considered to be numerically zero.

For a full list of interface functions to CLAPACK routines, see `Interface-CLAPACK.hpp` and its accompanying documentation. Interface functions are only provided for those CLAPACK functions, which are required by the quantitative finance algorithms discussed in this book. However, an experienced C++ programmer could easily adapt existing interface functions in order to access further CLAPACK functionality.

2.4 Numerical integration

Prices of derivative financial instruments can typically be represented as expectations, which in turn can typically be represented as (possibly multidimensional) integrals. The literature on numerical integration is vast and it is beyond the scope of this book to even attempt to present an overview. However, two methods are used in the implementation of quantitative finance models in this book, namely Gaussian quadrature for one–dimensional integrals and, for multidimensional integrals, cubature of vector–valued integrands over hypercubes. Gaussian quadrature is discussed below, while for cubature we restrict ourselves to interfacing with the C–code written by Steven G. Johnson.[22] The key application of the latter in the quantitative finance context is to evaluate the cumulative distribution function of the multivariate normal distribution.

2.4.1 Gaussian quadrature

The idea behind Gaussian quadrature is to approximate an integral by a weighted sum of the integrand evaluated at a chosen set of points (abscissas).

---

[22] Steven G. Johnson's code was downloaded from http://ab-initio.mit.edu/cubature/.

The weights w_j and abscissas x_j are chosen such that the approximation

$$\int_a^b W(x)f(x)dx \approx \sum_{j=0}^{N-1} w_j f(x_j) \qquad (2.54)$$

is exact if $f(x)$ is a polynomial of degree $2N-1$ or less, given a fixed integration interval (a,b) and known function $W(x)$, i.e. the approximation will work well if $f(x)$ is well–approximated by a polynomial. The weight function $W(x)$ and the integration interval (a,b) define a sequence of orthogonal polynomials $p_i(x)$. For a chosen degree N in (2.54), the abscissas x_j are given by the N roots of the orthogonal polynomial $p_N(x)$ of degree N. The weights w_j can then be determined such that (2.54) holds exactly for $f(x)$ equal to any of the first N orthogonal polynomials. For a more detailed overview, see Press, Teukolsky, Vetterling and Flannery (2007).

The abstract base class `GaussianQuadrature` implements (2.54) in the member function

```
double integrate(boost::function<double (double t)> f) const;
```

while there are derived classes `GaussHermite` and `GaussLaguerre` for different choices of $W(x)$, a and b, i.e. $W(x) = e^{-x^2}$, $-\infty < x < \infty$ for `GaussHermite` and $W(x) = e^{-x}$, $0 < x < \infty$ for `GaussLaguerre`. The weights w_j and abscissas x_j are appropriately initialised in the constructor in each case.

2.4.2 Cubature

Cubature refers to multidimensional numerical integration. The function

```
int cubature(boost::function<void
                        (const Array<double,1>&,
                         Array<double,1>&)> f,
           const Array<double,1>& xmin,
           const Array<double,1>& xmax,
           unsigned maxEval,
           double reqAbsError,
           double reqRelError,
           Array<double,1>& val,
           Array<double,1>& err);
```
$\qquad (2.55)$

declared in `InterfaceCubature.hpp`[23] serves as a C++ interface to the C–code of Steven G. Johnson, which implements adaptive multidimensional integration of vector–valued integrands over hypercubes.[24] For a function

$$f : \; \mathbb{R}^m \to \mathbb{R}^n$$

---

[23] Found in the folder `Include` on the website for this book.
[24] The code is based on the algorithms described in Genz and Malik (1980) and Berntsen, Espelid and Genz (1991).

it calculates the (in general vector–valued) integral

$$\int_{a_1}^{b_1} \int_{a_2}^{b_2} \cdots \int_{a_m}^{b_m} f(x) d^m x \qquad (2.56)$$

where x is an m-dimensional vector. f is passed to `cubature()` as a `boost::-function` object, representing a function, the first argument of which is a `const` reference to an `Array<double,1>` for x, and the second argument of which is a reference to an `Array<double,1>` to hold the output of $f(x)$, `xmin` and `xmax` are m-dimensional vectors representing the integration bounds a and b, respectively. `maxEval` specifies a maximum number of function evaluations. Integration stops when one of the following holds:

- `maxEval`, rounded up to an integer number of subregion evaluations, is reached

- the estimated error of the numerical integration is less than `reqAbsError`

- the estimated error of the numerical integration is less than `reqRelError` times the absolute value of the integral

If any of the three conditions is set to zero, it is ignored. `val` is an output `Array` containing the n-dimensional vector–valued result of the numerical integration, and `err` is an n-dimensional output `Array` of error estimates.

As suggested in the documentation of Johnson's routine, integration over infinite or semi-infinite intervals can be achieved by a transformation of variables. This transformation can be applied to each dimension of x separately, taking care to multiply the integrand by the corresponding Jacobian factor for each dimension being transformed. In one dimension, for a semi-infinite interval one sets $x = a + t/(1 - t)$, so that

$$\int_a^\infty f(x) dx = \int_0^1 f\left(a + \frac{t}{1-t}\right) \frac{1}{(1-t)^2} dt \qquad (2.57)$$

and

$$\int_{-\infty}^b f(x) dx = \int_{-b}^\infty f(-x) dx = \int_0^1 f\left(b - \frac{t}{1-t}\right) \frac{1}{(1-t)^2} dt \qquad (2.58)$$

For an infinite interval, set $x = t/(1 - t^2)$, so that

$$\int_{-\infty}^\infty f(x) dx = \int_{-1}^1 f\left(\frac{t}{1-t^2}\right) \frac{1+t^2}{(1-t^2)^2} dt \qquad (2.59)$$

2.4.3 The multivariate normal distribution

The multivariate normal distribution is important for a number of applications in quantitative finance, and its cumulative distribution function (CDF) is needed in particular to evaluate some of the analytical option pricing formulas in Chapters 4 and 8.

Denote by Σ the covariance matrix of a set of m random variables X_i, which are assumed to be jointly normally distributed. Not necessarily requiring Σ to be of full rank, the probability density function (PDF) of this joint distribution is then as given in van Perlo-ten Kleij (2004),

$$f(x) = (2\pi)^{-\frac{1}{2}k}(\det_k \Sigma)^{-\frac{1}{2}} \exp\left(-\frac{1}{2}(x-\mu)^\top \Sigma^+(x-\mu)\right), \quad x \in \mathbb{R}^m \quad (2.60)$$

where $\mu \in \mathbb{R}^m$ is the vector of the expectations $\mu_i = E[X_i]$, Σ^+ is the Moore/Penrose inverse of Σ, k is the rank of Σ, and $\det_k \Sigma$ is the product of the positive eigenvalues of Σ.

In general, the CDF is an m-dimensional integral of (2.60). However, note that the X_i can be represented by a linear combination of k independent univariate standard normal random variables Y_j: Denote by λ_j the k non-zero eigenvalues of Σ, and by Λ an $m \times k$ matrix, the columns of which are the corresponding eigenvectors of Σ. Then one can set[25]

$$X_i = \mu_i + \sum_{j=1}^{k} \sqrt{\lambda_j} Y_j \Lambda_{ij} \quad (2.61)$$

and the CDF becomes a k–dimensional integral. In particular, if $k = 1$,

$$P\{X_i \le c_i \ \forall \ 1 \le i \le m\} \quad (2.62)$$
$$= P\{\mu_i + \sqrt{\lambda_1} Y_1 \Lambda_{i1} \le c_i \ \forall \ 1 \le i \le m\}$$
$$= P\left\{Y_1 \le \frac{c_i - \mu_i}{\sqrt{\lambda_1}\Lambda_{i1}} \ \forall \ 1 \le i \le m\right\}$$
$$= P\left\{Y_1 \le \min_{1 \le i \le m} \frac{c_i - \mu_i}{\sqrt{\lambda_1}\Lambda_{i1}}\right\}$$

which can be evaluated using the univariate standard normal CDF.

Unfortunately, there is no general, fast and accurate algorithm for the CDF of an m–variate normal distribution. In fact, even for the bivariate and trivariate normal CDF, many available implementations tend to be problematic. West (2005) discusses this issue and provides improved code for normal CDFs up to dimension 3, based on Genz (2004).

Consequently, the class `MultivariateNormal` in `MultivariateNormal.hpp` on the CD–ROM implements the CDF in different ways depending on the dimension m. The constructor

```
MultivariateNormal(const Array<double,2>& covar);         (2.63)
```

---

[25] This relationship can also be used to generate realisations of the X_i from realisations of the independent standard normal variates Y_j for Monte Carlo simulation. This is implemented in

```
void MultivariateNormal::operator()(Array<double,1>& x);
```

which fills the output `Array` x with a random realisation of the X_i. Similarly,

```
void MultivariateNormal::transform_random_variables()(Array<double,1>& x);
```

uses (2.61) to transform a vector of independent standard normal random variates x into correlated random variates (also output in x) with covariances given by Σ.

takes the covariance matrix Σ as its argument; the expectations are assumed
to have been normalised to zero. The member function

```
double CDF(const Array<double,1>& d,
           unsigned long n = 1000000,                          (2.64)
           bool quasi = false);
```

calculates the CDF at d, i.e. it evaluates the integral

$$\int_{-\infty}^{d_1} \int_{-\infty}^{d_2} \cdots \int_{-\infty}^{d_m} f(x) d^m x \qquad (2.65)$$

with $f(x)$ given by (2.60). The methods used are:

1. If $k = 1$, the univariate normal CDF from the Boost library.

2. If $m = 2$, $k = 2$, the bivariate CDF as implemented by West (2005).

3. If $m = 3$, $k \geq 2$, the trivariate CDF as implemented by West (2005).

4. If $4 \leq m \leq 7$, $k \geq 2$, (2.65) is evaluated using the cubature() function
   (2.55). In this case, if $k < m$ the dimension of the numerical integration is
   larger than strictly necessary, but if one uses (2.61) to reduce the dimension
   of the integral, the integration is in general no longer over a hypercube.
   Thus (2.65) is used instead, as this fits easily into cubature().

5. If $m > 7$, $k \geq 2$, (2.65) is evaluated by Monte Carlo integration.

The parameter n of CDF() is used as the maxEval parameter of cubature() in
case 4, or as the number of Monte Carlo points in case 5. The boolean quasi
signals to use quasi-random Monte Carlo if true, pseudo-random Monte Carlo
if false.[26]

2.5 Optimisation and root search

One of the key tasks in the practical implementation of the models of quantitative finance discussed in subsequent chapters is *model calibration*, i.e. determining the model parameters in such a way as to obtain a fit to market
data. Typically, model prices for market instruments, e.g. swaps or options,
depend on the model parameters in a non-linear way. Model calibration involves finding the values of these parameters such that the distance between
model prices and those observed in the market is minimised, by some metric.
Thus calibration becomes a (possibly multidimensional) non-linear optimisation problem. If there is a sufficient number of parameters that an exact fit is
possible (and required), calibration becomes a root–finding problem.

While there are many commercial packages available that handle these two
tasks, there is no widely available open–source C++ library that provides the
optimisation and root search facilities needed for model calibration.[27] In this

---

[26] For a discussion of Monte Carlo integration, see Chapter 7.
[27] However, the GNU Scientific Library (GSL), found on the web at
http://www.gnu.org/software/gsl/, provides an excellent implementation in C of a
selection of algorithms.

section, we consider some algorithms suitable for the purposes of the subsequent chapters. They do not necessarily represent the most efficient solution in all cases, but they are relatively simple, easy to use and sufficient for present purposes. Each algorithm is implemented as a C++ template class. This avoids excessive (computationally costly) virtual function calls in the inner loops, while retaining the flexibility to apply the algorithm to any function with the appropriate interface. The presentation of the algorithms follows Press, Teukolsky, Vetterling and Flannery (2007), who also provide more sophisticated alternatives.

2.5.1 A simple bisection template

In the simplest case of a one–dimensional root search, suppose that we are looking for x such that

$$f(x) = y_0 \qquad (2.66)$$

for some continuous, one–dimensional function f. If there are x_1, x_2 such that $f(x_1) < y_0$ and $f(x_2) > y_0$, then there exists a solution to (2.66). If f is monotonic, the solution is unique.

A simple strategy to find the solution to (2.66) thus is to start with an interval $[a, b]$ in which the solution might lie. If $f(a)$ and $f(b)$ are either both less than or greater than y_0, expand the interval until $f(a)$ and $f(b)$ bracket y_0. Then there is a solution within the interval and we evaluate the function at a point in the interval, say the midpoint $(a + b)/2$. We then move either the lower or the upper interval boundary to the midpoint, such that the values of the function evaluated at the new interval boundaries still bracket the target value y_0. Continue to bisect the interval until (2.66) is met to the desired accuracy.

This algorithm is implemented as a template class in `Rootsearch.hpp`,[28] as part of the C++ library accompanying this book. The template parameters are

```
template <class F,class argtype,class rettype>
class Rootsearch
```
(2.67)

where F is a user–defined class representing the function f, i.e. F must provide the member operator

```
rettype operator()(argtype x)
```
(2.68)

The `Rootsearch` constructor is used to set up the root–finding problem. Table 2.1 lists the constructor arguments. Once the problem is set up, calling the member function `solve()` of `Rootsearch` returns a solution to (2.66), if it exists. If the maximum number of iterations is exceeded in search for a solution, `Rootsearch` will throw a `std::out_of_range` exception.

The file `RootsearchExample.cpp` gives a usage example for this template. Note that when the function is not monotonic, (2.66) may have multiple solutions. In this case, which solution is found depends on the starting point. This

---

[28] Found in the folder **Include** on the website for this book.

Type	Argument	Default value
F&	The function.	
rettype	Function target value y_0.	
argtype	Initial point in root search.	
argtype	Initial width of search interval for bisection algorithm.	
argtype	Lower bound for function argument (for functions not defined on $(-\infty, \infty)$).	
argtype	Upper bound for function argument (for functions not defined on $(-\infty, \infty)$).	
double	Desired accuracy.	1E-9
unsigned long	Maximum number of iterations.	100000

Table 2.1 *Rootsearch constructor arguments*

template is also used, for example, in Chapter 6 to calculate Black/Scholes implied volatility.

2.5.2 Newton/Raphson in multiple dimensions

Consider now equation (2.66) where both x and y_0 are n-dimensional vectors. In this case one typically needs to make use of some knowledge of the function f in order to find a solution. Newton/Raphson is arguably the simplest method, and one that works well if the starting point is sufficiently close to the solution and/or the function is sufficiently well–behaved with regard to monotonicity. Expanding f in a Taylor series, we have, in matrix notation,

$$f(x + \Delta x) = f(x) + J(x)\Delta x + \mathcal{O}(\Delta x^2) \qquad (2.69)$$

where x and Δx are n-dimensional vectors and J is the Jacobian matrix of partial derivatives, i.e.

$$J_{ij} = \frac{\partial f_i}{\partial x_j}$$

is the derivative of the i-th component of the vector–valued function f with respect to the j-th component of the argument vector x. Setting $f(x + \Delta x) = y_0$ in (2.69) and ignoring the higher order terms in the expansion, we have the linear system of equations

$$J(x)\Delta x = y_0 - f(x) \qquad (2.70)$$

which is then solved for Δx. x is updated by adding Δx and the process repeated until (2.66) is met to the desired accuracy.

This algorithm is implemented as a template class in NewtonRaphson.hpp. The template parameter is

```
template <class F> class NewtonRaphson
```
(2.71)

where F is a user–defined class representing the function f, i.e. F must provide a member returning the function value

```
Array<double,1> operator()(Array<double,1> x)
```
(2.72)

a member returning the Jacobian matrix

```
Array<double,2> Jacobian(Array<double,1> x)
```
(2.73)

and the member functions

```
int argdim() // The dimension of the argument vector.
int retdim() // The dimension of the result vector.
```
(2.74)

Note that this template is not as general as the one in the previous section: The function is explicitly assumed to be taking as its argument a Blitz++ Array of doubles and returning a Blitz++ Array of doubles. The NewtonRaphson constructor

```
NewtonRaphson(F& xf,double xeps,
              unsigned long xmaxit = 100000)
```
(2.75)

takes as its arguments the function object, the desired accuracy and the maximum number of iterations of the algorithm. Once the problem is set up, calling the member function

```
Array<double,1> solve(Array<double,1>& x,
                      const Array<double,1>& y0)
```
(2.76)

of NewtonRaphson returns a solution to (2.66), if found. Here x is the starting point of the root search and y0 is the target function value y_0. Note that the algorithm terminates if either the current point x no longer changes by more than the desired accuracy, or the target function value is achieved within the desired accuracy. If the maximum number of iterations is exceeded in search for a solution, NewtonRaphson will throw a std::runtime_error exception.

If implementing the Jacobian of the function f explicitly is too tedious, a class providing only the function value via the operator()() can be "decorated"[29] with a Jacobian calculated numerically by finite differences using the NumericalJacobian template also provided in NewtonRaphson.hpp.

The file NewtonRaphsonExample.cpp gives a usage example for this template as well as the template NumericalJacobian. Note that when the function is not monotonic, (2.66) may have multiple solutions. In this case, which solution is found depends on the starting point. This template is also used, for example, for fitting a term structure of interest rates to data (see Section 2.6.3, below).

2.5.3 Brent's line search and Powell's minimisation algorithm

Many financial problems involve non-linear optimisation in multiple dimensions. For example, calibrating models to market data might be done by min-

---

[29] This is an application of the "Decorator" object–oriented design pattern, described in the seminal book of Gamma, Helm, Johnson and Vlissides (1995), with the only modification that the additional functionality is not added dynamically at run-time, but rather at compile-time because for efficiency reasons we are working with templates rather than inheritance.

imising the sum of the squared differences between market and model prices, where the model prices are non-linear functions of the model parameters. This section introduces one relatively robust algorithm for non-linear optimisation, which is particularly easy to use, as it does not require derivatives of the objective function. The problem to be solved is[30]

$$\min_{x} f(x) \tag{2.77}$$

where

$$f : \mathbb{R}^d \to \mathbb{D}$$

is a function with a vector argument returning a scalar in a range $\mathbb{D}$ on the real line.

The algorithm is constructed by executing a succession of one-dimensional minimisations (line searches), thus as a first building block we need a routine which minimises $f(x)$ along a particular direction Δx through $\mathbb{R}^d$. If our current position in $\mathbb{R}^d$ is x_0, this one-dimensional problem can be written as

$$\min_{\lambda \in \mathbb{R}} f(x_0 + \lambda \Delta x) \tag{2.78}$$

This problem can be solved numerically using Brent's line search, which is based on parabolic interpolation. To shorten notation, set

$$g(\lambda) = f(x_0 + \lambda \Delta x) \tag{2.79}$$

Suppose that we have evaluated $g(\cdot)$ at three points λ_1, λ_2, λ_3. Then the abscissa λ of the minimum of the parabola through these points is given by

$$\lambda = \lambda_2 - \frac{1}{2} \frac{(\lambda_2 - \lambda_1)^2(g(\lambda_2) - g(\lambda_3)) - (\lambda_2 - \lambda_3)^2(g(\lambda_2) - g(\lambda_1))}{(\lambda_2 - \lambda_1)(g(\lambda_2) - g(\lambda_3)) - (\lambda_2 - \lambda_3)(g(\lambda_2) - g(\lambda_1))} \tag{2.80}$$

This always results in a valid value for λ, unless the three points lie on a straight line. The code must be made robust in this case, as for the case that the parabola through the three points has a maximum rather than a minimum. Suppose that it is known that the λ solving (2.78) lies in the interval (a, b) (because, say, $g(a) > g(c)$ and $g(b) > g(c)$ for some $c \in (a, b)$). The idea is to first calculate λ using (2.80), if possible. This λ is accepted as the next step in the line search only if $\lambda \in (a, b)$ *and*, in order to prevent non-convergent cycles in the algorithm, the change from the previous "best" λ is *less* than half the corresponding change in λ in the step before last. If the parabolic λ is not acceptable according to these criteria, a new λ is found by simple bisection. Thus (hopefully) fast convergence to a minimum is achieved by the parabolic interpolation, while the simple bisection steps ensure the robustness of the algorithm. This algorithm is implemented in the file Linesearch.hpp as

```
template <class F, class rettype>
class GeneralBrentLinesearch
```
(2.81)

---

[30] Note that given an algorithm to solve the minimisation problem, one could maximise a function $g(x)$ by setting $f(x) = -g(x)$.

in the namespace `quantfin::opt`. The class `F` represents the function $f(x)$ and must provide a

$$\text{rettype operator()(Array <rettype, 1>\& x)} \qquad (2.82)$$

where `rettype` is the type of the return value and the type of each element of the function argument x — typically, `rettype` will be set to `double`. `GeneralBrentLinesearch` provides a single

```
rettype operator()(F& f,
                   Array <rettype,1>& currpos,
                   const Array <rettype,1>& xdirection,
                   rettype eps,
                   unsigned long maxit);
```

where `currpos` represents the starting point x_0 in (2.78), `xdirection` the direction Δx of the line search, `eps` the numerical tolerance when identifying a minimum and `maxit` the maximum number of iterations allowed when searching for a minimum.

Now that we have a line search algorithm to minimise $f(x)$ along arbitrary directions in $\mathbb{R}^d$, we are in a position to solve the full multidimensional minimisation problem (2.77). Again, the basic idea is very simple: Successively minimise $f(x)$ along different directions in $\mathbb{R}^d$, repeating the process until a multidimensional minimum is reached. The main issue (and one that distinguishes different approaches to the problem) is to find a "good" choice of directions. In particular, one wants to avoid a situation in which gains made in the minimisation along one direction are in a sense lost in a subsequent minimisation along a different direction. Formally, this can be achieved as follows: If the algorithm has moved to a minimum along a direction u, then the gradient ∇f of f will necessarily be orthogonal to u in that point. If, in a move along a new direction v, this orthogonality is maintained, then any new point along v will still be a minimum along the direction u, and the minimisation along u will not have to be repeated. Ideally, one would want this condition to be maintained along all subsequent line search directions. Obviously, this is only possible if the Hessian matrix of second partial derivatives is constant, i.e. the function $f(x)$ can be written as a quadratic form

$$f(x) = c - bx + \frac{1}{2}xAx \quad , \quad b \in \mathbb{R}^d, A \in \mathbb{R}^{d \times d} \qquad (2.83)$$

in which case $\nabla f = Ax - b$, so along a new direction v the gradient ∇f changes with Av, and the orthogonality condition becomes $Av \perp u$, i.e.

$$u^\top A v = 0 \qquad (2.84)$$

Two vectors which satisfy this condition are said to be *conjugate*. Given a set of d linearly independent, conjugate directions, the minimisation algorithm will then necessarily converge after d line searches. In practice, of course, (2.83) holds only approximately (in the sense of a Taylor expansion of f around some point $x^{(0)}$), so repeated sets of d line searches are necessary.

Powell's method is a way to generate d mutually conjugate directions with-

1. Set the initial d line search directions $u^{(i)}$ to the basis vectors.

2. Set $x^{(0)}$ to the starting position.

3. For $i = 1$ to d, move from $x^{(i-1)}$ to the line search minimum in direction $u^{(i)}$ and save this point as $x^{(i)}$.

4. For $i = 1$ to $d - 1$, set $u^{(i)} = u^{(i+1)}$.

5. Set $u^{(d)} = x^{(d)} - x^{(0)}$.

6. Move from $x^{(d)}$ to the line search minimum in direction $u^{(d)}$. If this minimum is less than $f(x^{(0)})$, set $x^{(0)}$ to the location of this new minimum.

7. Go to 3.

Algorithm 2.1: Generating conjugate directions by Powell's method

out knowledge of the Hessian of f (nor the gradient, for that matter). Algorithm 2.1 describes the procedure.

After d iterations, this procedure results in a full set of conjugate directions, and would exactly minimise a quadratic form (2.83).

Press, Teukolsky, Vetterling and Flannery (2007) note that there is a problem, however: By throwing away $u^{(d)}$ in favour of $x^{(d)} - x^{(0)}$ in each iteration of Algorithm 2.1, one may end up with a set of directions which are linearly dependent, meaning that the search for a minimum is conducted only in a subspace of $\mathbb{R}^d$ and the true minimum is not found. They list several ways to prevent this, of which the simplest is to reinitialise the line search directions $u^{(i)}$ to the basis vectors after every d iterations of the algorithm.

The algorithm is implemented in the file Powell.hpp as

$$\text{template <class F, class rettype, class linesearch>} \atop \text{class GeneralPowell} \qquad (2.85)$$

in the namespace quantfin::opt. linesearch represents the one-dimensional line search algorithm, and will typically be given by

GeneralBrentLinesearch<F,rettype>

but this could be replaced by any class which appropriately defines operator() analogously to GeneralBrentLinesearch. As before, F represents the function $f(x)$ and must provide an operator (2.82). The usage of GeneralPowell is demonstrated in the file PowellDemo.cpp on the website.

In some situations, additional constraints may be imposed on the minimisation problem (2.77). Unlike in linear or quadratic optimisation, there is no algorithm for truly constrained optimisation in the general non-linear case, i.e. no algorithm in which the constraints are satisfied in every intermediate step. Commonly, one imposes a numerical penalty for any violation of the constraints, adding this penalty to the function value to be minimised. Hopefully, this then leads to a minimum of the objective function in a point in which all constraints are met.

In some cases, however, there are bounds on the function arguments, which must be respected in order for the function value to be well defined. Here, a logistic mapping of the function arguments can be used to transform a function $f(x)$

$$f : (a, b) \to \mathbb{D}$$

into an equivalent function $g(z)$

$$g : \mathbb{R} \to \mathbb{D}$$

by setting

$$x = a + \frac{b - a}{1 + e^{-z}} \tag{2.86}$$

$$\Leftrightarrow \quad z = \ln\left(\frac{x - a}{b - x}\right) \tag{2.87}$$

and $g(z) = f(x(z))$. A minimum of g in $z_{\min} \in \mathbb{R}$ will then correspond to a minimum of f in $x_{\min} = x(z_{\min}) \in [a, b]$. The same mapping procedure can be applied element–wise to multidimensional function arguments x, if the set of permissible x is given by a hypercube. This means that the permissible range of every component $x^{(i)}$ of x must be given by an interval (a_i, b_i) independently of all other components $x^{(j)}$, $i \neq j$, of x.

More generally, the permissible range of x may be given by a convex set $\mathbb{G}$, with the bounds on the individual components $x^{(i)}$ not independent of each other. In this case, the bounds on x can be imposed during each line search: Suppose $x_0 \in \mathbb{G}$. Suppose further that given a search direction Δx, there is a method that can determine a $\underline{\lambda}$ and a $\overline{\lambda}$ such that

$$\text{and} \quad \begin{array}{l} x_0 + \lambda \Delta x \in \mathbb{G} \quad \text{if } \lambda \in (\underline{\lambda}, \overline{\lambda}) \\ x_0 + \lambda \Delta x \notin \mathbb{G} \quad \text{if } \lambda \notin (\underline{\lambda}, \overline{\lambda}) \end{array} \tag{2.88}$$

Note that $\underline{\lambda}$ and $\overline{\lambda}$ must exist (though possibly $\underline{\lambda} = -\infty$ and/or $\overline{\lambda} = \infty$) as a consequence of the convexity of $\mathbb{G}$. Then we can restrict the line search in direction Δx to solve

$$\min_{\lambda \in (\underline{\lambda}, \overline{\lambda})} f(x_0 + \lambda \Delta x) \tag{2.89}$$

If $f(x_0 + \lambda \Delta x)$ is monotonic on $(\underline{\lambda}, \overline{\lambda})$, and thus does not have an interior minimum, then under the slightly more restrictive assumption of replacing (2.88) by

$$x_0 + \lambda \Delta x \in \mathbb{G} \text{ if and only if } \underline{\lambda} \leq \lambda \leq \overline{\lambda} \tag{2.90}$$

we can set the line search to return a minimum at $\underline{\lambda}$ or $\overline{\lambda}$, as the case may be.

This constrained line search is implemented via an optional argument of the `GeneralBrentLinesearch` constructor, a Boost function object

```
boost::function<void (Array<rettype,1>&,
                      const Array<rettype,1>&,            (2.91)
                      double&,double&)> xset_bounds;
```

This represents a function taking (in this order) as input arguments the current position x_0, the search direction Δx, and as output arguments references to the doubles $\underline{\lambda}$ and $\overline{\lambda}$, i.e. xset_bounds calculates $\underline{\lambda}$ and $\overline{\lambda}$ given x_0 and Δx. If the function pointer is NULL (the default), no bounds are imposed on λ.[31]

2.6 The term structure of interest rates

2.6.1 Fixed income terminology

The basic unifying concept of the terminology to be introduced in this section is the *zero coupon bond*, "zero bond" for short.

Definition 2.1 *The* zero coupon bond *price $B(t,T)$ is the value at time[32] t of a security paying 1 monetary unit at time T. T is called the* maturity *of the zero coupon bond; $T - t$ is the* time to maturity.

$B(t,T)$ can also be interpreted as a *discount factor*, i.e. the factor with which sure payoffs at time T must be multiplied in order to calculate their time t present value. While zero bonds are in general fictitious, meaning that there rarely exist liquid markets for these securities, they are the basic building blocks in terms of which the rest of the fixed income vocabulary may be unambiguously defined. Operating under the standing assumption that securities with the same future payoffs must have the same present value (law of one price), zero bonds can be seen as a canonical vector basis of the space of assets with deterministic payoffs. A *coupon bond*, for example, can be represented as the sum (or portfolio) of zero bonds:

Definition 2.2 *A* coupon bond $V(t, \mathbb{T}, k)$, *pays k at each coupon date $T_i \in \mathbb{T}$, $1 \leq i < n$, and pays $1 + k$ at the bond maturity T_n. The set $\mathbb{T} = \{T_0, T_1, \ldots, T_n\}$, with $T_i > T_{i-1}$ for all $0 < i \leq n$, is called the* tenor structure *of the coupon bond.*

Defining further $\eta(t)$ as the index of the next date in the tenor structure, i.e.

$$\eta(t) = \min\{0 < i \leq n | t < T_i\}, \tag{2.92}$$

the value $V(t, \mathbb{T}, k)$ at time t of the coupon bond can be calculated by taking the present value for each payment separately; thus the price of the coupon bond can be written as

$$V(t, \mathbb{T}, k) = B(t, T_n) + k \sum_{i=\eta(t)}^{n} B(t, T_i). \tag{2.93}$$

---

[31] For an example of imposing bounds on λ during an optimisation, see the calibration of a GramCharlier risk neutral distribution to market data in Section 6.1.

[32] Within the models of quantitative finance and their implementations, time is typically measured in years and fractions thereof.

Definition 2.3 *Depending on the way interest is compounded, interest rates are defined as follows:*

- *The simple or actuarial rate of interest* $r_a(t, T)$ *is given by*[33]

$$B(t, T) = (1 + (T - t)r_a(t, T))^{-1}.$$

- *The effective yield* $y_e(t, T)$ *is given by*

$$B(t, T) = (1 + y_e(t, T))^{-(T-t)}.$$

- *The continuously compounded yield* $y(t, T)$ *is given by*

$$B(t, T) = \exp\{-(T - t)y(t, T)\}.$$

Thus each of the above rates can easily be calculated given a zero coupon bond price and vice versa. For a given time t we can therefore represent the *term structure of interest rates* ("term structure" for short) as any of the mappings $T \to B(t, T)$, $T \to r_a(t, T)$, $T \to y_e(t, T)$ or $T \to y(t, T)$.

The rate of interest most often encountered in theoretical models of the term structure, is a purely mathematical construct:

Definition 2.4 *Assume logarithmic zero coupon bond prices are continuously differentiable with respect to maturity. The continuously compounded short rate (or "short rate") $r(t)$ is given by*

$$r(t) = \lim_{T \searrow t} y(t, T) = -\left.\frac{\partial \ln B(t, T)}{\partial T}\right|_{T \searrow t}.$$

In contrast to the rates in Definition 2.3, the short rate is valid only an infinitesimal period of time starting immediately. It is in terms of this rate that the theoretical *savings account* is defined, which can be interpreted as the value of an initial investment at time 0 of one monetary unit continuously reinvested ("rolled over") at the short rate:

Definition 2.5 *The value of the savings account $\beta(t, T)$ is given by*

$$\beta(t, T) = \exp\left\{\int_t^T r(s)ds\right\}.$$

Another instrument which figures prominently in subsequent chapters is the *forward contract*:

Definition 2.6 *A forward contract is the right and the obligation to buy (respectively to sell) an asset $S(T)$ at time T for the forward price $F(S, t, T)$, predetermined at time t, in such a manner that the time t value of the contract is zero.*

---

[33] Note that usually only rates for periods of up to one year are quoted in this manner.

Proposition 2.7 *The forward price $F(S, t, T)$ is given by*

$$F(S, t, T) = \frac{S(t)}{B(t, T)}.$$

PROOF: Consider the following strategy:

Time	t	T
Cashflow	$F(S, t, T)B(t, T)$	$-F(S, t, T)$
	$-S(t)$	$S(T)$

At time t we sell $F(S, t, T)$ zero coupon bonds $B(t, T)$ and buy the asset $S(t)$. When the zero coupon bonds mature in T, we have to pay $F(S, t, T)$ and the value of the asset we are holding is $S(T)$. Thus in T we are in the same position we would have been had we *gone long* a forward contract, i.e. entered the obligation to *buy* the asset in T for the forward price $F(S, t, T)$. Therefore, by the law of one price, if entering a forward contract results in no payment obligations at time t, the initial value of our strategy must also be zero, resulting in the proposed identity. $\square$

Forward interest rates or *forward rates* are defined by replacing the zero coupon bond price in 2.3 by a *forward bond price* $B(t, T_2)/B(t, T_1)$:

Definition 2.8 *Depending on the way interest is compounded, forward interest rates are defined as follows:*

- *The* simple *or* actuarial *forward rate of interest $f_a(t, T_1, T_2)$ is given by*

$$B(t, T_2)/B(t, T_1) = (1 + (T_2 - T_1)f_a(t, T_1, T_2))^{-1}.$$

- *The* effective *forward yield $f_e(t, T_1, T_2)$ is given by*

$$B(t, T_2)/B(t, T_1) = (1 + f_e(t, T_1, T_2))^{-(T_2 - T_1)}.$$

- *The* continuously compounded *forward yield $f(t, T_1, T_2)$ is given by*

$$B(t, T_2)/B(t, T_1) = \exp\{-(T_2 - T_1)f(t, T_1, T_2)\}.$$

The forward rate $f(t, T_1, T_2)$ can be interpreted as the rate of interest for money invested from T_1 to T_2, with the investment already being agreed upon in t. Furthermore, if logarithmic zero bond prices are continuously differentiable, we have

Definition 2.9 *The* instantaneous forward rate *$f(t, T)$ is given by*

$$f(t, T) = -\left.\frac{\partial \ln B(t, u)}{\partial u}\right|_{u=T}$$

or alternatively

$$B(t, T) = \exp\left\{-\int_t^T f(t, u)du\right\}.$$

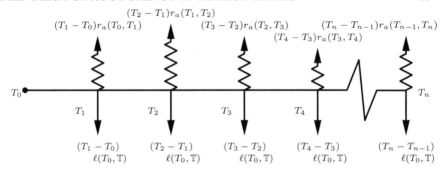

Figure 2.2 *Interest rate swap cashflows*

Note that $f(t,t) = r(t)$, the continuously compounded short rate. Given the mapping $T \to f(t,T)$, we have another alternative representation of the term structure of interest rates.

Generically, a *swap* is a contract to exchange one cashflow for another. Like forward contracts, swaps are typically set up in such a way that the net present value at initiation is zero. Also like forward contracts, the value of a swap can be represented in terms of more basic financial instruments. Thus swaps and forward contracts are simple examples of derivative financial instruments, or *derivatives* for short.

A plain vanilla[34] *interest rate swap* is a contract where one party pays the other a fixed interest on the swap's notional value, receiving a variable ("floating") rate of interest in return. Figure 2.2 schematically shows this exchange of cashflows. For example, the fixed interest payments could be made semi-annually, against variable payments determined by the current six–month London Interbank Offer Rate (LIBOR), an index of the rate of interest at which banks offer to lend funds in the interbank market. The fixed interest rate, also called *swap yield* or *swap rate*, is determined at contract inception such that the initial value of the swap is zero. For simplicity, we will confine ourselves to swaps settled *in arrears*, meaning that the "floating" payment at the end of a period is determined by the variable rate of interest at the beginning of that period. Therefore we can state

Proposition 2.10 *Let the notional amount of a swap be unity and fixed payments $\ell(T_0, \mathbb{T})$ occur at times $T_i \in \mathbb{T}$, $0 < i \leq n$, up to contract expiration in T_n. Then the swap yield $\ell(T_0, \mathbb{T})$ of a swap contracted in T_0 must equal the coupon on a coupon bond quoted at par, i.e. with price $V(T_0, \mathbb{T}, \ell(T_0, \mathbb{T})) = 1$.*

PROOF: Consider the position of the side paying the fixed rate $\ell(T_0, \mathbb{T})$ and receiving the variable interest payments, say three–month LIBOR. Note that our argument does not depend on how we choose the floating rate; we only

---

[34] In market jargon, the simplest version of a particular kind of derivative is often labelled "plain vanilla," supposedly because vanilla is the simplest flavour of ice cream.

need that the variable interest payments are adjusted to the "market rate" (for which we take LIBOR as a proxy) for each compounding period. Thus we would take six–month instead of three–month LIBOR if the floating payments occurred every six instead of three months. At contract inception at time T_0, let the side paying the fixed rate also buy the coupon bond $V(T_0, \mathbb{T}, \ell(T_0, \mathbb{T}))$. They can thus meet their obligations incurred in the swap contract by the coupon payments they receive from the bond, leaving them with a repayment of the notional amount 1 at maturity T_n. Furthermore, the party paying the fixed rate borrows one monetary unit from time T_0 to the time they receive the first variable interest payment from the swap. They can use this variable interest payment to pay the interest on the loan, borrowing 1 to repay the principal. They carry on this "roll–over" strategy up to expiration of the swap contract in T_n, where the remaining obligation of one monetary unit is met using the amount received from the maturing coupon bond. Thus all cashflows after time T_0 from swap, coupon bond and roll–over strategy cancel out, and therefore the cashflows in T_0, $-V(T_0, \mathbb{T}, \ell(T_0, \mathbb{T}))$ for the purchase of the coupon bond and $+1$ borrowed at the variable rate of interest, must also sum up to zero, yielding the proposed identity.   □

Remark 2.11 Using (2.93), the swap rate can be represented in terms of zero coupon bonds as

$$0 = (1 - B(T_0, T_n)) - \ell(T_0, \mathbb{T}) \sum_{i=1}^{n} (T_i - T_{i-1}) B(T_0, T_i)$$

$$\Leftrightarrow \quad \ell(T_0, \mathbb{T}) = \frac{1 - B(T_0, T_n)}{\sum_{i=1}^{N} (T_i - T_{i-1}) B(T_0, T_i)} \tag{2.94}$$

The sum

$$\sum_{i=1}^{N} (T_i - T_{i-1}) B(T_0, T_i)$$

is often called the *present value of a basis point* (PVBP), as it represents the dollar value of a change in the swap rate $\ell(T_0, \mathbb{T})$.

A *forward start swap* is a swap, which is contracted at some time $t < T_0$, i.e. before the beginning of the first accrual period $[T_0, T_1]$. The *forward swap rate* $\ell(t, \mathbb{T})$ is set at time t in such a way that the net present value of the contract is zero. In this case (2.94) becomes

$$\ell(t, \mathbb{T}) = \frac{B(t, T_0) - B(t, T_n)}{\sum_{i=1}^{N} (T_i - T_{i-1}) B(t, T_i)} \tag{2.95}$$

2.6.2 Term Structure Interpolation

Traditionally, continuous time term structure models assume that the term structure of interest rates is given for a continuum of maturities up to some time horizon[35], be it in the form of zero coupon bond prices or alternatively in terms of yields or instantaneous forward rates. This confronts anyone seeking to apply these models in practice with two problems: First, zero coupon bond prices or yields are usually not directly observable in the market, and neither are instantaneous forward rates. Second, the data that is available does not cover a continuum. Instead, one has to make do with coupon bond prices, swap yields or interest rate futures and specify a functional form for the interpolation between discrete data points.

This raises the question from which family of functions the interpolation rule should be chosen. The answer given to this question is usually based very much on pragmatic considerations, i.e. the problem is treated completely separately from any model for the evolution of interest rates. The only economic requirement that is always imposed is that the interpolated term structure should be in some sense consistent with the observed data. If there are only few data points available, such as in the swap or the futures markets, a perfect fit to the observations is usually required. In the coupon bond market, more data is available, but a perfect fit is either not possible or would yield highly unstable results: Even before any interpolation rule is applied, as Frachot (1994) notes[36], the fact that the coupon dates of different bonds usually do not coincide multiplies the number of zero coupon bond prices which have to be estimated, at the same time reducing the precision of the estimate. From a numerical point of view, the system of linear equations implied for zero bond prices by the relationship (2.93) is often close to a singularity. Therefore in this case statistical models incorporating error terms to explain deviations of the observed from theoretical prices are employed, resulting in a least squares estimate for the interpolated term structure.

For the pragmatic reasons of simplicity and robustness, linear interpolation is often applied, either on yields, on zero coupon bond prices or on logarithmic zero bond prices. Given the zero coupon bond prices $B(t, T_1)$ and $B(t, T_2)$, by loglinear interpolation we have for $T_1 \leq T_i \leq T_2$

$$\ln B(t, T_i) = \ln B(t, T_1) + \frac{T_i - T_1}{T_2 - T_1} \ln \left(\frac{B(t, T_2)}{B(t, T_1)} \right), \qquad (2.96)$$

implying for the instantaneous forward rate

$$-\frac{\partial \ln B(t, u)}{\partial u} \bigg|_{u=T_i} = \frac{\ln B(t, T_1) - \ln B(t, T_2)}{T_2 - T_1}$$

which is constant for all $T_i \in [T_1; T_2]$.

---

[35] See for example Hull and White (1990).
[36] See Frachot (1994), p. 24.

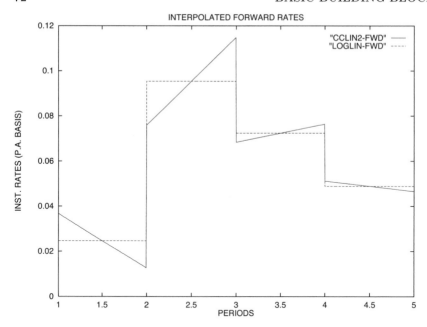

Figure 2.3 *Interpolated instantaneous forward rates*

Conversely, linear interpolation of continuously compounded yields gives

$$y(t, T_i) = \left(1 - \frac{T_i - T_1}{T_2 - T_1}\right) y(t, T_1) + \frac{T_i - T_1}{T_2 - T_1} y(t, T_2) \qquad (2.97)$$

$$\Leftrightarrow \quad \ln B(t, T_1) = \left(1 - \frac{T_i - T_1}{T_2 - T_1}\right) T_i \frac{\ln B(t, T_1)}{T_1} + \frac{T_i - T_1}{T_2 - T_1} T_i \frac{\ln B(t, T_2)}{T_2}$$

which leads to an instantaneous forward rate linear in T_i:

$$- \left. \frac{\partial \ln B(t, u)}{\partial u} \right|_{u = T_i} = - \frac{\ln B(t, T_1)}{T_1} + \frac{2T_i - T_1}{T_2 - T_1} \left(\frac{\ln B(t, T_1)}{T_1} - \frac{\ln B(t, T_2)}{T_2}\right).$$

The above results are demonstrated in Figure 2.3. The zero coupon bond prices for the maturities 1, 2, 3, 4 and 5 were set exogenously. Both methods lead to discontinuous forward rates at the interval boundaries, a fact which is difficult to reconcile with most term structure models.

More sophisticated methods for term structure interpolation were suggested by McCulloch (1971, 1975), using polynomial splines, and by Vasicek and Fong (1982), using exponential splines. The former approach has its origins in the Weierstrass Theorem, which states that a continuous function can be approximated to arbitrary precision over some interval by a polynomial defined over the same interval. However, using high degree polynomials results in approximations that may fluctuate wildly over a large range. This can be remedied by using approximating functions which are piecewise defined as low degree polynomials, the piecewise defined functions being called *splines*[37]. The spline approach can be likewise used to attain a perfect fit to swap yields or to estimate a term structure from a larger data set of zero coupon bond prices, the choice between perfect fit and least squares estimation depending on the number of data points available as compared to the number of free parameters. McCulloch (1971) used a quadratic spline function to estimate zero bond prices from U.S. treasury and corporate bond issues, while in McCulloch (1975) he employed a cubic spline constructed in such a manner that the spline ordinates, first and second derivatives are equal at the subsection boundaries. Thus forward rates are continuously differentiable.

As Shea (1985) notes, the main drawback of using polynomial splines to approximate the term structure of interest rates is that they "commonly yield estimates of forward interest rate structures that are unstable, fluctuate widely, and often drift off to very large, even negative, values."[38] In contrast, Vasicek and Fong (1982) assert that their interpolation method using piecewise exponential functions "has desirable asymptotic properties for long maturities, and exhibits both a sufficient flexibility to fit a wide variety of shapes of the term structure, and a sufficient robustness to produce stable forward rate curves."[39] Their approach is motivated by the assumption that discount functions are principally exponential decays and thus exponential splines should give a better fit than polynomials would. However, as Shea (1985) observes, when constructing splines we are concerned with *local* polynomial approximations to some (possibly exponential) function, so Vasicek and Fong's argument for preferring exponential over polynomial splines does not carry as much weight as it might appear, particularly in light of Shea's result that exponential splines exhibit the same instabilities as the polynomials do. This leads Shea (1985) to recommend the use of polynomials, for reasons of computational simplicity.

Adams and van Deventer (1994) focus on the requirement that in addition to giving a good fit to the observed data, an interpolation method should result in a forward rate curve that is as smooth as possible. They thus extend McCulloch's, respectively Vasicek/Fong's, spline approach by demanding maximum smoothness, formalised by the mathematically rigorous criterion of minimising

$$\int_0^T (f''(0,s))^2 ds$$

where the double prime denotes the second derivative of the instantaneous forward rate with respect to maturity. This is a common mathematical definition of smoothness used in engineering applications.[40] The following theorem,

---

[37] For an analysis of spline approximations of the term structure of interest rates, see Shea (1984).

[38] See Shea (1985), p. 320. He gives examples supporting this statement in Shea (1984).

[39] Vasicek and Fong (1982), p. 340.

[40] Adams and van Deventer (1994) note that the solution of the smoothest term structure problem depends on the particular smoothness criterion employed. They also give an alternative measure of smoothness which is invariant under rotation of the forward rate curve.

attributed to Oldrich Vasicek, gives the maximum smoothness term structure of all functional forms interpolating a set of observed zero coupon bond prices.

Theorem 2.12 *(Vasicek) The term structure $f(0,t), 0 \le t \le T$, of instantaneous forward rates that satisfies the maximum smoothness criterion*

$$\min \int_0^T (f''(0,s))^2 ds$$

while fitting the observed prices $B(0,t_1), B(0,t_2), \cdots, B(0,t_m)$ of zero coupon bonds with maturities $t_1, t_2, \cdots, t_m$ is a fourth–order spline with the cubic term absent given by

$$f(0,t) = c_i t^4 + b_i t + a_i \qquad for \quad t_{i-1} < t \le t_i, \quad with \quad i = 1, 2, \cdots, m+1$$

$$where \quad 0 = t_0 < t_1 < \cdots < t_m < t_{m+1} = T.$$

The coefficients $a_i, b_i, c_i, i = 1, 2, \cdots, m+1$ satisfy the equations

$$c_i t_i^4 + b_i t_i + a_i = c_{i+1} t_i^4 + b_{i+1} t_i + a_{i+1}, \quad i = 1, 2, \cdots, m$$

$$4 c_i t_i^3 + b_i = 4 c_{i+1} t_i^3 + b_{i+1}, \quad i = 1, 2, \cdots, m$$

$$\frac{1}{5} c_i (t_i^5 - t_{i-1}^5) + \frac{1}{2} b_i (t_i^2 - t_{i-1}^2) + a_i (t_i - t_{i-1}) = -\log\left(\frac{B(0,t_i)}{B(0,t_{i-1})}\right),$$

$$i = 1, 2, \cdots, m \quad (2.98)$$

and

$$c_{m+1} = 0$$

PROOF: See Adams and van Deventer (1994). $\qquad\qquad\qquad\qquad\qquad$ □

If the additional requirement is imposed that the derivative of the forward rate curve is zero at the time horizon

$$f'(0,T) = 0 \qquad \Leftrightarrow \qquad b_{m+1} = 0,$$

which is reasonable in the light of conventional wisdom about the asymptotic behaviour of the term structure, and the short rate is fixed to some known value, yielding

$$a_1 = r(0),$$

then the coefficients of the interpolating function are uniquely determined. Otherwise one can use the first order conditions resulting from the minimisation problem to cover the remaining degrees of freedom.[41] In any case, given zero coupon bond prices the coefficients are calculated by solving a system of linear equations.

As noted previously, however, market prices for zero coupon bonds are usually not available. To apply the interpolation method given swap yields $\ell(0,t_i)$,

---

[41] See Adams and van Deventer (1994), pp. 55–56.

we must replace the conditions (2.98) with[42]

$$1 = \exp\left\{-\int_0^{t_i} f(0,s)ds\right\} + \sum_{j=1}^{i} \ell(0,t_i) \cdot \exp\left\{-\int_0^{t_j} f(0,s)ds\right\},$$

$$i = 1, 2, \cdots, m \quad (2.99)$$

In this case determining the coefficients involves a fixed point problem which has to be solved numerically.

2.6.3 Implementing interest rate term structures

In the previous sections we have seen that the most basic representation of an interest rate term structure is as a set of zero coupon bond prices and the main implementation decision is how one interpolates between zero coupon bonds. All operations one would expect to be able to perform with an object representing a term structure of interest rates can be implemented given the zero coupon bonds and an interpolation scheme. Thus one can define an abstract base class `TermStructure`,[43] which determines the interface (i.e. the public member functions) common to all term structures and which implements everything that can be implemented without explicit knowledge of the interpolation scheme. The interpolation schemes themselves are implemented in the derived classes. `TermStructure` only contains one pure virtual function,

```
virtual double operator()(double t) const = 0;        (2.100)
```

which returns the (possibly interpolated) zero coupon bond price for the maturity t.

The term structure is represented internally by two protected member variables

```
Array <double,1> T;
Array <double,1> B;                                   (2.101)
```

representing a set of maturities and the associated zero coupon bond prices (normalised to a unit notional). Note that $B(0)$ is always set to 1, so if $T(0)$ corresponds to a date later than "today," the prices in the array B are actually time $T(0)$ forward zero coupon bond prices. Most of the member functions are straightforward and self-explanatory. For example,

```
double swap(const Array <double,1>& tenor) const;     (2.102)
```

calculates a (forward) swap rate from the (possibly interpolated) zero coupon bond prices by equation (2.95) in Section 2.6.1, where `tenor` contains the times (decimals in years) $t_0, t_1, \ldots, t_n$ defining the tenor of the swap (i.e. the swap pays in arrears at times t_i, $1 \le i \le n$, for accrual periods $[t_{i-1}, t_i]$). Listing 2.13 lists the member functions for accessing interest rates and discount factors given by a `TermStructure`.

---

[42] See Proposition 2.10.
[43] See `TermStructure.hpp`.

```
class TermStructure : public QFNestedVisitable {
public:
  /** The maturity t, time T(0) forward zero coupon bond price. */
  virtual double operator()(double t) const = 0;
  /** Continuously compounded forward yields for the accrual periods
      defined by the Array of dates xT. */
  Array<double,1> ccfwd(Array<double,1> xT) const;
  /// Get all delta-compounded forward rates for tenor structure xT.
  Array<double,1> simple_rate(const Array<double,1>& xT,double delta)
     const;
  /** Get the delta-compounded forward rate, observed today,
      for the forward date xT. */
  double simple_rate(double xT,double delta) const;
  /** Get all co-terminal delta-compounded forward swap rates */
  Array<double,1> swaps(const Array<double,1>& xT,double delta)
     const;
  /// Get forward swap rate, valid today.
  double swap(double T0,    ///< Start date of the forward swap.
              int n,        ///< Length of the swap in delta periods.
              double delta ///< Length of time between two payments.
              ) const;
  /// Get forward swap rate, valid today, for a given tenor.
  double swap(const Array<double,1>& tenor ///< Swap tenor structure
              ) const;
};
```

Listing 2.13: Member functions for accessing information in a `TermStructure`

The simplest concrete class derived from **TermStructure** is the class **FlatTermStructure**, with constructor

$$\text{FlatTermStructure(double lvl,}$$
$$\text{double tzero,} \qquad\qquad (2.103)$$
$$\text{double horizon);}$$

where `lvl` is the flat level of continuously compounded interest rates, `tzero` is the start and `horizon` is the end of the maturity time line. This class is mainly used for testing purposes. More practically useful term structure classes are classified by the way they can be fitted to the market, which is discussed in the next section.

2.6.4 Fitting the term structure to the market

Given a term structure of interest rates, the present value of any deterministic cashflow (i.e. a set of known future payments at known future dates) can be calculated. Conversely, one can fit a term structure of interest rates to a given set of market values of deterministic cashflows. This can always be done approximately (e.g. in a least–squares sense), and also exactly if there is no redundancy in the given set of cashflows, i.e. if none of the cashflows can be represented as a linear combination of the others.

Deterministic cashflows are implemented in the class `DeterministicCash-flow`. Its constructor is

```
DeterministicCashflow(const Array<double,1>& timeline,
                      const Array<double,1>& cashflow,
                      double marketval)
```

where `timeline` is an array of payment times and `cashflow` contains the corresponding payment amounts, while `marketval` is a given market value of the cashflow. Thus to create an instance of `DeterministicCashflow` representing a three–year coupon bond paying an annual coupon of 4% on a notional of 1, with market value at par, set

$$
\begin{aligned}
&\texttt{Array<double,1> timeline(3),cashflow(3);} \\
&\texttt{timeline = 1.0, 2.0, 3.0;} \\
&\texttt{cashflow = 0.04, 0.04, 1.04;} \\
&\texttt{DeterministicCashflow couponbond(timeline,cashflow,1.0);}
\end{aligned} \tag{2.104}
$$

`DeterministicCashflow` provides the member functions

$$
\begin{aligned}
&\texttt{double market\_value() const;} \\
&\texttt{void set\_market\_value(double v);}
\end{aligned} \tag{2.105}
$$

respectively to read and to modify the market value of the cashflow, and the member function

$$
\texttt{double NPV(const TermStructure\& ts) const;} \tag{2.106}
$$

to calculate the present value of a cashflow given the term structure `ts`. Thus the following sequence of statements sets the market value of the coupon bond to a price consistent with a term structure that is flat at 6% continuously compounded:

$$
\begin{aligned}
&\texttt{FlatTermStructure ts(0.06,0.0,10.0);} \\
&\texttt{couponbond.set\_market\_value(couponbond.NPV(ts));}
\end{aligned} \tag{2.107}
$$

Note that an approximate fit of the term structure to a given set of market values of deterministic cashflows (e.g. coupon bond prices or market rates) can be implemented at the level of the class `TermStructure` in a way that is very general, though not necessarily very efficient. This is done in the member function

$$
\begin{aligned}
&\texttt{virtual void} \\
&\quad\texttt{approximate\_fit(std::vector<DeterministicCashflow>} \\
&\qquad\qquad\qquad\texttt{cashflows,} \\
&\qquad\qquad\texttt{double eps = 1E-09);}
\end{aligned} \tag{2.108}
$$

implemented in the file `TermStructure.cpp`. The approximate fit is achieved by minimising the sum of the squared differences between the market values and the net present values of the cashflows, with the latter being calculated using the `TermStructure` for discounting. This objective function is implemented as the private nested class `TermStructure::fit_func`, which is then fed into the `GeneralPowell` multidimensional optimisation template. This method is extremely flexible, fitting any number of cashflows and achieving an exact fit when this is possible (or one of the possible exact fits when the

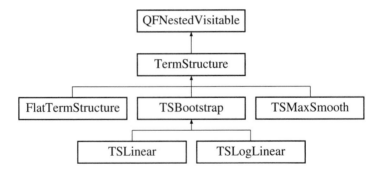

Figure 2.4 *The TermStructure class hierarchy*

exact fit is not unique). Consequently, there was no need to override this virtual member function in any of the derived classes in the class hierarchy in Figure 2.4.

However, given specific knowledge of the interpolation scheme, efficient calibration methods are available, especially when an exact fit is possible. For basic interpolation schemes,[44] which calculate $B(t, T)$ for $T_i < T < T_{i+1}$ using only $B(t, T_i)$ and $B(t, T_{i+1})$, an exact fit can be achieved by a bootstrap technique, fitting first the shortest zero coupon bond $B(t, T_1)$ to the market value of the shortest cashflow, then fitting $B(t, T_2)$ (given $B(t, T_1)$) to the second shortest cashflow, and so on. This requires only a one-dimensional root search in each step: In the i-th step, all future payments in the i-th cashflow that occur at or before T_{i-1} are discounted based on the (now given) discount factors $B(t, T_0) = 1$, $B(t, T_1)$, ..., $B(t, T_{i-1})$ and the interpolated values between them. The task, therefore, is to find a $B(t, T_i)$ such that the payments in the i-th cashflow, which occur between T_{i-1} and T_i, when discounted by $B(t, T_i)$ and any required interpolated discount factors, bring the total net present value of the cashflow to equal its market value. This is implemented as the member function

```
void bootstrap(std::vector<DeterministicCashflow>
                          cashflows,                        (2.109)
               double eps = 1E-12);
```

in `TSBootstrap`, the abstract base class of all term structures to which this method of fitting is applicable.

In the case of maximum smoothness interpolation, implemented in `TSMax-Smooth`, getting an exact fit is somewhat more involved. Fitting the spline coefficients to zero coupon bond prices in the constructor

---

[44] For example, log-linear interpolation of zero coupon bond prices (2.96) or linear interpolation of continuously compounded yields (2.97).

```
TSMaxSmooth(const Array<double,1>& xT, ///< Time line.
            const Array<double,1>& xB, ///< ZCB prices.
            double xrzero ///< Initial short rate.
           );
```

requires the solution of the system of linear equations given by Theorem 2.12, which one can implement using CLAPACK. For an exact fit to the market values of a set of more arbitrary cashflows, one must solve a system of non-linear equations. To derive these equations, note that after setting $b_{m+1} = 0$ and $c_{m+1} = 0$ in Theorem 2.12, one can eliminate the variables b_i, c_i, $i = 1, 2, \ldots, m + 1$ by expressing these in terms of the a_i, $i = 1, 2, \ldots, m + 1$. Furthermore, set $a_1 = r(0)$, the initial continuously compounded short rate. Each zero coupon bond price, and consequently the net present value of any cashflow, can then be calculated as a non-linear function of the variables a_i, $i = 2, \ldots, m + 1$. Given the market values of m different cashflows, the a_i can be found using the NewtonRaphson template.

Note that in each case where a TermStructure is exactly fitted to the market values of a set of cashflows, the time line of maturities of the zero coupon bonds making up the internal representation of the term structure has to be constructed from the end dates of the cashflows, i.e. if the cashflows are sorted from shortest to longest, then T(1) will be the last payment date of the first cashflow, and so on. (T(0) will be the time representing "today," thus typically $T(0) = 0$.)

## CHAPTER 3

Lattice models for option pricing

Lattice methods are among the most common and long–established methods for the pricing of derivative financial instruments, going back to the seminal work of Cox, Ross and Rubinstein (1979), who introduced the binomial lattice model[1] for option pricing. Almost all fundamental results of pricing by arbitrage arguments can be illustrated in this model, and lattice models are also effective numerical methods in their own right.

In this chapter, two broad classes of models will be considered: binomial models for a single underlying and models for the term structure of interest rates. The pricing of standard and exotic derivatives in these models is discussed. The common concepts linking all lattice models will be reflected in the C++ implementation, and many of these concepts will be picked up again in the finite difference schemes presented in Chapter 5.

3.1 Basic concepts of pricing by arbitrage

3.1.1 A simple example

Consider Figure 3.1. Suppose the Australian dollar is trading at 50 US cents today. Suppose the one-year interest rates are 4% and 5% in Australia and the US, respectively. Suppose we know that one year from now, one US dollar will be worth either AUD 2.20 or AUD 1.80, with either outcome equally likely.

What is the value today of an option to buy 100 USD for 200 AUD in one year's time? This question can be answered by a simple hedging argument. By taking positions ϕ_1 in AUD and ϕ_2 in USD at time t_0, we can replicate the option payoff at time t_1, choosing ϕ_1 and ϕ_2 in such a way that the value of the position at time t_1 matches the option payoff irrespective of whether the exchange rate moved up or down. Thus we require:

$$1.04\phi_1 + 2.31\phi_2 = 20$$
$$1.04\phi_1 + 1.89\phi_2 = 0$$

$$\Rightarrow \quad \phi_1 = -\frac{4500}{52} \quad \phi_2 = \frac{1000}{21}$$

I.e. borrow ϕ_1 AUD at 4% and invest ϕ_2 USD at 5%.

---

[1] Models of this type are often, somewhat misleadingly, called *binomial tree* models.

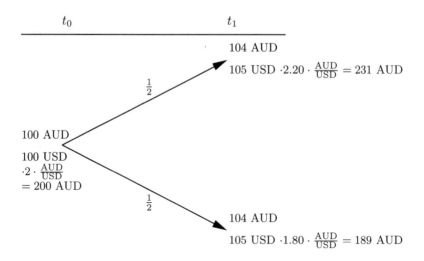

Figure 3.1 *The Australian perspective*

The initial value of the replicating portfolio is

$$\phi_1 + 2\phi_2 = 8.69963$$

This must also be the price of the option, because otherwise there is an *arbitrage opportunity*, i.e. an opportunity to make a riskless profit:

- If the option is trading at a value higher than this, we can make a riskless profit by selling the option and going long the replicating portfolio.
- If the option is trading at a value lower than this, we can make a riskless profit by buying the option and going short the replicating portfolio.

Note that any time t_1 payoff can be priced in this manner. In particular, we can define

Definition 3.1 *A* state price *is the value today of one dollar paid in one state of the world (only) tomorrow. An asset with such a payoff is also called an* Arrow/Debreu security, *after Arrow (1964) and Debreu (1959).*

Since the option pays 20 dollars in the "up" state and nothing in the "down" state and is worth 8.69963, the state price π_{up} of the "up" state must be

$$\pi_{\text{up}} = \frac{8.69963}{20} = 0.434982$$

Alternatively, one could derive this value by constructing a portfolio paying $1 in the "up" state and nothing in the "down" state.

To calculate the state price for the "down" state, consider the following. If we have one dollar in the "up" state and one dollar in the "down" state, we have one dollar for sure at time t_1. The value of this at time t_0 must be equal

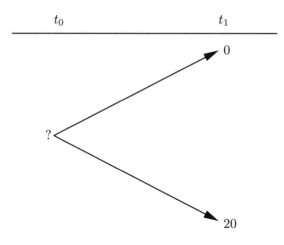

Figure 3.2 *A put option payoff*

to the discount factor:

$$\pi_{\text{up}} + \pi_{\text{down}} = \frac{1}{1.04}$$
$$\Rightarrow \quad \pi_{\text{down}} = 0.526557$$

Alternatively, one could derive this value by constructing a portfolio paying $1 in the "down" state and nothing in the "up" state.

Given the state prices, we can easily value other contingent payoffs. For example, what is the value of an option to sell 100 USD for 200 AUD at time t_1 (Figure 3.2)? Using the previously calculated state prices, we get

$$0 \cdot \pi_{\text{up}} + 20 \cdot \pi_{\text{down}} = 10.5311$$

Because the payoff of an Arrow/Debreu security is positive in one state and zero in all others, state prices must always be positive in order to avoid arbitrage. Furthermore, we have already noted that their sum is equal to the discount factor. If we *normalise state prices* by dividing by the discount factor,

$$p_{\text{up}} = \frac{\pi_{\text{up}}}{\pi_{\text{up}} + \pi_{\text{down}}} = 0.452381 \qquad p_{\text{down}} = \frac{\pi_{\text{down}}}{\pi_{\text{up}} + \pi_{\text{down}}} = 0.547619$$

the value of the call option can be calculated as

$$8.69963 = \frac{1}{1.04}(p_{\text{up}} \cdot 20 + p_{\text{down}} \cdot 0)$$

Normalising the state prices thus results in numbers, which can be interpreted as probabilities (they lie between zero and one and they sum to one), and the time t_0 value of any time t_1 payoff must equal the *discounted expected payoff* under these probabilities.

However, this relationship does not only apply to derivatives; rather, the

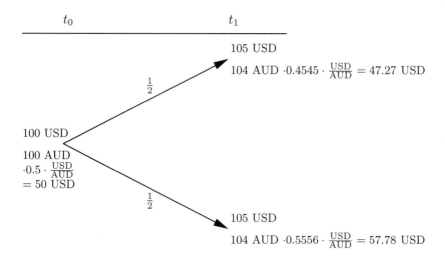

Figure 3.3 *The American perspective*

current value of any asset must equal its discounted expected payoff. For example, the current value of 100 USD is

$$200 = \frac{1}{1.04}(p_{\text{up}} \cdot 231 + p_{\text{down}} \cdot 189)$$

Valuing an asset by its discounted expected payoff is considered *risk neutral* in the sense of utility theory.[2] This leads to the following

Definition 3.2 *Normalised state prices are called* risk neutral probabilities.

It is important to note that the risk neutral probabilities are determined by market prices, *not* by the "real world" probabilities by which events occur. They are "risk neutral probabilities" only in the sense that these are the probabilities by which a risk neutral investor (if he or she did exist) would have to calculate expectations in order to arrive at the arbitrage–free price. However, the arbitrage–free price does not need risk neutrality to justify it; the argument of a riskless profit is much more fundamental and compelling. In particular, recall that neither the "real world" probabilities nor investors' preferences entered into the calculation of the option price.

Let us now consider the previous example from the American perspective. Then, the call option becomes a put, with payoffs expressed in US dollars as

---

[2] In contrast, a *risk averse* investor would value a risky payoff at less than its discounted expected value. Note, however, that the concepts of risk neutrality and risk aversion refer to expectations calculated under either some subjective probabilities assumed by the investor, or the objective, "real world" probabilities under which events occur — not the artificial probabilities derived by normalising state prices.

in Figure 3.3. So what is the value in t_0 of an option to sell 200 AUD for 100 USD at t_1?

As before, by taking positions ϕ_1 in AUD and ϕ_2 in USD, we can replicate the option.

$$0.4727\phi_1 + 1.05\phi_2 = 100 - 200 \cdot 0.4545 = 9.09$$
$$0.5778\phi_1 + 1.05\phi_2 = 0$$
$$\Rightarrow \quad \phi_1 = -\frac{4500}{52} \quad \phi_2 = \frac{1000}{21}$$

The positions in AUD and USD are the same as when replicating the option from the Australian perspective. The initial value of the replicating portfolio is

$$0.5\phi_1 + \phi_2 = 4.34982 \text{ USD} = 8.69963 \text{ AUD}$$

Thus, unsurprisingly, the value of the option is the same no matter which perspective we take.

However, the USD state prices are

$$\tilde{\pi}_{\text{up}} = \frac{4.34982}{9.09} = 0.478480$$
$$\tilde{\pi}_{\text{down}} = \frac{1}{1.05} - \tilde{\pi}_{\text{up}} = 0.473901$$

and the risk neutral probabilities are

$$\tilde{p}_{\text{up}} = 1.05 \cdot \tilde{\pi}_{\text{up}} = 0.502404 \qquad \tilde{p}_{\text{down}} = 1.05 \cdot \tilde{\pi}_{\text{down}} = 0.497596$$

Thus the USD risk neutral probabilities differ from the AUD risk neutral probabilities. This is because the risk neutral probabilities are associated with the asset in terms of which we measure value: For AUD, that asset is the AUD money market account, worth 1 AUD at t_0 and 1.04 AUD at t_1, or 0.50 USD at t_0 and either 0.4727 USD or 0.5778 USD at t_1. For USD, that asset is the USD money market account, worth 1 USD at t_0 and 1.05 USD at t_1, or 2 AUD at t_0 and either 2.31 AUD or 1.89 AUD at t_1.

The asset in which we choose to measure value is called the *numeraire*. Under the risk neutral probabilities associated with a particular numeraire, the current price of all assets is given by the expected value of the respective payoffs, discounted by the numeraire asset. Any asset with strictly positive payoffs is a valid numeraire. Typically, however, when speaking of risk neutral probabilities without explicit mention of the numeraire, it is assumed that the numeraire is the savings account in domestic currency.

3.1.2 The general single period case

Suppose now that there are N possible states of the world at time t_1. Furthermore, let there be M different assets available in the market and denote by S_{ij} the time t_1 price of the i-th asset in state j, and by S_i the time t_0 price of the i-th asset. Denote by $\phi = \{\phi^{(1)}, \ldots, \phi^{(M)}\}$ a *portfolio* of these assets, where $\phi^{(i)}$ is the number of units of asset i in the portfolio. The time t_0 value of the portfolio is then given by

$$V_0(\phi) = \sum_{i=1}^{M} \phi^{(i)} S_i$$

and its time t_1 value $V_j(\phi)$ in state j is

$$V_j(\phi) = \sum_{i=1}^{M} \phi^{(i)} S_{ij}$$

In this setting we can state

Definition 3.3 *An* arbitrage opportunity *is a portfolio* $\phi \in \mathbb{R}^M$, *such that*

$$V_0(\phi) < 0 \qquad \text{and, for all states } j \qquad V_j(\phi) \geq 0 \qquad (3.1)$$

(3.1) is sometimes called *strong arbitrage*, whereas *weak arbitrage* would be a portfolio ϕ, such that

$$V_0(\phi) \leq 0 \qquad \text{and, for all states } j \qquad V_j(\phi) \geq 0 \qquad (3.2)$$

with $V_j(\phi) > 0$ for at least one state j.

In the following, without loss of generality, we will take S_1 to be the numeraire.

Theorem 3.4 *The absence of arbitrage is equivalent to the existence of a risk neutral probability measure.*

PROOF: To show that the existence of a risk neutral probability measure implies the absence of arbitrage, let p_j denote the risk neutral probability of the state j. Then the time t_0 value of any portfolio ϕ is given by its expected discounted payoff, i.e.

$$\frac{V_0(\phi)}{S_1} = \sum_{j=1}^{N} p_j \sum_{i=1}^{M} \phi^{(i)} \frac{S_{ij}}{S_{1j}} = \sum_{j=1}^{N} p_j \frac{V_j(\phi)}{S_{1j}}$$

If $V_j(\phi) \geq 0$ for all j, this implies $V_0(\phi) \geq 0$, because $p_j \geq 0$ for all j. Thus there is no strong arbitrage opportunity. If furthermore all states have positive probability $(p_j > 0)$, then there is no weak arbitrage opportunity.

To show that the absence of arbitrage implies the existence of a risk neutral probability measure, we need the following

Lemma 3.5 (Minkowski/Farkas–Lemma) *Let S be a real $M \times N$ matrix and $\mathcal{A} = \{\phi \in \mathbb{R}^M | S^\top \phi \geq 0\}$. Then there exists a vector $\lambda \in \mathbb{R}^M$ with*

$$\lambda^\top \phi \geq 0 \quad \forall \, \phi \in \mathcal{A}, \qquad (3.3)$$

if and only if there exists a non-negative vector $\pi \in I\!\!R^N$ with

$$\lambda^\top = \pi^\top S^\top. \tag{3.4}$$

In the current setting of Theorem 3.4, S is the matrix of the future payoffs of the assets, i.e.

$$S = \begin{pmatrix} S_{11} & S_{12} & \cdots & S_{1N} \\ S_{21} & S_{22} & \cdots & S_{2N} \\ \vdots & \vdots & \ddots & \vdots \\ S_{M1} & S_{M2} & \cdots & S_{MN} \end{pmatrix}$$

and thus $\mathcal{A}$ can be interpreted as the set of all portfolios with non-negative payoffs. No arbitrage then requires that there is price vector λ (where $\lambda_i = S_i$ is the time t_0 price of the i–th asset) with

$$\lambda^\top \phi \geq 0 \quad \forall \, \phi \in \mathcal{A},$$

Lemma 3.5 then implies the existence of a non-negative vector π satisfying (3.4), which can be interpreted as a vector of state prices. Normalising state prices as in Section 3.1.1, we get a risk neutral probability measure, i.e.

$$p_j = \frac{\pi_j}{\sum_{j=1}^N \pi_j}$$

$\square$

Definition 3.6
- *A vector of payoffs $Z = \{Z_1, \ldots, Z_N\}$ is called* attainable, *if there exists a portfolio ϕ, such that for all states j, $V_j(\phi) = Z_j$.*
- *This portfolio is called the* replicating portfolio *(or hedge portfolio) for the payoff Z.*
- *If the payoff of the i–th asset can be replicated by a portfolio of the remaining assets, then the i–th asset is called* redundant.
- *If all possible payoffs $Z \in I\!\!R^N$ are attainable, the market is* complete.

Clearly, for a market to be complete, we must have $M \geq N$. If there are no redundant securities, then $M \leq N$.

Example 3.7 *An arbitrage opportunity?* Consider the three price processes in Figure 3.4. Note that the first asset is obviously riskless with an interest rate of zero, thus in this case the state prices are identical to the risk neutral probabilities. Any two of the three assets complete the market, and we could use any two assets to derive the state prices. Using the first and second asset, we have

$$\begin{aligned} 100\phi_1 + 75\phi_2 &= 1 \\ 100\phi_1 + 25\phi_2 &= 0 \end{aligned}$$

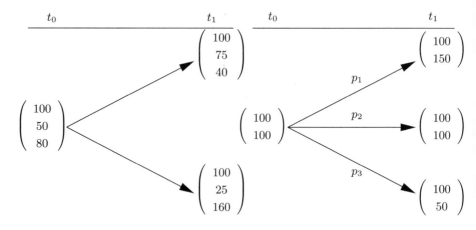

Figure 3.4 *An arbitrage opportunity?*   Figure 3.5 *An incomplete market*

which yields

$$\Rightarrow \qquad \phi_1 = -0.005 \qquad \phi_2 = 0.02$$
$$\Rightarrow \qquad p_{\text{up}} = \pi_{\text{up}} = 100\phi_1 + 50\phi_2 = 0.5$$
$$\Rightarrow \qquad p_{\text{down}} = 0.5$$

However, the current value of the third asset does not equal its discounted expected future value:

$$80 \neq 40p_{\text{up}} + 160p_{\text{down}} = 100$$

Thus we can make a riskless profit by *buying* the third asset and going *short* the replicating portfolio, which we can calculate by

$$100\phi_1 + 75\phi_2 = 40$$
$$100\phi_1 + 25\phi_2 = 160$$

resulting in

$$\Rightarrow \qquad \phi_1 = 2.2 \qquad \phi_2 = -2.4$$

It is easily verified that if we buy one unit of the third asset, sell 2.2 units of the first asset and buy 2.4 units of the second, all future payoffs cancel and we make an immediate arbitrage profit of 20.

Note that in this example, no risk neutral probability measure exists, as any probability measure derived using two of the three asset price processes will be inconsistent with the price process of the remaining asset.

Example 3.8 *An incomplete market.* Consider the two price processes in Figure 3.5. Trying to price by replication a call option with strike 80 on the second asset leads to the system of equations

$$100\phi_1 + 150\phi_2 = 70$$
$$100\phi_1 + 100\phi_2 = 20$$
$$100\phi_1 + 50\phi_2 = 0$$

which cannot be solved, thus the option cannot be replicated and the market is incomplete.

A valid set of risk neutral probabilities (taking the first asset as numeraire) would be

$$p_1 = \frac{1}{3} \qquad p_2 = \frac{1}{3} \qquad p_2 = \frac{1}{3}$$

as this ensures that the time t_0 value of all assets equals their discounted expected payoff. However, any set of p_1, p_2, p_3 satisfying

$$100 = 100(p_1 + p_2 + p_3) \qquad (3.5)$$
$$100 = 150p_1 + 100p_2 + 50p_3 \qquad (3.6)$$

would also constitute a valid risk neutral probability measure, e.g. $p_1 = \frac{1}{4}, p_2 = \frac{1}{2}, p_3 = \frac{1}{4}$. In this case, there is an infinite number of risk neutral probability measures and a range of arbitrage–free prices for the option — *the risk neutral probabilities are unique only if the market is complete.*

To determine the arbitrage–free price bounds on the option, we solve the equations (3.5)–(3.6) in terms of one of the variables, say p_1:

$$p_3 = p_1 \qquad p_2 = 1 - 2p_1$$

Then the price of the option must satisfy

$$C = 70p_1 + 20p_2 + 0p_3$$
$$= 70p_1 + 20(1 - 2p_1) = 20 + 30p_1 \qquad (3.7)$$

Thus we have the minimum arbitrage–free price of 20 for $p_1 = 0$ and the maximum arbitrage–free price of 50 for $p_1 = 1$.

If the option price lies outside these bounds, there is an arbitrage opportunity. Suppose that the option price is 10, i.e. the option is too cheap. Consequently, we can make a riskless profit by buying the option and taking an offsetting position in the two original assets. To work out the arbitrage strategy, substitute $C = 10$ into (3.7), yielding $p_1 = p_3 = -1/3$. Since the risk neutral probabilities (and therefore the state prices) for states 1 and 3 are negative, any portfolio with positive payoffs on these states (and zero in state 2) represents an arbitrage opportunity. Denote by ϕ_3 the position in the option and solve

$$100\phi_1 + 150\phi_2 + 70\phi_3 = 1$$
$$100\phi_1 + 100\phi_2 + 20\phi_3 = 0$$
$$100\phi_1 + 50\phi_2 + 0\phi_3 = 1$$

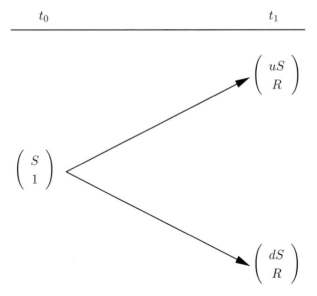

Figure 3.6 *A one–period binomial model*

resulting in
$$\phi_1 = \frac{1}{30} \qquad \phi_2 = -\frac{7}{150} \qquad \phi_3 = \frac{1}{15}$$
This portfolio has a non-negative future payoff and an initial price of
$$100\phi_1 + 100\phi_2 + 10\phi_3 = -\frac{2}{3}$$
i.e. setting up the portfolio yields an immediate risk–free profit of 0.67.

3.2 Hedging and arbitrage–free pricing in a binomial lattice model

3.2.1 The one–period case

Figure 3.6 shows a typical one–period binomial model. The risky asset has an initial value of S, which can move either upwards by a multiplicative factor of u or downwards by a factor of d. The riskless asset is represented by the accumulation factor R, where $R = 1 + r$ when r is the riskless interest rate with simple compounding.

Options are priced by the same hedging argument as in Section 3.1. Consider a derivative which pays f_{up} in the "up" state and f_{down} in the "down" state. The replicating portfolio (ϕ_1, ϕ_2) is then determined by the equations
$$\begin{aligned}
\phi_1 uS + \phi_2 R &= f_{\text{up}} \\
\phi_1 dS + \phi_2 R &= f_{\text{down}}
\end{aligned}$$

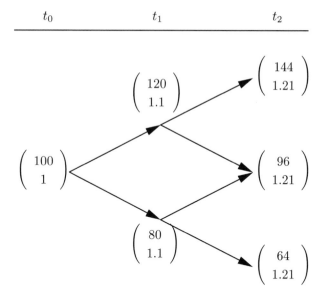

Figure 3.7 *A two–period example*

$$\Rightarrow \quad \phi_1 = \frac{f_{\text{up}} - f_{\text{down}}}{(u-d)S} \qquad \phi_2 = \frac{u f_{\text{down}} - d f_{\text{up}}}{(u-d)R}$$

and the price of the derivative equal the initial investment in the replicating portfolio, $\phi_1 S + \phi_2$.

State prices can be calculated by setting $f_{\text{up}} = 1$ and $f_{\text{down}} = 0$ to get π_{up}:

$$\pi_{\text{up}} = \frac{1}{(u-d)} - \frac{d}{(u-d)R} = \frac{R-d}{(u-d)R}$$

and similarly

$$\pi_{\text{down}} = \frac{u-R}{(u-d)R}$$

The risk neutral probabilities are therefore

$$p_{\text{up}} = \frac{\pi_{\text{up}}}{\pi_{\text{up}} + \pi_{\text{down}}} = R \cdot \pi_{\text{up}} = \frac{R-d}{u-d} \qquad (3.8)$$

$$p_{\text{down}} = R \cdot \pi_{\text{down}} = \frac{u-R}{u-d}$$

3.2.2 Binomial lattices

The one–period case is easily extended to multiple periods. Consider the two–period example in Figure 3.7. The model parameters are $S = 100$, $r = 10\%$, $u = 1.2$, and $d = 0.8$. Note that since $Sud = Sdu$, the lattice *recombines*.

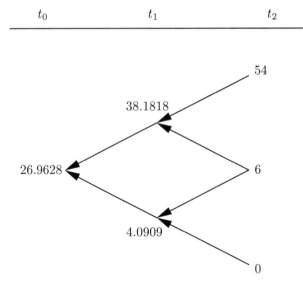

Figure 3.8 *Rolling a payoff back through the lattice*

Computationally, this is an important feature of the model, since it means that the number of nodes after the i-th period is $i + 1$, so the total number of nodes in an n–period lattice is $(n + 1)(n + 2)/2$. If the binomial paths did not recombine,[3] the number of nodes would grow exponentially, severely limiting the number of periods which can be computed in a reasonable amount of time.

To price an option to buy the underlying asset for a strike of 90 at the end of the second period, one can follow these steps to roll the discounted expected payoff backwards through the lattice as in Figure 3.8:

1. Calculate the payoffs at expiry: In the uppermost state after two periods, the payoff is $144 - 90 = 54$, in the middle state $96 - 90 = 6$ and in the lowest state the option expires out of the money, so the payoff is zero.

2. To calculate expectations, the risk neutral probabilities are needed:
$$p_{up} = \frac{R - d}{u - d} = \frac{1.1 - 0.8}{1.2 - 0.8} = \frac{3}{4}$$

3. We can treat the two–period lattice as consisting of three one–period lattices, one in the first period and two in the second. To roll the value of the option at the end of the second period (i.e. the option payoff at maturity) back to the beginning of the second period/end of the first period, we calculate discounted expected values. For the uppermost node at the end of

---

[3] In this case one could correctly speak of a *binomial tree*.

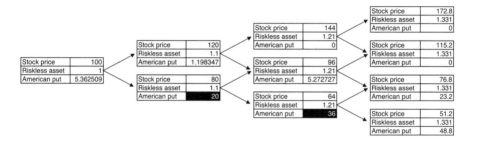

Figure 3.9 *Pricing an American put option in a binomial lattice*

the first period, we have

$$\frac{1}{1.1}\left(\frac{3}{4}\cdot 54 + \frac{1}{4}\cdot 6\right) = 38.1818$$

and for the lowermost node

$$\frac{1}{1.1}\left(\frac{3}{4}\cdot 6 + \frac{1}{4}\cdot 0\right) = 4.0909$$

4. Rolling the thus calculated option values back from the end of the first period to the beginning, we have

$$\frac{1}{1.1}\left(\frac{3}{4}\cdot 38.1818 + \frac{1}{4}\cdot 4.0909\right) = 26.9628$$

which is the option price that we wished to calculate.

Alternatively, one could calculate the option price directly via the discounted expected value:

$$\frac{1}{1.21}\left(54\cdot\left(\frac{3}{4}\right)^2 + 6\cdot 2\cdot\frac{3}{4}\cdot\frac{1}{4} + 0\cdot\left(\frac{1}{4}\right)^2\right) = 26.9628 \qquad (3.9)$$

where the middle terminal node (payoff of 6) is weighted with a factor of 2 because two paths (up–down or down–up) lead to this payoff.

The rollback procedure described above is easily generalised to an arbitrary number of periods, as is the direct valuation along the lines of (3.9). For the latter, see Section 3.2.3, below. The former is mainly used to value (typically path–dependent) options, for which no direct valuation formula is available, such as the American put.[4] Since the holder has a right to exercise an American put at any time during the life of the option, we must check whether it is optimal to exercise the option early. At each node the (rolled back) option price represents the value of holding on to the option (the *continuation* value).

---

[4] In the absence of dividends, the Merton (1973a) result holds for American call options, i.e. it is not rational to exercise an American call prematurely. Thus an American call has the same price as the corresponding European call.

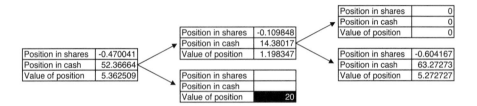

Position in shares	0
Position in cash	0
Value of position	0

Position in shares	-0.109848
Position in cash	14.38017
Value of position	1.198347

Position in shares	-0.470041
Position in cash	52.36664
Value of position	5.362509

Position in shares	0
Position in cash	0
Value of position	0

Position in shares	-0.604167
Position in cash	63.27273
Value of position	5.272727

Position in shares	
Position in cash	
Value of position	20

Figure 3.10 *Hedging an American put option in a binomial lattice*

If the payoff resulting from exercising the option at this node is greater than the continuation value, we must substitute this payoff for the rolled back option price before continuing to roll back through the lattice. Figure 3.9 gives an example of pricing an American put with exercise price 100 in a four–period lattice with the same model parameters as in the previous example. Highlighted in black are the two nodes in which the value of immediate exercise is greater than the continuation value.

Similarly, one can generalise the single–period binomial hedge to multiple periods. To this end, we state

Definition 3.9 *A trading strategy is called* self-financing *if, after it is set up initially, it does not require any inflow or outflow of funds from the hedging portfolio until the conclusion of the strategy.*

In the multiperiod binomial lattice, a self-financing dynamic hedging strategy can be constructed by considering the multiperiod case as a sequence of single–period binomial steps. At the beginning of each step, the positions in the underlying and the riskless asset must be chosen in such a way that at the end of the period the value of the hedge portfolio equals the value of the option regardless of whether an up–move or a down–move occurred. By constructing the option price lattice by the rollback procedure described above, one ensures that the value of the option at the end of a period equals the value of the hedge portfolio required for the next period. Thus at each node between the initial set–up of the hedge portfolio and the expiry (or possibly early exercise) of the option, the hedge portfolio is rebalanced by shifting funds between the position in the underlying and the riskless asset, but no additional funds are invested into or extracted from the portfolio.

Figure 3.10 illustrates the hedging strategy for the American put option in Figure 3.9, giving the rebalanced position in the underlying asset and in cash in each node of the lattice. The value of the hedge position matches the option price in each node, and the option payoff in the terminal nodes. Note that the hedging strategy terminates when the option is exercised early, and thus the node after two down moves of the stock price (with stock price equal to 64) does not figure in the strategy.

3.2.3 The binomial option pricing formula

Generalising the direct calculation of the discounted expected value in equation (3.9) to a binomial lattice with n time steps, we get the binomial option pricing formula of Cox, Ross and Rubinstein (1979) (CRR):

Proposition 3.10 *In a binomial lattice model with n time steps, in each of which the price of the underlying asset S can either move up by a factor of u or down by a factor of d, and in which the one–period accumulation factor is R, the value of a European call option with strike K expiring in n periods is given by*

$$C = \frac{1}{R^n} \sum_{i=a}^{n} \binom{n}{i} p^i (1-p)^{n-i} (u^i d^{n-i} S - K) \tag{3.10}$$

and that of a European put option by

$$P = \frac{1}{R^n} \sum_{i=0}^{a-1} \binom{n}{i} p^i (1-p)^{n-i} (K - u^i d^{n-i} S) \tag{3.11}$$

where a is the minimum number of "up" moves such that

$$u^a d^{n-a} S > K$$

The formulas (3.10) and (3.11) can be proven using the backward induction argument of the previous section. Note that

$$\binom{n}{i} = \frac{n!}{i!(n-i)!}$$

is the so-called *binomial coefficient*, giving the number of paths, which have exactly i "up" moves and $n-i$ "down" moves.

We can also take another view of the CRR formulas. Set

$$\tilde{p} = p \cdot \frac{u}{R}$$

Then

$$\begin{aligned}
1 - \tilde{p} &= 1 - \frac{R-d}{u-d} \cdot \frac{u}{R} = \frac{(u-d)R - (R-d)u}{(u-d)R} \\
&= \frac{(u-R)d}{(u-d)R} = (1-p) \cdot \frac{d}{R}
\end{aligned}$$

and we can write (3.10) as

$$C = S \sum_{i=a}^{n} \binom{n}{i} \tilde{p}^i (1-\tilde{p})^{n-i} - \frac{K}{R^n} \sum_{i=a}^{n} \binom{n}{i} p^i (1-p)^{n-i} \tag{3.12}$$

It can be shown that $\tilde{p}, (1 - \tilde{p})$ are the risk–neutral probabilities associated with taking S as the numeraire. Thus (3.12) represents the call option price as the difference between

- the present value of the underlying asset S times the risk neutral probability (using S as the numeraire) that the option will be exercised and

- the present value of the strike price times the risk neutral probability (using the riskless asset as the numeraire) that the option will be exercised.

Remark 3.11 For a fixed time horizon T, Cox, Ross and Rubinstein (1979) set the period length $\Delta t = T/n$ and

$$u = e^{\sigma\sqrt{\Delta t}} \qquad d = e^{-\sigma\sqrt{\Delta t}} \qquad R = e^{r\Delta t} \tag{3.13}$$

where σ is the per annum volatility of the underlying asset and r is the continuously compounded interest rate. If S_0 is the initial stock price, the stock price at time T is given by

$$u^i d^{n-i} S_0 = \exp\{\ln S_0 + i \ln u + (n-i)\ln d\} \tag{3.14}$$

where i is the (random) number of "up" moves in the binomial model. We have

$$i \ln u + (n-i)\ln d = 2i\sigma\sqrt{\Delta t} - n\sigma\sqrt{\Delta t} \tag{3.15}$$

and, using the properties of the binomial distribution,

$$E[2i\sigma\sqrt{\Delta t} - n\sigma\sqrt{\Delta t}] = 2\sigma\sqrt{T}\sqrt{np} - \sigma\sqrt{T}\sqrt{n} \tag{3.16}$$

$$\text{Var}[2i\sigma\sqrt{\Delta t}] = 4\sigma^2 Tp(1-p) \tag{3.17}$$

By the Central Limit Theorem,[5] as n tends to infinity, the distribution of $2i\sigma\sqrt{\Delta t}$ will approach a normal distribution with matching mean and variance, i.e.

$$\lim_{n\to\infty} E[2\sigma\sqrt{\Delta t}i - n\sigma\sqrt{\Delta t}]$$

$$= \lim_{n\to\infty} \sigma\sqrt{T}\sqrt{n}\frac{2(e^{r\frac{T}{n}} - e^{-\sigma\sqrt{\frac{T}{n}}}) - e^{\sigma\sqrt{\frac{T}{n}}} + e^{-\sigma\sqrt{\frac{T}{n}}}}{e^{\sigma\sqrt{\frac{T}{n}}} - e^{-\sigma\sqrt{\frac{T}{n}}}}$$

$$= \lim_{\Delta t\to 0} \sigma T\frac{2e^{r\Delta t} - e^{\sigma\sqrt{\Delta t}} - e^{-\sigma\sqrt{\Delta t}}}{\sqrt{\Delta t}(e^{\sigma\sqrt{\Delta t}} - e^{-\sigma\sqrt{\Delta t}})} \tag{3.18}$$

---

[5] See e.g. Kendall and Stuart (1969).

and using L'Hopital's Rule[6]

$$= \sigma T \lim_{\Delta t \to 0} \frac{2re^{r\Delta t} + (e^{-\sigma\sqrt{\Delta t}} - e^{\sigma\sqrt{\Delta t}})\frac{1}{2}\sigma\Delta t^{-\frac{1}{2}}}{\frac{1}{2}\Delta t^{-\frac{1}{2}}(e^{\sigma\sqrt{\Delta t}} - e^{-\sigma\sqrt{\Delta t}}) + \sqrt{\Delta t}(e^{\sigma\sqrt{\Delta t}} - e^{-\sigma\sqrt{\Delta t}})\frac{1}{2}\sigma\Delta t^{-\frac{1}{2}}}$$

(3.19)

Furthermore, again using L'Hopital's Rule, we can write

$$\lim_{\Delta t \to 0} \frac{e^{\sigma\sqrt{\Delta t}} - e^{-\sigma\sqrt{\Delta t}}}{\sqrt{\Delta t}} = \lim_{\Delta t \to 0} \frac{(e^{\sigma\sqrt{\Delta t}} + e^{-\sigma\sqrt{\Delta t}})\frac{1}{2}\sigma\Delta t^{-\frac{1}{2}}}{\frac{1}{2}\Delta t^{-\frac{1}{2}}} = 2\sigma \quad (3.20)$$

so using the properties of limits, (3.18) becomes

$$\lim_{n \to \infty} E[2i\sigma\sqrt{\Delta t} - n\sigma\sqrt{\Delta t}] = \sigma T \frac{2r - \sigma^2}{\sigma + \sigma} = rT - \frac{1}{2}\sigma^2 T \quad (3.21)$$

Similarly, under $\tilde{p}$

$$\lim_{n \to \infty} \tilde{E}[2i\sigma\sqrt{\Delta t} - n\sigma\sqrt{\Delta t}] = rT + \frac{1}{2}\sigma^2 T \quad (3.22)$$

and under p as well as $\tilde{p}$

$$\lim_{n \to \infty} \text{Var}[2i\sigma\sqrt{\Delta t}] = \sigma^2 T \quad (3.23)$$

Thus

$$\lim_{n \to \infty} \sum_{i=a}^{n} \binom{n}{i} \tilde{p}^i (1 - \tilde{p})^{n-i} = N(h_1) \quad (3.24)$$

$$\lim_{n \to \infty} \sum_{i=a}^{n} \binom{n}{i} p^i (1 - p)^{n-i} = N(h_2) \quad (3.25)$$

where $N(\cdot)$ is the cumulative distribution function of the standard normal distribution and

$$h_{1,2} = \frac{\ln \frac{S}{K} + (r \pm \frac{1}{2}\sigma^2)T}{\sigma\sqrt{T}} \quad (3.26)$$

So (3.12) becomes

$$C = SN(h_1) - Ke^{-rT}N(h_2) \quad (3.27)$$

Thus the limit of the Cox, Ross and Rubinstein (1979) option pricing formula is the well–known Black and Scholes (1973) formula. In fact, when one considers not just the terminal distribution, but the entire binomial price process, it

---

[6] For the present calculation, we use the version of L'Hopital's Rule which states that if f and g are differentiable functions on an open interval containing x_0, with $g'(x) \neq 0$ for all x in the interval, and if

$$\lim_{x \to x_0} f(x) = \lim_{x \to x_0} g(x) = 0 \quad \text{and} \quad \lim_{x \to x_0} \frac{f'(x)}{g'(x)} = c,$$

then

$$\lim_{x \to x_0} \frac{f(x)}{g(x)} = c.$$

can be shown that by the *Functional Central Limit Theorem*[7] the price process converges to the geometric Brownian motion underlying the Black/Scholes model discussed in Chapter 4. In this sense, the binomial lattice represents one possible numerical method to calculate, say, American put option prices in the Black/Scholes model.

3.2.4 Dividend–paying assets in the binomial model

Many real–life assets pay the holder an income stream over time; e.g. stocks pay dividends, instruments in a foreign currency earn interest and holding a commodity can incur a storage cost (generating a negative income). Stock indices are portfolios of the component stocks, and their dividend payments are typically aggregated into a *dividend yield* assumed to be paid continuously.

In order to maintain market completeness (permitting us to price options by arbitrage arguments), we assume that any cashflows generated by the underlying asset are deterministic. For this, there are three common modelling choices:

1. A dividend yield paid continuously (expressed in terms of either simple or continuous compounding).

2. Proportional dividends paid at discrete points in time.

3. Fixed dividends paid at discrete points in time.

It is relatively straightforward to deal with dividend yields in the existing model framework. Given a divided yield y_d, define

$$D = 1 + \Delta t y_d \qquad (3.28)$$

for simple compounding, or

$$D = e^{y_d \Delta t} \qquad (3.29)$$

for continuous compounding, for each period of length Δt. Then, if dividends are reinvested, the value of one share with price S at the beginning of a period becomes either uSD or dSD at the end of the period. The absence of arbitrage requires

$$S \quad = \quad \frac{1}{R}(puSD + (1-p)dSD) \qquad (3.30)$$

$$\Longleftrightarrow \quad p \quad = \quad \frac{\frac{R}{D} - d}{u - d} \qquad (3.31)$$

i.e. dividend yields can be incorporated into the existing binomial framework by simply adjusting the risk neutral probabilities.

Suppose now that instead of receiving a dividend each period, discrete dividends are received only at set points in time. Maintaining the absence of arbitrage in the model, with the risk–neutral probabilities unchanged, this means that the share price drops by the amount of the dividend per share

---

[7] Also called Donsker's Invariance Principle — see e.g. Theorem 4.20 of Karatzas and Shreve (1991).

when the stock goes ex–dividend. This means that a recombining lattice can only be constructed for the case of dividends proportional to the share price: Given a dividend δ_i at time t_i, consider

$$(S(t_{i-1})u)(1 - \delta_i)d = (S(t_{i-1})d)(1 - \delta_i)u \qquad (3.32)$$

i.e. the left hand side, representing an "up" move in the stock price from time t_{i-1} to t_i and a "down" move from t_i to t_{i+1}, is equal to the right hand side, representing a "down" move followed by an "up" move. Thus the lattice recombines. This property is lost if the dividend is given as a fixed absolute amount, i.e.

$$(S(t_{i-1})u - \delta_i)d \neq (S(t_{i-1})d - \delta_i)u \qquad (3.33)$$

This substantially increases the number of nodes required for a given refinement of the time line. Therefore it is best to deal with this case using other methods.

3.3 Defining a general lattice model interface

As described in Section 3.2.2, the general method to price a contingent claim such as an option in a binomial lattice model is to calculate the payoff in each state at maturity and then "roll back" this payoff through the lattice by calculating discounted expected values in each time step, possibly applying additional conditions along the way (such as for example the early exercise condition in the case of an American option). In Listing 3.1, we define a class BinomialLattice[8], which represents a generic binomial lattice model and supplies public member functions performing these tasks.

The constructor takes the parameters needed to set up the model: An object encapsulating relevant characteristics of the underlying asset S, the constant continuously compounded rate of interest r, the time horizon T and the number of time points N (this includes the initial time point $t_0 = 0$, so the number of time *steps* is $n = N - 1$). Since by Remark 3.11 the binomial lattice can be seen as a discretisation of a Black and Scholes (1973) model, the underlying asset is represented by an instance of the class BlackScholesAsset[9], which can be queried for properties such as the time zero ("today") asset value and the asset's (deterministic) volatility σ.

The member function apply_payoff(i,f) applies the payoff function f to the underlying in each node in period i. The payoff function is represented as a boost::function object taking a double argument and returning a double.[10] The payoff can then be "rolled back" from the i–th period to period 0 using the member function rollback() and the time zero option value can be read using the member function result(). In Listing 3.2, a European call option

---

[8] Found in the file Binomial.hpp in the folder Include on the website for this book.
[9] Declared in the file BlackScholesAsset.hpp in the folder Include on the website for this book.
[10] For an overview of the features of the Boost library used in this book, see Section 2.2.

```
class BinomialLattice {
protected: /* ... */
public:
  BinomialLattice(const BlackScholesAsset& xS, ///< Underlying asset.
                  double xr,                    ///< Risk-free interest.
                  double xT,                    ///< Time horizon.
                  int xN                        ///< Number of time steps.
                  );
  void set_CoxRossRubinstein();
  void set_JarrowRudd();
  void set_Tian();
  void set_LeisenReimer(double K);
  /// Calculate value of the underlying asset at a given grid point.
  inline double underlying(int i, ///< Time index
                           int j  ///< State index
                           ) const;
  /// Calculate all values of the underlying for a given time slice.
  inline void underlying(int i,Array<double,1>& un) const;
  /// Apply the (payoff) function f to values for a given time slice.
  inline void apply_payoff(int i,boost::function<double (double)> f);
  /// Roll back (via discounted expected values)
  /// from a given time to a given time.
  inline void rollback(int from,int to);
  /// Rollback with early exercise (or similar) condition
  /// represented by the function f.
  inline void rollback(int from,int to,
                   boost::function<double (double,double)> f);
  /// Access the (time zero) result.
  inline double result() const;
};
```

Listing 3.1: Public member functions of `BinomialLattice`

is priced in this way. Note that the payoff is given by an instance of the class `Payoff`, defined in `Payoff.hpp`, which overloads

$$\text{double Payoff::operator()(double S)} \qquad (3.34)$$

to return the option payoff as a function of S, the price of the underlying asset at maturity.

For American options, or options with a similar path dependence, the payoff at expiry must be rolled back through the lattice using the overloaded member function `rollback(from,to,f)`, where f is now a `boost::function` object taking two `double` arguments (the continuation value and the value of the underlying asset) and returning a `double`. In Listing 3.2, this is used to price an American put and a down–and–out barrier call option. In the case of the American put, f is supplied by `EarlyExercise` and returns the greater of the continuation value and the payoff of early exercise. A down–and–out barrier

European call

```
ConstVol vol(sgm);
BlackScholesAsset stock(&vol,initial_stockprice);
BinomialLattice btree(stock,r,maturity,N);
Payoff call(strike);
boost::function<double (double)> f;
f = boost::bind(std::mem_fun(&Payoff::operator()),&call,_1);
btree.apply_payoff(N-1,f);
btree.rollback(N-1,0);
cout << "Binomial call (CRR): " << btree.result() << endl;
```

American put

```
ConstVol vol(sgm);
BlackScholesAsset stock(&vol,initial_stockprice);
BinomialLattice btree(stock,r,maturity,N);
Payoff put(strike,-1);
boost::function<double (double)> f;
f = boost::bind(std::mem_fun(&Payoff::operator()),&put,_1);
EarlyExercise amput(put);
boost::function<double (double,double)> g;
g = boost::bind(boost::mem_fn(&EarlyExercise::operator()),
                         &amput,_1,_2);
btree.apply_payoff(N-1,f);
btree.rollback(N-1,0,g);
cout << "Binomial American put (CRR): " << btree.result() << endl;
```

Down–and–out barrier call

```
ConstVol vol(sgm);
BlackScholesAsset stock(&vol,initial_stockprice);
BinomialLattice btree(stock,r,maturity,N);
Payoff call(strike);
boost::function<double (double)> f;
f = boost::bind(std::mem_fun(&Payoff::operator()),&call,_1);
KnockOut down_and_out(barrier);
boost::function<double (double,double)> g;
g = boost::bind(boost::mem_fn(&KnockOut::operator()),
                         &down_and_out,_1,_2);
btree.apply_payoff(N-1,f);
btree.rollback(N-1,0,g);
cout << "Binomial down and out call (CRR): "
     << btree.result() << endl;
```

Listing 3.2: Pricing options in a `BinomialLattice`

call is an option, which has a standard call payoff at expiry if and only if the price of the underlying did not fall below a given barrier during the life of the option — the payoff is zero otherwise. In this case f is supplied by

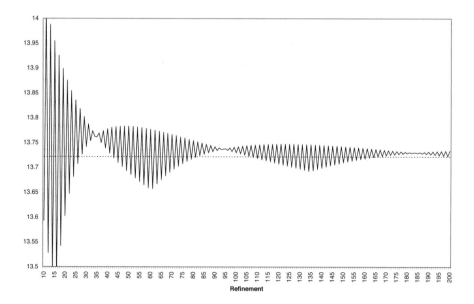

Figure 3.11 *Convergence of a binomial call option price (solid line) to the continuous limit (dashed line). Model: Cox/Ross/Rubinstein with initial stock price $S = 100$, strike $K = 110$, interest rate $r = 0.05$, maturity $T = 1.5$, volatility $\sigma = 0.3$.*

KnockOut and returns zero if the barrier is breached, and the continuation value otherwise.[11]

3.4 Implementing binomial lattice models

The member functions `apply_payoff()`, `rollback()` and `result()` can be implemented generically for all of the binomial lattice models with constant transition probabilities p and up and down factors u and d.

Four models commonly cited in the literature differ only in the way p, u and d are determined. This is implemented via the member functions set_CoxRoss-Rubinstein(), set_JarrowRudd(), set_Tian() and set_LeisenReimer(), which set p, u and d to the values proposed by Cox, Ross and Rubinstein (1979), Jarrow and Rudd (1983), Tian (1993) and Leisen and Reimer (1996), respectively.[12] Note that in each case, the binomial lattice converges to the Black and Scholes (1973) model as n tends to infinity. In addition to the

---

[11] The classes `EarlyExercise` and `KnockOut` are defined in `Payoff.hpp`. The full listings for European call, American put and down–and–out barrier call can be found on the website in the files `BinomialCall.cpp`, `BinomialAmPut.cpp` and `BinomialDownOut.cpp`, respectively.

[12] As an alternative design, one could have created an abstract base class defining the interface for binomial lattice models, from which a concrete class for each of the specific models is derived.

Cox/Ross/Rubinstein values given in Remark 3.11, we have the approach of Jarrow and Rudd (1983) with

$$u = \exp(r - \frac{1}{2}\sigma^2)\Delta t + \sigma\sqrt{\Delta t} \tag{3.35}$$

$$d = \exp(r - \frac{1}{2}\sigma^2)\Delta t - \sigma\sqrt{\Delta t} \tag{3.36}$$

where p is set to 0.5. The u and d are chosen in this manner in order to ensure that the first two moments of the logarithmic returns in the binomial lattice of any refinement match those of the continuous limit exactly. In this case the model is strictly speaking arbitrage–free only in the limit, because for any $\Delta t > 0$, the no-arbitrage condition (3.8), on the risk neutral probabilities holds only approximately.

Alternatively, following Tian (1993), we can set

$$v = e^{\sigma^2 \Delta t} \tag{3.37}$$

$$u = \frac{1}{2}ve^{r\Delta t}(v + 1 + \sqrt{v^2 + 2v - 3}) \tag{3.38}$$

$$d = \frac{1}{2}ve^{r\Delta t}(v + 1 - \sqrt{v^2 + 2v - 3}) \tag{3.39}$$

and p is calculated using (3.8). As in the method of Jarrow and Rudd, Tian aims to match moments of the discrete distribution from the binomial lattice to the continuous limit. In either case, the arguments of Remark 3.11 can be applied to show that the models converge to Black/Scholes as $\Delta t \to 0$. However, as illustrated in Figure 3.11 for the Cox/Ross/Rubinstein model, in all these models there are considerable oscillations and wave patterns along the way, and there is no guarantee that the option price calculated using an $n + 1$ or $n + 2$ step binomial lattice is closer to the continuous limit than the price calculated using an n–step lattice.

This phenomenon was studied in detail by Leisen and Reimer (1996); it is due to the way the probability mass is concentrated on node points in the binomial lattice. They propose an alternative method of setting u and d in such a way that equalities in (3.24) and (3.25) already hold exactly for a given refinement n and strike K, rather than only in the limit. As a result, in their approach the European option price converges its continuous limit with order 2, as opposed to the order of convergence 1 of the models presented above.[13] Furthermore, the convergence is monotonic.

The key observation of Leisen and Reimer (1996) is that the convergence of the binomial lattice price to the Black/Scholes limit is driven by the con-

---

[13] A sequence of n–step lattices used to calculate an option price is said to converge with order $\rho > 0$ if there exists a constant $\kappa > 0$ such that the difference between the continuous limit and the value calculated using an n–step lattice is at most $\kappa n^{-\rho}$ for all $n \in \mathbb{N}$.

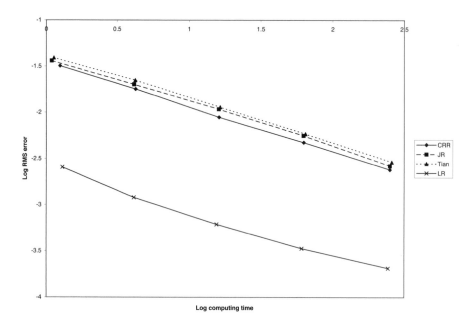

Figure 3.12 *Log/log plot of the root mean squared error of the American put option price using the methods of Cox/Ross/Rubinstein (CRR), Jarrow/Rudd (JR), Tian and Leisen/Reimer (LR).*

vergence

$$\sum_{i=a}^{n}\binom{n}{i}p^i(1-p)^{n-i}\underset{n\to\infty}{\longrightarrow}N(h) \tag{3.40}$$

$$\underbrace{\phantom{\sum_{i=a}^{n}\binom{n}{i}p^i(1-p)^{n-i}}}_{=:\Phi(a;n,p)}$$

when $\Phi(a; n, p)$ is the complementary binomial distribution function. They therefore set $\tilde{p}$ such that $\Phi(a; n, \tilde{p}) = N(h_1)$ and p such that $\Phi(a; n, p) = N(h_2)$. In our implementation, this is achieved using a Peizer and Pratt (1968) method 2 inversion,[14] setting

$$p(h) = 0.5 \mp \sqrt{0.25 - 0.25\exp\left\{-\left(\frac{h}{n+\frac{1}{3}+\frac{0.1}{n+1}}\right)^2\left(n+\frac{1}{6}\right)\right\}} \tag{3.41}$$

and $a = (n-1)/2$. Given p and $\tilde{p}$, one can then use the no-arbitrage conditions to calculate

$$u = R\frac{\tilde{p}}{p} \quad \text{and} \quad d = \frac{R - pu}{1 - p}$$

Note that the probabilities are chosen as a function of h_1 and h_2, and are

---

[14] Leisen and Reimer (1996) also mention alternative methods.

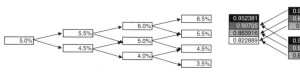

Figure 3.13 *A simple interest rate lattice...*

Figure 3.14 *...and the corresponding zero coupon bond prices for $p = 0.5$.*

therefore specific to the contract being priced. Consequently, the convergence results of Leisen and Reimer (1996) do not readily apply to options other than European calls and puts; in fact, they show that in their approach, as in previous models, American put option prices converge with order 1. However, there is evidence, reproduced here in Figure 3.12, that although the order of convergence is the same, the initial error in the Leisen/Reimer lattice is lower, leading to considerable higher accuracy at the same computational effort.[15]

3.5 Lattice models for the term structure of interest rates

3.5.1 General construction

Suppose now that we wish to model the evolution of interest rates as a random process. Naively, we might construct a binomial lattice as in Figure 3.13 for the single-period interest rate with simple compounding. The question then arises how one should choose the risk-neutral probabilities in order to preclude arbitrage. It can be seen by the following argument that is a non-trivial issue: Via the relationship

$$B_{i,j}(t_i, t_{i+1}) = (1 + (t_{i+1} - t_i)r_{ij})^{-1} \qquad (3.42)$$

we can calculate the one-period zero coupon bond price at each node j in each period i. Furthermore, given the risk–neutral transition probabilities p, one can calculate the two-period zero coupon bond price at each node j in period i from the one-period zero coupon bond prices in the next period via

$$B_{i,j}(t_i, t_{i+2}) = \underbrace{B_{i,j}(t_i, t_{i+1})}_{\text{discounting}} \underbrace{(pB_{i+1,j+1}(t_{i+1}, t_{i+2}) + (1 - p)B_{i+1,j}(t_{i+1}, t_{i+2}))}_{\text{expected future value}}$$

$$(3.43)$$

---

[15] As in Leisen and Reimer (1996), Figure 3.12 plots the logarithm of the root mean squared error (RMSE) of the American put option prices in 2500 random scenarios calculated using the different binomial models against the logarithm of the required computing time. The "correct" American put option price for each scenario was calculated using the Cox/Ross/Rubinstein model at a very high refinement of 20,000 time steps (the maximum number of time steps used in Figure 3.12 is 800).

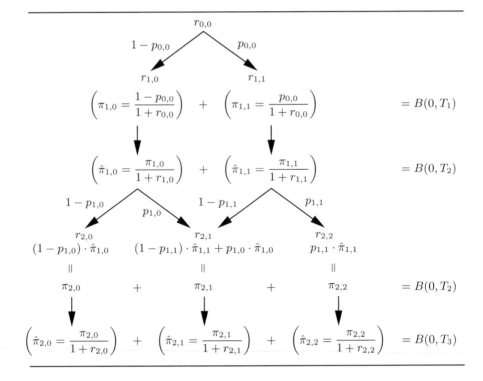

Figure 3.15 *The forward substitution technique*

"Rolling back" all available bond prices through the lattice in this way, as illustrated for $p = 0.5$ in Figure 3.14, in an n–step lattice we get all initial ("time zero") zero coupon bond prices $B_{0,0}(t_0, t_1), ..., B_{0,0}(t_0, t_{n+1})$. The absence of arbitrage requires that the prices thus calculated match the corresponding prices observed in the market. The key difference in constructing arbitrage-free interest rate models, as opposed to models of a stock, a commodity or an exchange rate, lies thus in the fact that the absence of arbitrage must be enforced not just on the evolution of one asset, but on an entire term structure of zero coupon bond prices. In a binomial lattice model, the most generic way of achieving this is to construct the model by a *forward substitution technique*.[16] From this perspective, it also becomes evident what modelling freedom remains after the arbitrage constraints are satisfied.

Consider Figure 3.15. The initial single–period interest rate $r_{0,0}$ is set to

---

[16] This technique first appeared in the literature in Jamshidian (1991). The presentation here follows Sandmann and Schlögl (1996).

match the current observed single–period discount factor, i.e.

$$r_{0,0} = \frac{1}{t_1 - t_0} \left(\frac{1}{B(0, t_1)} - 1 \right) \tag{3.44}$$

The state prices for the two nodes at time t_1 are then given by

$$\pi_{1,1} = \frac{p_{0,0}}{1 + (t_1 - t_0)r_{0,0}} \quad \text{and} \quad \pi_{1,0} = \frac{1 - p_{0,0}}{1 + (t_1 - t_0)r_{0,0}} \tag{3.45}$$

and the condition $\pi_{1,1} + \pi_{1,0} = B(0, t_1)$ is trivially satisfied for any choice of transition probability $p_{0,0}$. Suppose now that for time t_1, the single–period interest rate is $r_{1,1}$ for the "up" state and $r_{1,0}$ for the "down" state. Then the one–period zero coupon bond prices are $(1 + (t_2 - t_1)r_{1,1})^{-1}$ and $(1 + (t_2 - t_1)r_{1,0})^{-1}$, respectively. This is a bond maturing in t_2, priced at time t_1, and we can calculate the time t_0 value of this bond by weighting its possible time t_1 values with the corresponding state prices, i.e.

$$B(0, t_2) = \hat{\pi}_{1,1} + \hat{\pi}_{1,0} \tag{3.46}$$

with

$$\hat{\pi}_{1,j} = \frac{\pi_{1,j}}{1 + (t_2 - t_1)r_{1,j}} \quad j = 0, 1$$

$B(0, t_2)$ is an initial discount factor observed in the market, and (3.46) must be satisfied in order to avoid arbitrage. Thus, in general, we have the freedom to either fix $r_{1,0}$ and $r_{1,1}$ exogenously, in which case (3.46) will determine the risk–neutral probability $p_{0,0}$, or we could exogenously set the risk-neutral probability, in which case $r_{1,0}$ and $r_{1,1}$ must be chosen such that (3.46) is satisfied.

In the next step, the state prices (as viewed from t_0) for the three possible states at time t_2 can be calculated using the previously determined $\hat{\pi}_{1,j}$ and the transition probabilities $p_{1,j}$ (which may, in general, be state dependent):

$$\begin{aligned} \pi_{2,0} &= (1 - p_{1,0})\hat{\pi}_{1,0} \\ \pi_{2,1} &= (1 - p_{1,1})\hat{\pi}_{1,1} + p_{1,0}\hat{\pi}_{1,0} \\ \pi_{2,2} &= p_{1,1}\hat{\pi}_{1,1} \end{aligned}$$

Analogously to (3.46), we have for the initial zero coupon bond maturing at time t_3,

$$B(0, t_3) = \sum_{j=0}^{2} \hat{\pi}_{2,j}, \qquad \hat{\pi}_{2,j} = \frac{\pi_{2,j}}{1 + (t_3 - t_2)r_{2,j}}$$

imposing an arbitrage constraint on the choice of the $p_{1,j}$ and $r_{2,j}$. The further construction of the model then proceeds by forward induction, i.e.

$$\pi_{i,j} = \begin{cases} (1 - p_{i-1,0})\hat{\pi}_{i-1,0} & \text{if } j = 0 \\ (1 - p_{i-1,j})\hat{\pi}_{i-1,j} + p_{i-1,j-1}\hat{\pi}_{i-1,j-1} & \text{if } 0 < j < i \\ p_{i-1,j-1}\hat{\pi}_{i-1,j-1} & \text{if } j = i \end{cases} \tag{3.47}$$

$$\hat{\pi}_{i,j} = \frac{\pi_{i,j}}{1 + (t_{i+1} - t_i)r_{i,j}}$$

with the $p_{i-1,j}$ and $r_{i,j}$ chosen to satisfy

$$B(0, t_{i+1}) = \sum_{j=0}^{i} \hat{\pi}_{i,j} \qquad (3.48)$$

and guaranteeing the absence of arbitrage vis–à–vis the discount factors given by the market.

Remark 3.12 It is worthwhile to take a closer look at the relationship of the risk–neutral transition probabilities and state prices in the present framework. Note that when "rolling back" the value of an asset by calculating its discounted expected value as in equation (3.43), the discount factor $B_{i,j}(t_i, t_{i+1})$ depends on the state j. Viewed from time t_0, the value of a contingent claim paying $X(t_n)$ at time t_n could thus be calculated as the expected discounted value

$$X(t_0) = E\left[X(t_n) \prod_{i=0}^{n-1} B(t_i, t_{i+1})\right] \qquad (3.49)$$

where the expectation is taken under the risk–neutral probabilities.

Defining the *discrete rollover savings account*, i.e. the value of initially \$1 invested at the single–period rate, with the proceeds including interest reinvested each period, as

$$A(t_k) = \prod_{i=0}^{k-1} (1 + r_{i,.}(t_{i+1} - t_i)) = \prod_{i=0}^{k-1} B(t_i, t_{i+1})^{-1} \qquad (3.50)$$

(3.49) can be written as

$$\frac{X(t_0)}{A(t_0)} = E\left[\frac{X(t_n)}{A(t_n)}\right]$$

i.e. the discrete rollover savings account is the numeraire asset associated with the risk–neutral probabilities in the current framework. Because A evolves randomly with the changing single–period interest rate, the discounting remains inside the expectation and (3.49) must be evaluated by backward induction ("rolling back") through the binomial lattice. For assets and contingent claims without path–dependent features, such as European options or the zero coupon bonds themselves, this is unnecessarily tedious: Suppose a contingent claim pays $X_j(t_n)$ in state j at time t_n. Given the state prices $\pi_{n,j}$, this can be valued directly as

$$X(t_0) = \sum_{j=0}^{n} \pi_{n,j} X_j(t, n) \qquad (3.51)$$

Furthermore, as in Section 3.1, note that the state prices are positive and their sum is equal to the discount factor,

$$B(t_0, t_n) = \sum_{j=0}^{n} \pi_{n,j} \qquad (3.52)$$

so the normalised state prices

$$\bar{p}_{n,j} := \frac{\pi_{n,j}}{B(t_0, t_n)} \tag{3.53}$$

can be interpreted as probabilities and we can write

$$X(t_0) = B(t_0, t_n) \sum_{j=0}^{n} \bar{p}_{n,j} X_j(t, n)$$

$$\Leftrightarrow \quad \frac{X(t_0)}{B(t_0, t_n)} = E_{t_n}\left[\frac{X(t_n)}{B(t_n, t_n)}\right]$$

where the expectation $E_{t_n}[\cdot]$ is taken under the probabilities $\bar{p}_{n,j}$, which are the risk–neutral probabilities associated with taking the zero coupon bond maturing in t_n as the numeraire. Since asset prices divided by a zero coupon bond price are forward prices, one also refers to the probabilities $\bar{p}_{n,j}$ as probabilities under the *time t_n forward measure*.[17] Implicitly, this way of simplifying the calculation was already used in the construction of the binomial lattice framework, i.e. (3.48) is nothing other than

$$B(t_0, t_{i+1}) = \sum_{j=0}^{i} \pi_{i,j} B_j(t_i, t_{i+1}) = B(t_0, t_i) E_{t_i}[B(t_i, t_{i+1})] \tag{3.54}$$

3.5.2 Concrete models

The first arbitrage–free term structure model to appear in the literature was Ho and Lee (1986). In this model, the risk–neutral transition probabilities are set exogenously and are constant in time and states. Moving from one time step to the next, all zero coupon bonds are perturbed in the same way around their (one–step) forward value, i.e. set

$$B_{j+1}(t_{i+1}, t_{l+1}) = \frac{B_j(t_i, t_{l+1})}{B_j(t_i, t_{i+1})} \psi(l - i) \tag{3.55}$$

$$B_j(t_{i+1}, t_{l+1}) = \frac{B_j(t_i, t_{l+1})}{B_j(t_i, t_{i+1})} \psi^*(l - i) \tag{3.56}$$

with

$$\psi(k) = \frac{1}{p + (1-p)\delta^k} \tag{3.57}$$

$$\psi^*(k) = \frac{\delta^k}{p + (1-p)\delta^k} = \psi(k)\delta^k \tag{3.58}$$

where $\delta > 0$ is an additional model parameter; the further δ departs from 1, the greater the resulting volatility of bond prices and interest rates. Note that

---

[17] In the binomial lattice models discussed in Section 3.2, interest rates were deterministic, so normalised state prices were identical to the original risk–neutral measure.

the choice of the perturbation function ψ and ψ^* ensures that

$$B_j(t_i, t_{l+1}) = B_j(t_i, t_{i+1})(pB_{j+1}(t_{i+1}, t_{l+1}) + (1-p)B_j(t_{i+1}, t_{l+1})) \quad (3.59)$$

Thus the model is arbitrage free by construction and satisfies the constraint (3.48). However, the particular bond price dynamics postulated in (3.55)-(3.58) permit zero coupon bond prices greater than 1 (and thus negative interest rates).[18] The model also lacks flexibility in that it does not allow for time–dependent volatilities, which are needed in order to calibrate to at–the–money caplet prices observed in the market.[19]

Independently of each other, Black, Derman and Toy (1990) and Sandmann and Sondermann (1989, 1991) proposed a binomial term structure model, which does not generate negative interest rates and can easily be adapted to time–dependent volatilities. This is the model defined and implemented in TSBinomial.hpp and TSBinomial.cpp.[20] Again, the risk–neutral transition probabilities are set exogenously and typically (though not necessarily) constant in time and states — our implementation simply sets them to $p = 0.5$. As an exogenous risk measure the local standard deviation (volatility) $\sigma_{i,j}$ of the logarithmic single–period interest rate is specified.[21] For node j at time t_i, we therefore have

$$\sigma_{i,j}^2 = \mathrm{Var}[\ln r_{i+1,\cdot}|r_{i,j}] = p_{i,j}(1-p_{i,j})\left(\ln \frac{r_{i+1,j+1}}{r_{i+1,j}}\right)^2$$

$$\Leftrightarrow \quad r_{i+1,j+1} = r_{i+1,j}\exp\left\{\frac{\sigma_{i,j}}{\sqrt{p_{i,j}(1-p_{i,j})}}\right\} > r_{i+1,j}$$

Thus all possible realisations of the single-period rate at time t_{i+1} can be expressed in terms of the lowest possible rate $r_{i+1,0}$:

$$r_{i+1,j} = r_{i+1,0}\prod_{k=0}^{j-1}\exp\left\{\frac{\sigma_{i,k}}{\sqrt{p_{i,k}(1-p_{i,k})}}\right\} \quad \forall 1 \le j \le i+1 \quad (3.60)$$

In our implementation, with σ time–, but not state–dependent, this simplifies to

$$r_{i+1,j} = r_{i+1,0}\exp\left\{\frac{j\sigma(t_i)}{\sqrt{p(1-p)}}\right\} \quad (3.61)$$

---

[18] In the continuous–time limit, the Ho/Lee model is a one–factor Gaussian Heath, Jarrow and Morton (1992) model without mean reversion (see Sommer (1994)). The properties of such a model are discussed in Chapter 8.

[19] One could argue that the binomial lattice models discussed and implemented in Section 3.2 suffer from a similar drawback. However, for binomial lattices for equities and similar assets, a binomial lattice can be constructed fitting observed option prices both at and away from the money, see e.g. Derman and Kani (1994).

[20] Found, respectively, in the folders Include and Source on the website for this book.

[21] Note that the single–period interest rate is still given in terms of simple compounding. As shown by Sandmann and Sondermann (1997), this ensures stable limit behaviour of the model, which would not be the case if σ represented the volatility of the logarithmic continuously compounded rate.

```
class TSBinomialLatticeNode {
private:
    double        Q; ///< State price of this node
                  ///< as viewed from the initial node.
    double        r; ///< Short rate realisation in this node.
public:
    TSLogLinear* ts; ///< Term structure realisation in this node.
    inline TSBinomialLatticeNode();
    inline double state_price() const { return Q; };
    inline double& state_price() { return Q; };
    inline double short_rate() const { return r; };
    inline double& short_rate() { return r; };
};
```

Listing 3.3: The class `TSBinomialLatticeNode`

Substituting (3.60) or (3.61) into (3.48) when constructing the model by forward induction results in a one–dimensional fixed point problem to determine the lowest interest rate $r_{i+1,0}$ such that the model price for $B(t_0, t_{i+2})$ matches the observed market price. Expressing $B(t_0, t_{i+2})$ in terms of $r_{i+1,0}$, (3.48) becomes

$$B(t_0, t_{i+2}) = \sum_{j=0}^{i+1} \pi_{i+1,j} \left(1 + (t_{i+2} - t_{i+1}) r_{i+1,0} \exp\left\{ \frac{j\sigma(t_i)}{\sqrt{p(1-p)}} \right\} \right)^{-1}$$

(3.62)

For $r_{i+1,0} = 0$,

$$B(t_0, t_{i+2}) = \sum_{j=0}^{i+1} \pi_{i+1,j} = B(t_0, t_{i+1})$$

and for $r_{i+1,0} \to \infty$, $B(t_0, t_{i+2})$ will tend to zero. Furthermore, $B(t_0, t_{i+2})$ as given by (3.62) is monotonic in $r_{i+1,0}$. Thus if the market $B(t_0, t_{i+1}) > B(t_0, t_{i+2})$ (which follows if forward rates are positive) and $B(t_0, t_{i+2}) > 0$ (which follows from the absence of arbitrage), then there is a unique $r_{i+1,0}$ such that the model price given by (3.62) matches the market price, and it can be found by a simple one–dimensional root search.

In the class `TSBinomialMethod`, the binomial lattice is represented as a

`std::vector<std::vector<TSBinomialLatticeNode*>*>`

i.e. as a vector of pointers to vectors of node objects `TSBinomialLatticeNode`, given in Listing 3.3. Each node is characterised by its state price π_{ij} and the single–period interest rate r_{ij} in this node. The `TSBinomialMethod` constructor takes as arguments a reference to a `TermStructure` object representing the initial term structure of interest rates observed in the market, the volatility of the logarithmic single–period interest rates (either as a `double`, representing constant per annum volatility, or as an `Array` of doubles, with one volatility for each time period in the lattice), and lastly an `Array` of doubles repre-

senting the discretisation of the time line in the lattice. Note that there is
no requirement for the time steps to be of equal length. The constructor first
calls the private member function `allocateLattice()`, which allocates the
memory for the lattice nodes and initialises the first node (state price equal
to 1 and single–period interest rate equal to the corresponding rate in the
initial term structure). Subsequently, the constructor calls `createLattice()`,
given in Listing 3.4, which constructs the model by inductively calculating
the $\pi_{i,j}$ and $r_{i,j}$ as described above. Here, the root search to find the $r_{i+1,0}$ to
satisfy (3.62) is implemented by applying the `Rootsearch` template described
in Chapter 2 to the private member function `bond_r(double r)`, which evaluates the right hand side of (3.62) for the "current period" for $r_{i+1,0} = r$.
The `TSBinomialMethod` object uses the data member `current_period` to
keep track of the "current period" in the forward construction of the lattice.
Calling `rs.solve()` on the `Rootsearch` instance `rs` returns the $r_{i+1,0}$, which
sets the right hand side of (3.62) equal to the initial zero coupon bond price
$B(t_0, t_{i+2})$.

Because the model allows for time-dependent volatilities, it can also be calibrated to a sequence of market prices of European interest rate derivatives,
the payoff of which depends on the single–period interest rate. Typically, these
will be at–the–money caplets.[22] The member function `calibrate()` takes a
`std::vector` of pointers to `TSEuropeanInstrument` objects as its argument
and proceeds as follows. First, the calibration instruments are sorted with respect to increasing maturity. Then, the volatility calibration is conducted inductively, commencing with the shortest maturity: In a manner similar to the
calibration of the model to the initial zero coupon bond prices, a root search is
set up using the private member function `price_current_instrument(double
sgm)`. Given that the volatilities are already calibrated, say, up to time t_i, and
the maturity of the current instrument is $t_k > t_i$, the root search attempts
to find a constant σ such that if we set $\sigma(t_l) = \sigma$ for all $i \leq l < k$, the
model price of the instrument equals its market price. Note that the lat-
tice must be reconstructed from t_i to t_k (by calling `createLattice(i,k)` in
`price_current_instrument()`) every time there is a change in $\sigma(t_l)$, $i \leq l <
k$.

The full set of zero coupon bond prices in each node, if required, must be
calculated by backward induction as in equation (3.43). This is done by calling the public member function `rollbackTermstructures()`, which assigns
pointers to valid `TermStructure` objects[23] to the `ts` data member of each
node.

---

[22] The instruments do not necessarily have to be caplets, nor do they have to be at the
money. The member function `calibrate()` will also accept a sequence of away–from–the–
money floorlets, for example. However, there can be at most one calibration instrument for
each time step in the lattice — time–dependent volatilities alone will generally not allow
calibration to, say, caplets of different strikes for the same maturity. Also, the current
implementation requires the payoff of the calibration instruments to be contingent only
on the single–period interest rate.

[23] In the current implementation, these are hardwired to be of the type `TSLoglinear`, but

```
void TSBinomialMethod::createLattice(int from,int to)
{
  int i;
  std::vector<TSBinomialLatticeNode*>::iterator j,k;
  rfactor = exp(sgm/std::sqrt(p*(1-p)));
  double lowest_rate = node(0,0)->short_rate();
  boost::function<double (double)> f;
  // Define functor for root search.
  f = boost::bind(std::mem_fun(&TSBinomialMethod::bond_r),this,_1);
  for (i=from;i<to;i++) {
    // Calculate state prices for new nodes.
    double prev_part = 0.0;
    k = nodes[i+1]->begin();
    for (j=nodes[i]->begin();j!=nodes[i]->end();j++) {
      double new_part =
        (*j)->state_price()/(1.0+dt(i)*(*j)->short_rate());
      (*k)->state_price() = (1.0-p)*prev_part + p*new_part;
      k++;
      prev_part = new_part; }
    (*k)->state_price() = (1.0-p)*prev_part;
    // Calculate lowest interest rate.
    set_current_period(i+1);
    Rootsearch<boost::function<double (double)>,double,double>
      rs(f,initial_ts(T(i+2)),lowest_rate*0.8,lowest_rate*0.4,0.0);
    double r = lowest_rate = rs.solve();
    // Set short rates.
    for (j=nodes[current_period]->begin();
         j!=nodes[current_period]->end();j++) {
      (*j)->short_rate() = r;
      r *= rfactor(i); }
}
```

Listing 3.4: Creating the binomial lattice for a term structure model.

The design of the interface routines for pricing contingent claims with TSBinomialMethod parallels that of BinomialLattice. We have the public member functions.

```
void apply_payoff(int i,
        boost::function<double (constTermStructure&)> f)
```

where the payoff is calculated based on the term structures realised at time t_i, and

```
void rollback(int from,int to)
```

and

this could be replaced by any class derived from TermStructure, which can be constructed from zero coupon bond prices.

```
void
rollback(int from,int to,
    boost::function<double(double,const TermStructure &)> f)
```

to roll the payoff back through the lattice, with or without an early exercise condition, or alternatively

```
void rollback(int from)
```

to price path–independent payoffs directly using the state prices. The example program TSBinomialExample.cpp on the website for this book shows how to use these functions to price an interest rate floor, a barrier cap, a European and a Bermudan swaption.

## CHAPTER 4

The Black/Scholes world

Financial engineers have at their disposal a set of mathematical "power tools," which came into standard use for option pricing in the wake of the seminal papers of Black and Scholes (1973) and Merton (1973b), though these techniques were known well previously to mathematicians, theorists and practitioners in the physical sciences. These methods will be discussed and used here mainly within the original Black/Scholes framework, which for a large range of applications still remains the dominant paradigm. However, it should be noted that they are equally applicable, albeit often at a considerably higher level of technical complexity, in the many extensions, generalisations and modifications of the model that have since been proposed.

The key assumption of the Black/Scholes model is that the underlying asset(s) exhibit deterministic proportional volatility. Whenever this assumption holds, Black/Scholes-type derivative pricing formulae follow.[1] This feature is cast into an object-oriented framework and will be one of the building blocks used in subsequent chapters. In addition to standard call and put options, the present chapter also covers a selection of exotic options with closed form pricing formulae in the Black/Scholes framework.

4.1 Martingales

Consider a *stochastic process* X, defined on a probability space $(\Omega, \mathcal{F}, P)$ as a collection of random variables $X_t(\omega)$, with $t \in T$ and $\omega \in \Omega$. T is a set of indices (discrete or continuous) typically interpreted as time. The ω are elements of some space Ω, typically interpreted as the set of possible "states of the world." P is a probability measure on this space, assigning probabilities $P(A)$ to all $A \in \mathcal{F}$, where $\mathcal{F}$ is a σ–field representing all the events (i.e. subsets of Ω) which are measurable (i.e. to which probabilities can be assigned).

In the context of the mathematics of stochastic processes, information at any particular point in time t is represented by the σ–field $\mathcal{F}_t$.

Definition 4.1 *A σ–field $\mathcal{F}$ on Ω is a set of subsets of Ω satisfying*

1. $\Omega \in \mathcal{F}$, $\emptyset \in \mathcal{F}$, where $\emptyset$ denotes the empty set.

2. If $A \in \mathcal{F}$, then $A^{\complement} \in \mathcal{F}$, where $A^{\complement}$ denotes the complement of A, i.e. the set of all elements in Ω which are not in A.

---

[1] For a succinct treatment of the significance of this assumption as a special case of a more general framework leading to Black/Scholes–type formulae, see Rady (1997).

3. If $A_1, A_2, \ldots \in \mathcal{F}$, then

$$\bigcap_{i=1}^{\infty} A_i \in \mathcal{F} \qquad and \qquad \bigcup_{i=1}^{\infty} A_i \in \mathcal{F}$$

i.e. any finite or infinite intersections or unions of sets in $\mathcal{F}$ are also in $\mathcal{F}$.

We assume that we do not forget any information as time progresses, so $\mathcal{F}_s \subseteq \mathcal{F}_t$ for $s \leq t$. A collection $\{\mathcal{F}_t\}_{t \in T}$ of σ–fields on Ω satisfying this property is called a *filtration*.

$\mathcal{F}_t$ can be interpreted as the set of statements of which it is known at time t whether or not they are true. This motivates why we require $\mathcal{F}_t$ to be a σ–field:

1. Ω represents the statement which is trivially known to be true, e.g., "A die roll will be either 1, 2, 3, 4, 5, or 6." $\emptyset$ represents the negation of that statement.

2. If it is known whether or not a statement is true, it is also known whether or not the negation of that statement is true.

3. If it is known whether several (separate) statements are true or not, then it is also known whether or not a statement formed by a conjunction of these statements with "or" (union) or "and" (intersection) is true.

The information contained in the knowledge of the value that a random variable Y has taken is represented by the σ–field $\sigma(Y)$ *generated* by Y. This is defined as the smallest σ–field containing all sets $\{\omega | a < Y(\omega) \leq b\}$ for all $-\infty < a < b < \infty$.

For a stochastic process X, if the value of the random variable X_t is observed at time t (or earlier), this can be expressed as $\sigma(X_t) \subseteq \mathcal{F}_t$. In this case, X is said to be *adapted* to the filtration $\{\mathcal{F}_t\}_{t \in T}$.

The best predictor for the future value of a random variable Y is given by its expected value, denoted $E[Y]$. The best predictor of Y given the information contained in the σ–field $\mathcal{F}$ is given by the *conditional expectation*. This conditional expectation is defined as the random variable $E[Y|\mathcal{F}]$, which does not contain more information than $\mathcal{F}$, i.e. $\sigma(E[Y|\mathcal{F}]) \subseteq \mathcal{F}$, and which satisfies the relation $E[Y\mathbb{I}_A] = E[E[Y|\mathcal{F}]\mathbb{I}_A]$ for $A \in \mathcal{F}$. $\mathbb{I}_A$ denotes the indicator function, i.e. $\mathbb{I}_A(\omega) = 1$ if $\omega \in A$ and $\mathbb{I}_A(\omega) = 0$ otherwise. $E[Y|\mathcal{F}]$ is thus a coarser version of the random variable Y in the sense that $E[Y|\mathcal{F}](\omega)$ is constant for all ω in an "elementary" set $A \in \mathcal{F}$, i.e. such $A \in \mathcal{F}$ that there is no $B \in \mathcal{F}$ with $B \subset A$. Note that if $\mathcal{F}$ contains all information on Y, i.e. $\sigma(Y) \subseteq \mathcal{F}$, then $E[Y|\mathcal{F}] = Y$.

Given the above concepts, one can now formally state:

Definition 4.2 *A stochastic process $\{M_t\}_{t \in T}$ on $(\Omega, \mathcal{F}, P)$ is called a martingale with respect to a filtration $\{\mathcal{F}_t \subseteq \mathcal{F}\}_{t \in T}$ and the probability measure P if*

(i) M_t is adapted to $\{\mathcal{F}_t\}_{t \in T}$

(ii) $E[|M_t|] < \infty$ *for all* $t \in T$

(iii) $E[M_t|\mathcal{F}_s] = M_s$ *for* $s \leq t$

If condition (iii) is replaced by $E[M_t|\mathcal{F}_s] \leq M_s$ $(E[M_t|\mathcal{F}_s] \geq M_s)$, the process M is said to be a supermartingale (submartingale). From a more technical viewpoint, for the purposes of many applications the conditions in the above definition can be relaxed to some extent, in which case one speaks of a local martingale (see for example Protter (1990) for a rigorous definition of this concept).

In financial engineering applications, asset price processes are commonly modelled as *semimartingales*. These are processes which can be decomposed into the sum of a local martingale and an adapted process of finite variation, where the latter represents the drift of the asset price process.

4.2 Option pricing in continuous time

In models for option pricing, the stochastic process of the underlying asset price is typically driven by Brownian motion (also called Wiener process, denoted $W(t)$). This is a stochastic process starting at 0, with continuous sample paths and independent, stationary, Gaussian increments. This is to say that for all $t > s$, the increments $W(t) - W(s)$ of standard Brownian motion are normally distributed with mean zero and variance $t - s$, and are independent of $W(s)$. If one considers the asset price over time to be subjected to an aggregate of many small influences too complex to model individually, the functional central limit theorem[2] suggests Brownian motion as a natural way of modelling the uncertainty of the price process. Furthermore, Brownian motion satisfies the Markov property that for all $u > t$, the distribution of $W(u)$ given $W(t)$ is independent of all $W(s)$ with $s < t$. In other words, all information, present $W(t)$ and past $W(s)$, relevant for predicting the future value $W(u)$, is contained in the current value $W(t)$. This is consistent with the efficient markets hypothesis[3] that all information about an asset is aggregated into its current market price.

Suppose that the dynamics of the asset $X(t)$ underlying an option are given by a semimartingale, i.e. by some trend plus a random term. If this random term is given by Brownian motion, then in a continuous–time framework one can write, in differential notation,

$$dX(t) = \mu(t, X(t))dt + \sigma(t, X(t))dW(t) \tag{4.1}$$

where $\mu(t, X(t))$ is the drift (or trend) and $\sigma(t, X(t))$ is the volatility term, by which the infinitesimal increments $dW(t)$ of standard Brownian motion are scaled. Brownian motion is of unbounded variation on every interval and thus

---

[2] Also called Donsker's Invariance Principle — see e.g. Theorem 4.20 of Karatzas and Shreve (1991).

[3] For a review of the efficient markets hypothesis, see Fama (1970).

the integral

$$\int_a^b f(t, X(t)) \mathrm{d}W(t) \tag{4.2}$$

cannot be defined in the classical (Lebesgue/Stieltjes) manner. To deal with stochastic differential equations such as (4.1), Itô calculus (introduced by K. Itô in 1944) is required. The fundamental result, on which this calculus is based, is the Itô lemma:[4]

Theorem 4.3 *Let $f : [0, T] \times \mathbb{R} \to \mathbb{R}$ be a function continuously differentiable in its first argument and twice continuously differentiable in its second argument. Then the dynamics of $Y(t) = f(t, X(t))$ are given by*

$$\mathrm{d}f(t, X(t)) = \partial_1 f(t, X(t)) \mathrm{d}t + \partial_2 f(t, X(t)) \mathrm{d}X(t) + \frac{1}{2} \partial_{2,2} f(t, X(t)) \mathrm{d}\langle X \rangle(t)$$

where ∂_i denotes the derivative with respect to the i–th argument and $\mathrm{d}\langle X \rangle(t) = (\mathrm{d}X(t))^2$ is the quadratic variation of X, calculated according to the rules $(\mathrm{d}t)^2 = 0$, $\mathrm{d}t \cdot \mathrm{d}W(t) = 0$ and $(\mathrm{d}W(t))^2 = \mathrm{d}t$.

Setting $\mu(t, X(t)) \equiv \mu$ and $\sigma(t, X(t)) \equiv \sigma$, the solution to (4.1) becomes

$$X(t) = X(0) + \mu t + \sigma W(t) \tag{4.3}$$

which is commonly termed generalised Brownian motion. In the Black/Scholes framework, proportional returns of a stock $S(t)$ are modelled by a trend plus a random term, i.e.

$$\mathrm{d}S(t) = S(t)(\mu \mathrm{d}t + \sigma \mathrm{d}W(t)) \tag{4.4}$$

Setting $\mathrm{d}X(t) \equiv \mathrm{d}W(t)$ and $f(t, X(t)) = e^{(\mu - \frac{1}{2}\sigma^2)t + \sigma X(t)}$ in the Itô lemma, one easily verifies that the geometric Brownian motion

$$S(t) = S(0) \exp \left\{ \left(\mu - \frac{1}{2}\sigma^2 \right) t + \sigma W(t) \right\} \tag{4.5}$$

solves (4.4).

4.2.1 The Black/Scholes partial differential equation

To value an option, Black and Scholes (1973) use an arbitrage pricing approach. Since the stock price is the only source of uncertainty in this framework, suppose that a European call option paying

$$C(T, S(T)) = \max(0, S(T) - K) \tag{4.6}$$

at time T has a time $t \leq T$ price $C(t, S(t))$, which is a function of time and the current stock price. Consider a portfolio $V(t, S(t))$ consisting of a long position in the call and a short position in

$$\Delta = \partial_2 C(t, S(t)) \tag{4.7}$$

---

[4] This is a special case of the multidimensional Itô formula, given for example in Revuz and Yor (1994), p.139.

shares. Applying Theorem 4.3 to $V(t, S(t)) = C(t, S(t)) - \Delta S(t)$, the stochastic terms cancel and

$$dV(t, S(t)) = \left(\partial_1 C(t, S(t)) + \frac{1}{2} \partial_{2,2} C(t, S(t)) S(t)^2 \sigma^2 \right) dt \qquad (4.8)$$

Thus, the portfolio is locally riskless and in order to avoid arbitrage must earn a return equal to the riskfree rate of interest r. This results in the Black/Scholes partial differential equation (PDE)

$$\partial_1 C(t, S(t)) + \frac{1}{2} \partial_{2,2} C(t, S(t)) S(t)^2 \sigma^2 = r(C(t, S(t)) - \partial_2 C(t, S(t)) S(t)) \quad (4.9)$$

This PDE must hold for any derivative depending solely on S. The type of derivative determines the terminal and boundary conditions under which (4.9) must be solved, e.g. (4.6) for a European call option. For some terminal/boundary conditions and some specifications of the underlying dynamics, (4.9) can be solved explicitly, e.g. for a European call or put option in the original Black/Scholes framework; in other cases one must resort to numerical methods, such as the finite difference schemes covered in Chapter 5.

4.2.2 The probabilistic approach

Alternatively, one can price an option by calculating the expected discounted payoff under the risk–neutral probabilities. The equivalence between the two approaches is established by the Feynman/Kac formula,[5] which states that given the dynamics (4.1), under some technical conditions on $\mu(\cdot, \cdot)$ and $\sigma(\cdot, \cdot)$, the solution $f(t, x)$ to the PDE

$$\partial_1 f(t, x) + \mu(t, x) \partial_2 f(t, x) + \frac{1}{2} \sigma^2(t, x) \partial_{2,2} f(t, x) = 0 \qquad (4.10)$$

with the terminal condition $f(T, x) = g(x)$, can be written as

$$f(t, x) = E[g(X(T)) | X(t) = x] \qquad (4.11)$$

Setting $f(t, x) = e^{(T-t)r} C(t, x)$, $\mu(t, x) = xr$ and $\sigma(t, x) = x\sigma$, (4.10) is equivalent to (4.9) and therefore

$$C(t, S(t)) = e^{-(T-t)r} E[C(T, S(T)) | S(t)] \qquad (4.12)$$

where $S(t)$ evolves according to (4.4) with $\mu = r$, i.e. the expectation is calculated using a probability measure under which the drift of the underlying asset is equal to the riskfree rate. This is called the risk–neutral or spot martingale measure, as the price process of any traded asset discounted by the riskfree rate must be a martingale to avoid the possibility of arbitrage. The continuously compounded savings account $\beta(t) = \exp\{rt\}$ is the numeraire associated with this martingale measure.

Again, different derivatives can be priced by evaluating different payoffs

---

[5] See Karatzas and Shreve (1991), p. 268.

inside the expectation. When (4.9) cannot be solved explicitly, this representation suggests Monte Carlo simulation as an alternative numerical method (see Chapter 7). In cases where it is possible to find a closed form solution, one may favour probabilistic methods to solve (4.12) instead of (4.9). Then it is often convenient to change probability measure to another, equivalent measure under which asset prices discounted by different numeraire are martingales. This is the continuous–time analogue of the probability measures associated with different numeraires in the discrete–time setting of Chapter 3. Recall in particular the interpretation of the Cox/Ross/Rubinstein binomial option pricing formula (3.12) in terms of the probability of exercise under different probability measures. In fact, in the Black/Scholes framework considered here, we have continuous–time analogues of all the results derived in the case of one period and a finite state space in Section 3.1.2:[6] Firstly, the market is complete in the Harrison and Pliska (1981, 1983) sense, that is, the terminal value of every contingent claim can be replicated with a self–financing portfolio strategy. Secondly, the absence of arbitrage is equivalent to the existence of a unique equivalent martingale measure for each valid choice of numeraire, leading to unique arbitrage–free prices for derivative financial instruments. A valid numeraire is any asset or self-financing portfolio strategy, the value of which is a strictly positive semimartingale.

Girsanov's Theorem[7] establishes the link between equivalent probability measures. In what follows below, let E_P denote the expectation operator under the measure P. A probability measure Q is said to be equivalent to a probability measure P on a σ–field $\mathcal{F}$ if the two measures agree on null sets, i.e. for any $A \in \mathcal{F}$, $Q(A) = 0$ implies $P(A) = 0$ and vice versa.

Theorem 4.4 *Let W be a standard d–dimensional Brownian motion on a probability space $(\Omega, \mathcal{F}, P)$ with filtration $\{\mathcal{F}_t\}_{t \in [0,\infty)}$. Assume λ to be an adapted real–valued d–dimensional process such that the process*

$$Z(t) := \exp \left\{ \sum_{i=1}^{d} \int_0^t \lambda_i(s)\, dW_i(s) - \frac{1}{2} \int_0^t \|\lambda(s)\|^2 ds \right\} \qquad (4.13)$$

is a martingale under the probability measure P. Define a process

$$\tilde{W} = \{\tilde{W}(t) = (\tilde{W}_1(t), \dots, \tilde{W}_d(t)), \mathcal{F}_t, t \in [0, \infty)\}$$

by

$$\tilde{W}_i(t) := W_i(t) - \int_0^t \lambda_i(s)\, ds, \qquad 1 \le i \le d, \ t \in [0, \infty). \qquad (4.14)$$

Then for each fixed $T \in [0, \infty)$, the process $\{\tilde{W}_t, \mathcal{F}_t, 0 \le t \le T\}$ is a standard d–dimensional Brownian motion on $(\Omega, \mathcal{F}_T, \tilde{P}_T)$, where $\tilde{P}_T$ is a probability measure equivalent to P on $\mathcal{F}_T$, defined by

$$\tilde{P}_T(A) := E_P[\mathbb{I}_A Z(T)] \qquad \text{for all } A \in \mathcal{F}_T. \qquad (4.15)$$

---

[6] For a rigorous treatment, see e.g. Musiela and Rutkowski (1997).
[7] See Karatzas and Shreve (1991), p. 191.

The probability measure $\tilde{P}_T$ can equivalently be defined by its Radon/Nikodým derivative

$$\frac{\mathrm{d}\tilde{P}_T}{\mathrm{d}P} = Z(T), \qquad P\text{-a.s.} \tag{4.16}$$

When evaluating conditional expectations under a change of measure, a version of Bayes' theorem holds:

Theorem 4.5 *Let $\tilde{P}_T$ have a Radon/Nikodým derivative with respect to P given by (4.16) and consider a random variable on the probability space $(\Omega, \mathcal{F}_T, P)$, integrable with respect to P and $\tilde{P}_T$. Furthermore, let the σ–field $\mathcal{F}_t$ be a subset of $\mathcal{F}_T$. Then*

$$E_{\tilde{P}_T}[X(T)|\mathcal{F}_t] = \frac{E_P[X(T)Z(T)|\mathcal{F}_t]}{E_P[Z(T)|\mathcal{F}_t]} \tag{4.17}$$

The key result, which facilitates the derivation of option pricing formulae by a change of measure/change of numeraire, is the following

Theorem 4.6 (Geman, El Karoui and Rochet (1995)) *Let S_0 be a non–dividend paying numeraire and Q_0 the corresponding martingale measure. Let S_1 be another strictly positive asset price process. Then there exists an equivalent probability measure Q_1 defined by its Radon/Nikodým derivative with respect to Q_0*

$$\left.\frac{dQ_1}{dQ_0}\right|_{\mathcal{F}_T} = \frac{S_0(0)S_1(T)}{S_1(0)S_0(T)}$$

such that

i) The basic securities prices with respect to the numeraire S_1 are Q_1–martingales.

ii) If a contingent claim has a fair price (given by the expectation of the discounted value process of the replicating self–financing portfolio strategy) under (S_0, Q_0), then it has the same fair price under (S_1, Q_1) and the replicating self–financing portfolio strategy ("hedging strategy" for short) is the same.

To illustrate this technique, consider the derivation of the Black/Scholes formula applied to currency options. Let the continuously compounded savings accounts for domestic and foreign currency be given by

$$\beta(t) = e^{rt}, \qquad \tilde{\beta}(t) = e^{\tilde{r}t} \tag{4.18}$$

Under the domestic risk neutral measure Q, the spot exchange rate X in units of domestic currency per unit of foreign currency is assumed to follow a geometric Brownian motion

$$X(t) = X(0)\exp\left\{\mu t - \frac{1}{2}\sigma_X^2 t + \sigma_X W(t)\right\}. \tag{4.19}$$

The foreign savings account $\tilde{\beta}$ is an asset with a price in foreign currency, and $X(t)\tilde{\beta}(t)$ is an asset with a price in domestic currency. Therefore

$$\frac{X(t)e^{\tilde{r}t}}{e^{rt}} = X(0)e^{(\mu+\tilde{r}-r)t}e^{\sigma_X W(t)-\frac{1}{2}\sigma_X^2 t} \tag{4.20}$$

must be a martingale under Q.[8] So we must have $\mu = r - \tilde{r}$. Consider a currency call option, which at time T pays $\max(0, X(T) - K)$. At time zero, we can price this as

$$E_Q[e^{-rT}\max(0, X(T) - K)] =$$
$$e^{-rT}E_Q[X(T)\mathbb{I}_{\{X(T)>K\}}] - e^{-rT}KE_Q[\mathbb{I}_{\{X(T)>K\}}]. \tag{4.21}$$

From (4.19) it follows that

$$E[\mathbb{I}_{\{X(T)>K\}}] = N\left(\frac{\ln\frac{X(0)}{K} + (r - \tilde{r})T - \frac{1}{2}\sigma_X^2 T}{|\sigma_X|\sqrt{T}}\right) \tag{4.22}$$

where $N(\cdot)$ is the cumulative distribution function of the standard normal distribution.

To evaluate the first expectation on the right–hand side of (4.21), change to the foreign risk neutral measure $\tilde{Q}$, with numeraire $\tilde{\beta}(t) = e^{\tilde{r}t}$:

$$\frac{d\tilde{Q}}{dQ} = \frac{X(T)\tilde{\beta}(T)\beta(0)}{X(0)\tilde{\beta}(0)\beta(T)} = \exp\left\{-\frac{1}{2}\sigma_X^2 T + \sigma_X W(T)\right\}. \tag{4.23}$$

And therefore, by the Girsanov theorem, $d\tilde{W}(t) = dW(t) - \sigma_X dt$. Under $\tilde{Q}$,

$$dX(t) = X(t)((r - \tilde{r})dt + \sigma_X^2 dt + \sigma_X d\tilde{W}), \tag{4.24}$$

yielding

$$e^{-rT}E_Q[X(T)\mathbb{I}_{\{X(T)>K\}}] \tag{4.25}$$

$$= e^{-rT}E_{\tilde{Q}}\left[X(T)\mathbb{I}_{\{X(T)>K\}}\frac{X(0)\tilde{\beta}(0)\beta(T)}{X(T)\tilde{\beta}(T)\beta(0)}\right] \tag{4.26}$$

$$= \frac{\beta(0)}{\beta(T)}\frac{X(0)\tilde{\beta}(0)\beta(T)}{\tilde{\beta}(T)\beta(0)}\tilde{Q}\{X(T) > K\} \tag{4.27}$$

$$= X(0)e^{-\tilde{r}T}N\left(\frac{\ln\frac{X(0)}{K} + (r - \tilde{r} + \frac{1}{2}\sigma_X^2)T}{\sigma_X\sqrt{T}}\right) \tag{4.28}$$

Thus the Black/Scholes option price is[9]

$$X(0)e^{-\tilde{r}T}N(h_1) - Ke^{-rT}N(h_2) \tag{4.29}$$

---

[8] Note that the spot exchange rate $X(t)$ itself is *not* an asset, so discounting the exchange rate alone by the numeraire does not result in a martingale.

[9] This version of the Black/Scholes formula was first derived by Garman and Kohlhagen (1983).

with

$$h_{1,2} = \frac{\ln \frac{X(0)}{K} + (r - \tilde{r})T \pm \frac{1}{2}\sigma_X^2 T}{|\sigma_X|\sqrt{T}} \qquad (4.30)$$

Similarly, one can use the change of measure technique to derive the classical Black/Scholes formula for equity options. However, the currency option example is particularly illustrative, since — as in Chapter 3 — the numeraire change can be interpreted as a change from measuring value in domestic currency to measuring value in foreign currency. Note that $N(h_1)$ is the probability under the foreign risk–neutral measure that the option will end in the money, while $N(h_2)$ is the probability under the domestic risk–neutral measure that the option will end in the money.

Finally, consider two arbitrary assets X and Y with deterministic, time-dependent proportional volatility and paying continuous dividend yields d_x and d_y. The most straightforward way to extend the model to the simultaneous stochastic dynamics of several (arbitrarily correlated) assets is to reinterpret σ as a vector and W as a standard d-dimensional Brownian motion (as in Theorem 4.4). Any multiplication between vectors is then to be interpreted as a sum product and

$$\int_s^t \sigma(u)dW(u) = \sum_{i=1}^d \int_s^t \sigma^{(i)}(u)dW^{(i)}(u) \qquad (4.31)$$

where the superscript (i) denotes the i-th component of the volatility vector and of the d-dimensional Brownian motion, respectively. With this reinterpretation of σ and W, the previous derivations remain unchanged (i.e. they are equally valid for vector–valued σ and W). As there is only one vector–valued Brownian motion driving all asset price processes, there is no need to explicitly deal with correlation, greatly simplifying notation. However, any desired correlation can be achieved[10] by an appropriate choice of volatility vectors: Denote by $\sigma_X(t)$ and $\sigma_Y(t)$ the vector–valued volatility functions of the assets X and Y, respectively. The instantaneous correlation of the asset dynamics is given by

$$\rho_{XY}(t) = \frac{\sigma_X(t)\sigma_Y(t)}{\sqrt{\sigma_X^2(t)}\sqrt{\sigma_Y^2(t)}} \qquad (4.32)$$

and the correlation between $\ln X(T)$ and $\ln Y(T)$, viewed from time t ("now"), is given by

$$\text{Corr}[\ln X(T), \ln Y(T)] = \frac{\int_t^T \sigma_x(s)\sigma_y(s)ds}{\sqrt{\int_t^T \sigma_x^2(s)ds \int_t^T \sigma_y^2(s)ds}} \qquad (4.33)$$

The value process of the selffinancing strategy, which invests in X at time zero and reinvests all dividends, is $X(t)e^{d_x t}$.[11] Analogously to (4.20), the absence

---

[10] Thus the approach chosen here is equivalent to a formulation of the model in terms of one–dimensional, but correlated, driving Brownian motions.

[11] If one maintains the previous interpretation of X as an exchange rate, this selffinancing

of arbitrage requires that under the domestic spot martingale measure,

$$dX(t) = X(t)((r - d_x)dt + \sigma_x(t)dW(t)) \qquad (4.34)$$
$$dY(t) = Y(t)((r - d_y)dt + \sigma_y(t)dW(t)) \qquad (4.35)$$

and by Itô's Lemma

$$X(t) = X(0)\exp\left\{(r - d_x)t + \int_0^t \sigma_x(s)dW(s) - \frac{1}{2}\int_0^t \sigma_x^2(s)ds\right\} \qquad (4.36)$$

$$Y(t) = Y(0)\exp\left\{(r - d_y)t + \int_0^t \sigma_y(s)dW(s) - \frac{1}{2}\int_0^t \sigma_y^2(s)ds\right\} \qquad (4.37)$$

Again using the change–of–measure technique, one can prove[12]

Theorem 4.7 *Under the above assumptions, consider an option to exchange X for K units of Y at the maturity date T, i.e. a European option with payoff $[X(T) - KY(T)]^+$.*

1. The time $0 \le t \le T$ value of the option is given by

$$X(t)e^{-d_x(T-t)}\mathcal{N}(h_1(t)) - KY(t)e^{-d_y(T-t)}\mathcal{N}(h_2(t)) \qquad (4.38)$$

with

$$h_{1,2}(t) = \frac{\ln\frac{X(t)}{KY(t)} + (d_y - d_x)(T - t) \pm \frac{1}{2}\int_t^T (\sigma_x(s) - \sigma_y(s))^2 ds}{\sqrt{\int_t^T (\sigma_x(s) - \sigma(s))^2 ds}} \qquad (4.39)$$

2. The hedge portfolio for this option in terms of X and Y is given by

$$e^{-d_x(T-t)}\mathcal{N}(h_1(t)) \quad \text{units of } X$$
$$e^{-d_y(T-t)}\mathcal{N}(h_2(t))K \quad \text{units of } Y$$

This theorem covers all standard European options as special cases. For example, the currency call option formula (4.29) is obtained by setting $d_x = \tilde{r}$, $\sigma_x(t) = \sigma_x$ constant, $Y(t) = 1$, $d_y = r$ and $\sigma_y(t) = 0$. The formula for a put option on Y with strike $\hat{K}$ results, if one sets $X(t) = \hat{K}$, $d_x = r$, $\sigma_x = 0$ and $K = 1$.

Theorem 4.7 allows for continuous dividend yields. The other two cases of modelling dividends introduced in Section 3.2.4 are also easily integrated into the formula: Since the call option holder does not receive and the put option holder does not have to give up any dividends until the option is exercised, one can simply adjust the current asset prices $X(t)$ and $Y(t)$ by subtracting the present value of the (known) future dividends, which will be paid before

---

strategy corresponds to investing in the foreign savings account $e^{\tilde{r}t}$, resulting in the value process $X(t)e^{\tilde{r}t}$, as before.

[12] This theorem is a slight adaptation (by allowing for continuous dividend yields) of Theorem 3.1 of Frey and Sommer (1996), who provide a proof. It is a generalisation of the formula given by Margrabe (1978) for an option to exchange one asset for another.

the maturity of the European option.[13] Suppose that the asset X pays a fixed dividend of δ_x at time $t < \tau < T$. Then the call option becomes

$$(X(t) - \delta_x e^{-r(\tau-t)})\mathcal{N}(h_1) - Ke^{-r(T-t)}\mathcal{N}(h_2) \qquad (4.40)$$

with

$$h_{1,2} = \frac{\ln\frac{X(t)-\delta_x e^{-r(\tau-t)}}{K} + r(T-t) \pm \frac{1}{2}\int_t^T \sigma_x^2(s)ds}{\sqrt{\int_t^T \sigma_x^2(s)ds}} \qquad (4.41)$$

If one instead makes the modelling assumption that X pays a *proportional* dividend of δ_x some time τ between t and T, i.e. the amount paid as a dividend will be $\delta_x X(\tau)$, then the present value of this dividend is $\delta_x X(t)$ (regardless of τ), and the call option formula becomes

$$(1-\delta_x)X(t)\mathcal{N}(h_1) - Ke^{-r(T-t)}\mathcal{N}(h_2) \qquad (4.42)$$

with

$$h_{1,2} = \frac{\ln\frac{(1-\delta_x)X(t)}{K} + r(T-t) \pm \int_t^T \sigma_x^2(s)ds}{\sqrt{\int_t^T \sigma_x^2(s)ds}} \qquad (4.43)$$

4.2.3 Time–varying interest rates and dividend yields

Theorem 4.7 is easily extended to deterministically time–varying interest rates and dividend yields. First of all, one should note that the distinction between interest rates and dividend yields is purely semantic in the context of the present model: E.g. when using Theorem 4.7 to price a call option on foreign currency, we set the dividend yield of X to the foreign interest rate and the dividend yield of Y to the domestic interest rate. The exchange rate $X(t)$ gives the value in domestic currency at time t of a position in one unit of foreign currency, and foreign currency generates income at the foreign rate of interest, while the "strike" Y is paid in domestic currency, a position in which generates income at the domestic rate of interest. Making the domestic and foreign continuously compounded spot interest rates r and $\tilde{r}$ deterministically time–varying, one unit of domestic currency invested at time t accumulates to

$$\exp\left\{\int_t^T r(s)ds\right\}$$

at time T, while the value in domestic currency at time T of one unit of foreign currency invested at time t is

$$X(T)\exp\left\{\int_t^T \tilde{r}(s)ds\right\}$$

---

[13] In fact, $X(t)$ is multiplied by $e^{-d_x(T-t)}$ and $Y(t)$ is multiplied by $e^{-d_y(T-t)}$ in Theorem 4.7 for exactly this reason.

As long as r and $\tilde{r}$ are deterministic,[14] we can replace $(T - t)d$ in Theorem 4.7 by $\int_t^T r(s)ds$ to obtain the currency call option formula

$$X(t)\exp\left\{-\int_t^T \tilde{r}(s)ds\right\}\mathcal{N}(h_1(t)) - K\exp\left\{-\int_t^T r(s)ds\right\}\mathcal{N}(h_2(t)) \quad (4.44)$$

with

$$h_{1,2}(t) = \frac{\ln\frac{X(t)}{K} + \int_t^T(r(s) - \tilde{r}(s))ds \pm \frac{1}{2}\int_t^T \sigma_X^2(s)ds}{\sqrt{\int_t^T \sigma_X^2(s)ds}} \quad (4.45)$$

or in terms of the zero coupon bond prices (discount factors)

$$B(t,T) = \exp\left\{-\int_t^T r(s)ds\right\} \quad (4.46)$$

one can write (4.44) as

$$B(t,T)\left(\frac{X(t)\tilde{B}(t,T)}{B(t,T)}\mathcal{N}(h_1(t)) - K\mathcal{N}(h_2(t))\right) \quad (4.47)$$

with

$$h_{1,2}(t) = \frac{\ln\frac{X(t)\tilde{B}(t,T)}{B(t,T)} \pm \frac{1}{2}\int_t^T \sigma_X^2(s)ds}{\sqrt{\int_t^T \sigma_X^2(s)ds}} \quad (4.48)$$

Written in this form, this is the Black (1976) formula for a currency call option.

4.3 Exotic options with closed form solutions

An option to exchange one asset for another, the option priced in Theorem 4.7 would commonly be considered an "exotic" option, as opposed to the standard "plain vanilla" European calls and puts. However, this case also illustrated how standard options are special cases of a more general, "exotic" payoff structure. Skipper and Buchen (2003) took this to the next level with their "Quintessential Option Pricing Formula."[15] It covers a very large class of derivatives, the payoff of which may depend on several underlying assets and several *event dates*, e.g. discrete barrier or lookback monitoring dates. For derivatives with continuous monitoring, the work of Buchen and Konstandatos[16] points the way toward a similarly comprehensive framework.[17] In

---

[14] Chapter 8 shows how to extend this to stochastic interest rates.

[15] See also Buchen (2012).

[16] See Buchen (2001), Konstandatos (2004, 2008) and Buchen and Konstandatos (2005).

[17] As will be seen in what follows, the formulae for the case of discrete "event dates" can be written to allow for a non-flat term structure of interest rates and time-varying (but still deterministic) volatility. For the continuous monitoring case, this is not straightforward, so for this case we will follow the standard literature on closed form solutions for exotic option (see e.g. Zhang (1998)) and restrict ourselves to constant interest rates and constant volatility.

either case, a generalisation of binary options[18] is used to form the building blocks from which the exotic options are assembled. This leads intuitively to a highly flexible approach to implementing exotic option pricing formulae.

4.3.1 M–Binaries and the "Quintessential Option Pricing Formula"

Consider a set of assets X_i, $i = 1, \ldots, N$, with dividend yield d_{X_i}. As discussed in Section 4.2.3, we generalise the previously used dynamics (4.36) by allowing for deterministically time–varying interest rates and dividend yields.[19] In analogy to zero coupon bond prices, define the "dividend discount factors"

$$D_i(t, T) = \exp\left\{ - \int_t^T d_{X_i}(s)ds \right\} \qquad (4.49)$$

Under the risk neutral measure, the assets follow the dynamics

$$X_i(t) = \frac{X_i(0)D_i(0,t)}{B(0,t)} \exp\left\{ \int_0^t \sigma_{X_i}(s)dW(s) - \frac{1}{2} \int_0^t \sigma_{X_i}^2(s)ds \right\} \qquad (4.50)$$

Skipper and Buchen (2003) define an *M–Binary* as a derivative financial instrument, the payoff of which depends on the prices of a subset of the assets X_i observed at a set of discrete times T_k, $k = 1, \ldots, M$, and can be represented as

$$V_{T_M} = \underbrace{\left(\prod_{j=1}^n (X_{\bar{i}(j)}(T_{\bar{k}(j)}))^{\alpha_j} \right)}_{\text{payoff amount}} \underbrace{\prod_{l=1}^m \mathbb{I}\left\{ S_{l,l} \prod_{j=1}^n (X_{\bar{i}(j)}(T_{\bar{k}(j)}))^{A_{l,j}} > S_{l,l} a_l \right\}}_{\text{payoff indicator}}$$

$$(4.51)$$

The M–Binary is a European derivative in the sense that payoff is assumed to occur at time T_M. n is the *payoff dimension*, i.e. the number of (asset, time) combinations, on which the payoff of the M–Binary depends. Thus $1 \leq n \leq N \cdot M$ and the index mapping functions $\bar{i}(\cdot)$ and $\bar{k}(\cdot)$ serve to select (asset, time) combinations from the $\mathbb{X} \times \mathbb{T}$ set of possible combinations. m is the *exercise dimension*, i.e. the number of indicator functions of events determining whether a payoff occurs or not. The payoff amount must be representable as a product of powers α_j of asset prices.[20] Similarly, the indicator functions condition on the product of (possibly different) powers $A_{j,l}$ of asset prices being greater than some constants a_l, where this inequality can be reversed by setting $S_{l,l} = -1$ (instead of $S_{l,l} = 1$ for the "greater than" case). The products of powers of lognormally distributed asset prices are also

---

[18] These derivatives are also known as *digital options*.

[19] Skipper and Buchen (2003) assume constant r, d and σ, but their results are immediately applicable to the case of time–dependent interest rates, dividend yields and volatilities. Chapter 8 shows how to integrate stochastic interest rates into the present setting.

[20] Thus a notable case excluded by this representation is the arithmetic average or *Asian* option. This option does not have a closed form solution; it is priced numerically in Chapter 7.

lognormal, allowing one to calculate the time $t < T_1$ value of the M–Binary as

$$V_t = B(t, T_M)E[V_{T_M}] = \beta \mathcal{N}_m(Sh; SCS) \prod_{j=1}^{n} (X_{\bar{i}(j)}(t))^{\alpha_j} \qquad (4.52)$$

where $\mathcal{N}_m(\cdot; SCS)$ is the cumulative distribution function of the m–variate normal distribution with mean 0 and covariance matrix SCS, and

$$\beta = B(t, T) \exp\left\{\alpha^\top \mu + \frac{1}{2}\alpha^\top \Gamma \alpha\right\} \qquad \text{scalar}$$

$$h = D^{-1}(\ell + A(\mu + \Gamma\alpha)) \qquad \text{vector}$$

$$C = D^{-1}(A\Gamma A^\top)D^{-1} \qquad \text{matrix}$$

A and S are matrices and α is a vector as per (4.51), and μ is an n–dimensional vector with

$$\mu_j = -\ln \frac{B(t, T_{\bar{k}(j)})}{D_{\bar{i}(j)}(t, T_{\bar{k}(j)})} - \frac{1}{2}\int_t^{T_{\bar{k}(j)}} \sigma^2_{X_{\bar{i}(j)}}(s)ds \qquad (4.53)$$

Γ is the $n \times n$ covariance matrix of logarithmic asset prices for the n (asset,time) combinations, i.e.

$$\Gamma_{jl} = \int_t^{\min(T_{\bar{k}(j)}, T_{\bar{k}(l)})} \sigma_{X_{\bar{i}(j)}}(s)\sigma_{X_{\bar{i}(l)}}(s)ds \qquad (4.54)$$

Furthermore, ℓ is an m–dimensional vector with

$$\ell_l = \left(\sum_{j=1}^{n} A_{jl} \ln X_{\bar{i}(j)}(t)\right) - \ln a_l \qquad (4.55)$$

Lastly, D is an $m \times m$ diagonal matrix given by

$$D^2 = \text{diag}(A\Gamma A^\top) \qquad (4.56)$$

4.3.2 M–Binaries with exercise dimension one

In (4.52), the dimension of the required normal distribution is the *exercise dimension*, i.e. the number of indicator functions multiplied together to form the "payoff indicator." For a substantial number of common exotic payoffs, this dimension is one, and thus only a univariate normal distribution is required to price these options.

The simplest M–Binary is a cash–or–nothing option with payoff

$$\mathbb{I}\{SX_1(T_1) > a_1\} \qquad (4.57)$$

i.e. we have $N = M = n = m = 1$ and

$$\alpha = 0 \quad A = 1 \quad a = a_1 \quad \bar{i}(1) = 1 \quad \bar{k}(1) = 1$$

where $S = 1$ or -1 depending on whether we have an "up" or "down" type

binary. Following the notation used by Buchen and Konstandatos (2005), we
write the time t value of (4.57) as

$$B_{a_1}^S(X_1(t), \tau) = B(t, t+\tau)E[\mathbb{I}\{SX_1(T_1) > a_1\}] \qquad (4.58)$$

$$= B(t, T_1)\mathcal{N}\left(S\frac{\ln\frac{X_1(t)D_1(t,T_1)}{a_1 B(t,T_1)} - \frac{1}{2}\int_t^{T_1}\sigma_{X_1}^2(s)ds}{\sqrt{\int_t^{T_1}\sigma_{X_1}^2(s)ds}}\right)$$

with $\tau = T_1 - t$ the time to maturity.

The values for the asset–or–nothing option with payoff

$$X_1(T_1)\mathbb{I}\{SX_1(T_1) > a_1\} \qquad (4.59)$$

are the same, except that here $\alpha = 1$. Thus the time t value of (4.59) is

$$A_{a_1}^S(X_1(t), \tau) =$$

$$X_1(t)D_1(t, T_1)\mathcal{N}\left(S\frac{\ln\frac{X_1(t)D_1(t,T_1)}{a_1 B(t,T_1)} + \frac{1}{2}\int_t^{T_1}\sigma_{X_1}^2(s)ds}{\sqrt{\int_t^{T_1}\sigma_{X_1}^2(s)ds}}\right) \qquad (4.60)$$

A long position in one unit of an "up" (4.59) and a short position of a_1 units
of an "up" (4.57) results in the payoff of a European call option on X_1 with
strike a_1, i.e.

$$X_1(T_1)\mathbb{I}\{SX_1(T_1) > a_1\} - a_1\mathbb{I}\{SX_1(T_1) > a_1\} \qquad (4.61)$$

and applying (4.52) results in the Black/Scholes formula in the current setting,
i.e.

$$X_1(t)D_1(t, T_1)\mathcal{N}(h') - B(t, T_1)a_1\mathcal{N}(h)$$

with

$$h' = \frac{\ln\frac{X_1(t)D_1(t,T_1)}{a_1 B(t,T_1)} + \frac{1}{2}\int_t^{T_1}\sigma_{X_1}^2(s)ds}{\sqrt{\int_t^{T_1}\sigma_{X_1}^2(s)ds}}$$

$$h = h' - \sqrt{\int_t^{T_1}\sigma_{X_1}^2(s)ds}$$

Setting $S = -1$ and reversing the long and short positions results in the corresponding put. For an option to obtain K_1 units of the asset X_1 in exchange
for K_2 units of the asset X_2, define the two M–Binaries $V^{(1)}$ and $V^{(2)}$ by
setting $N = n = 2$, $M = m = 1$, $\bar{\imath}(i) = i$, $\bar{k}(k) = 1$ and

$$\alpha^{(1)} = \begin{pmatrix}1\\0\end{pmatrix} \qquad S^{(1)} = 1 \qquad A^{(1)} = (1\ -1) \qquad a^{(1)} = \frac{K_2}{K_1}$$

$$\alpha^{(2)} = \begin{pmatrix}0\\1\end{pmatrix} \qquad S^{(2)} = 1 \qquad A^{(2)} = (1\ -1) \qquad a^{(2)} = \frac{K_2}{K_1}$$

Using (4.52) to calculate the value of

$$K_1 V_t^{(1)} - K_2 V_t^{(2)}$$

yields the exchange option pricing formula of Theorem 4.7.

Options on the discrete geometric mean. These derivatives give the holder the right to exchange the geometric mean of the asset price observed at discrete points in time T_k, i.e.

$$\bar{X}_{\text{Geo}}(T_{\bar{N}}) = \sqrt[\bar{N}]{\prod_{k=1}^{\bar{N}} X(T_k)} \tag{4.62}$$

for either a cash amount (the "fixed strike" case) or the value of the underlying asset at maturity (the "floating strike" case). The time $T_{\bar{N}}$ payoff is thus given by

$$\left(\prod_{k=1}^{\bar{N}} X(T_k)^{1/\bar{N}} - K \right) \mathbb{I} \left\{ \prod_{k=1}^{\bar{N}} X(T_k)^{1/\bar{N}} > K \right\} \tag{4.63}$$

and

$$\left(X(T_{\bar{N}}) - K \prod_{k=1}^{\bar{N}} X(T_k)^{1/\bar{N}} \right) \mathbb{I} \left\{ X(T_{\bar{N}}) > K \prod_{k=1}^{\bar{N}} X(T_k)^{1/\bar{N}} \right\} \tag{4.64}$$

respectively. If the option already has been in existence for a while and the current geometric mean is $\bar{X}_{\text{Geo}}(t)$ based on $\hat{N}$ observations, then, supposing that there are still $\bar{N}$ observations to follow, (4.62) becomes

$$\bar{X}_{\text{Geo}}(T_{\bar{N}}) = (\bar{X}_{\text{Geo}}(t))^{\frac{\hat{N}}{\hat{N}+\bar{N}}} \prod_{k=1}^{\bar{N}} X(T_k)^{\frac{1}{\hat{N}+\bar{N}}} \tag{4.65}$$

and the payoff expressions change accordingly. For this latter, more general case, we write the price of a fixed strike geometric average option as

$$(\bar{X}_{\text{Geo}}(t))^{\frac{\hat{N}}{\hat{N}+\bar{N}}} V_t^{(1)} - K V_t^{(2)} \tag{4.66}$$

and the floating strike geometric average option as

$$V_t^{(1)} - K (\bar{X}_{\text{Geo}}(t))^{\frac{\hat{N}}{\hat{N}+\bar{N}}} V_t^{(2)} \tag{4.67}$$

where the inputs for (4.52) to calculate $V_t^{(1)}$ and $V_t^{(2)}$ in each of the two cases are given in Table 4.1.

Forward start options. These are options expiring at some time T_2, the exercise price of which is set equal to the observed asset price $X_1(T_1)$ at some (pre-determined) future date $T_1 < T_2$. The time T_2 payoff of a forward start call is thus

$$\max(X_1(T_2) - X_1(T_1), 0)$$

and the Skipper/Buchen inputs are $N = 1$, $M = n = 2$, $m = 1$, $\bar{\imath}(i) = 1$, $\bar{k}(k) = k$,

$$S = 1 \qquad A = (-1 \ 1) \qquad a = 1$$

	Fixed strike		Floating strike	
# of assets $N =$	1		1	
Payoff dimension $n =$	$\bar{N}$		$\bar{N}$	
# of observation times $M =$	$\bar{N}$		$\bar{N}$	
Exercise dimension $m =$	1		1	
Asset index mapping $\bar{\imath}(i) =$	1		1	
Time index mapping $\bar{k}(k) =$	k		k	
	$V^{(1)}$	$V^{(2)}$	$V^{(1)}$	$V^{(2)}$
Payoff powers α	$\frac{1}{\bar{N}+\tilde{N}}$	0	$\begin{aligned}\alpha_{\bar{N}} &= 1\\ \alpha_{j\neq\bar{N}} &= 0\end{aligned}$	$\frac{1}{\bar{N}+\tilde{N}}$
Sign matrix S	1	1	1	1
Indicator powers A	$\frac{1}{\bar{N}+\tilde{N}}$	$\frac{1}{\bar{N}+N}$	$\begin{aligned}A_{1,\bar{N}} &= 1 - \frac{1}{\bar{N}+\tilde{N}}\\ A_{1,j\neq\bar{N}} &= -\frac{1}{\bar{N}+\tilde{N}}\end{aligned}$	$\begin{aligned}A_{1,\bar{N}} &= 1 - \frac{1}{\bar{N}+\tilde{N}}\\ A_{1,j\neq\bar{N}} &= -\frac{1}{\bar{N}+\tilde{N}}\end{aligned}$
Thresholds a	$K(\bar{X}_{\text{Geo}}(t))^{\frac{-\tilde{N}}{\bar{N}+\tilde{N}}}$	$K(\bar{X}_{\text{Geo}}(t))^{\frac{-\tilde{N}}{\bar{N}+\tilde{N}}}$	$K(\bar{X}_{\text{Geo}}(t))^{\frac{\tilde{N}}{\bar{N}+\tilde{N}}}$	$K(\bar{X}_{\text{Geo}}(t))^{\frac{\tilde{N}}{\bar{N}+\tilde{N}}}$

Table 4.1 Inputs to calculate the price of fixed and floating strike discretely monitored geometric average options using the Skipper/Buchen formula.

and the payoff powers are, respectively,

$$\alpha^{(1)} = \begin{pmatrix} 0 \\ 1 \end{pmatrix} \quad \text{and} \quad \alpha^{(2)} = \begin{pmatrix} 1 \\ 0 \end{pmatrix}$$

for M–Binaries $V^{(1)}$ and $V^{(2)}$ in the time t option price given by

$$S(V_t^{(1)} - V_t^{(2)})$$

For the forward start put, all inputs remain the same, with the exception of $S = -1$.

4.3.3 M–Binaries with higher exercise dimensions

Discretely monitored barrier options. The payoff indicator functions for these options condition on whether the underlying asset price breached a pre-set barrier at any of the monitoring times over the life of the option. For example, the payoff of a down–and–out call option on asset X_1 with strike K and barrier H can be written as

$$(X(T_M) - K)\mathbb{I}\{X(T_M) > K\} \prod_{k=1}^{M} \mathbb{I}\{X(T_k) > H\} \qquad (4.68)$$

i.e. the option has the same payoff as a European call option if and only if the asset price was above the barrier at all monitoring times (otherwise the option is "knocked out" and worthless). The exercise dimension of this option is thus $m = M + 1$.[21]

Conversely, if the option pays only if the asset price was below the barrier at all monitoring times, one speaks of an up–and–out option. Since both the up–and–out and down–and–out could either be a call or a put, we have a total of four different "out" options, the Skipper/Buchen inputs for which are given in Tables 4.2 and 4.3.

Alternatively, instead of being "knocked out," an option could be "knocked in," i.e. the option has the same payoff as a European option if and only if the barrier *was* breached at at least one of the monitoring times. This condition does not fit the M–Binary framework. However, the observation that the sum of a knock–out option and the corresponding knock–in option is equal to the corresponding standard option, e.g.

down–and–out call + down–and–in call = standard European call

allows the knock–in options to be priced as the difference of the standard option and a knock–out option, thus allowing us to price all eight possible barrier options.

---

[21] Strictly speaking, this then requires a very high dimensional multivariate normal distribution to evaluate, but this can be reduced by focusing only on the most significant eigenvalues of the covariance matrix (see Section 4.4). In the limit, for continuously monitored barriers, the resulting formula can be expressed in terms of the cumulative distribution function of the univariate standard normal distribution (see Section 4.3.5).

	Call		Put	
# of assets $N =$	1		1	
Payoff dimension $n =$	$M+1$		$M+1$	
# of observation times $M =$	M		M	
Exercise dimension $m =$	$M+1$		$M+1$	
Asset index mapping $\bar{\imath}(i) =$	1		1	
Time index mapping $\bar{k}(k) =$	$\begin{cases} k & \text{if } k \leq M \\ M & \text{if } k = M+1 \end{cases}$		$\begin{cases} k & \text{if } k \leq M \\ M & \text{if } k = M+1 \end{cases}$	
	$V^{(1)}$	$V^{(2)}$	$V^{(1)}$	$V^{(2)}$
Payoff powers α	$\alpha_{M+1} = 1$ $\alpha_{j \neq M+1} = 0$	$\alpha_j = 0$	$\alpha_{M+1} = 1$ $\alpha_{j \neq M+1} = 0$	$\alpha_j = 0$
Sign matrix S	$S_{j,j} = 1$	$S_{j,j} = 1$	$S_{j,j} = \begin{cases} 1 & \text{if } j \leq M \\ -1 & \text{if } j = M+1 \end{cases}$	$S_{j,j} = \begin{cases} 1 & \text{if } j \leq M \\ -1 & \text{if } j = M+1 \end{cases}$
Indicator powers A	$A_{j,l} = 1$	$A_{j,l} = 1$	$A_{j,l} = 1$	$A_{j,l} = 1$
Thresholds a	$a_{M+1} = K$ $a_{j \neq M+1} = H$	$a_{M+1} = K$ $a_{j \neq M+1} = H$	$a_{M+1} = K$ $a_{j \neq M+1} = H$	$a_{M+1} = K$ $a_{j \neq M+1} = H$

Table 4.2 *Inputs to calculate the price of discretely monitored down–and–out barrier options using the Skipper/Buchen formula. Option values are given by $V^{(1)} - KV^{(2)}$ for calls and $KV^{(2)} - V^{(1)}$ for puts.*

	Call		Put	
# of assets $N =$	1		1	
Payoff dimension $n =$	$M+1$		$M+1$	
# of observation times $M =$	M		M	
Exercise dimension $m =$	$M+1$		$M+1$	
Asset index mapping $\bar{\imath}(i) =$	1		1	
Time index mapping $\bar{k}(k) =$	$\begin{cases} k & \text{if } k \leq M \\ M & \text{if } k = M+1 \end{cases}$		$\begin{cases} k & \text{if } k \leq M \\ M & \text{if } k = M+1 \end{cases}$	
	$V^{(1)}$	$V^{(2)}$	$V^{(1)}$	$V^{(2)}$
Payoff powers α	$\alpha_{M+1} = 1$ $\alpha_{j \neq M+1} = 0$	$\alpha_j = 0$	$\alpha_{M+1} = 1$ $\alpha_{j \neq M+1} = 0$	$\alpha_j = 0$
Sign matrix S	$S_{j,j} = \begin{cases} -1 & \text{if } j \leq M \\ 1 & \text{if } j = M+1 \end{cases}$	$S_{j,j} = \begin{cases} -1 & \text{if } j \leq M \\ 1 & \text{if } j = M+1 \end{cases}$	$S_{j,j} = -1$	$S_{j,j} = -1$
Indicator powers A	$A_{j,l} = 1$	$A_{j,l} = 1$	$A_{j,l} = 1$	$A_{j,l} = 1$
Thresholds a	$a_{M+1} = K$ $a_{j \neq M+1} = H$	$a_{M+1} = K$ $a_{j \neq M+1} = H$	$a_{M+1} = K$ $a_{j \neq M+1} = H$	$a_{M+1} = K$ $a_{j \neq M+1} = H$

Table 4.3 Inputs to calculate the price of discretely monitored up-and-out barrier options using the Skipper/Buchen formula. Option values are given by $V^{(1)} - KV^{(2)}$ for calls and $KV^{(2)} - V^{(1)}$ for puts.

	$V^{(1)}$	$V^{(2)}$	$V^{(3)}$
# of assets $N =$	1	1	1
Payoff dimension $n =$	2	2	1
# of observation times $M =$	2	2	1
Exercise dimension $m =$	2	2	1
Asset index mapping $\bar{\imath}(i) =$	1	1	1
Time index mapping $\bar{k}(k) =$	k	k	1
Payoff powers α	$\binom{0}{1}$	$\binom{0}{0}$	0
Sign matrix S	$\begin{pmatrix} 1 & 0 \\ 0 & 1 \end{pmatrix}$	$\begin{pmatrix} 1 & 0 \\ 0 & 1 \end{pmatrix}$	1
Indicator powers A	$\begin{pmatrix} 1 & 0 \\ 0 & 1 \end{pmatrix}$	$\begin{pmatrix} 1 & 0 \\ 0 & 1 \end{pmatrix}$	1
Thresholds a	$\binom{a_1}{K_2}$	$\binom{a_1}{K_2}$	a_1

Table 4.4 *Inputs to calculate the price of a call–on–call compound option using the Skipper/Buchen formula.*

Compound options. These are options on options, e.g. a call–on–call compound option gives the holder the right to buy a call option at time T, for a cash amount K_1, where that call option on some asset X_1 matures at time $T_2 > T_1$ and has strike K_2. The holder will exercise the right to buy the T_2 option if at time T_1 this option is worth more than K_1. In order to express the compound option in terms of M–Binaries, this condition must be expressed in terms of an inequality involving $X_1(T_1)$. Since the time T_1 value of the T_2 option on X_1 is monotonic in $X_1(T_1)$, we can find (by a simple root search) an a_1 such that the holder will exercise the compound option for any $X_1(T_1) > a_1$. The time T_2 payoff of the call–on–call compound option then becomes[22]

$$(X_1(T_2) - K_2)\mathbb{I}\{X_1(T_2) > K_2\}\mathbb{I}\{X_1(T_1) > a_1\}-$$
$$\frac{K}{B(T_1, T_2)}\mathbb{I}\{X_1(T_1) > a_1\} \quad (4.69)$$

Thus the compound option consists of three M–Binaries, the first two of which involve a bivariate normal distribution. Table 4.4 lists the Skipper/Buchen inputs for this option. The other three first–order[23] compound options differ only in the sign matrices S and in how the three M–Binaries $V^{(1)}$, $V^{(2)}$ and $V^{(3)}$ are assembled to form the option price. In order to calculate the time

---

[22] The time T_1 cash payment K_1 is taken to time T_2 by dividing by the time T_1 price of a zero coupon bond maturing in T_2.

[23] One could take this to higher orders by considering options on compound options, etc.

$t < T_1$ value of these options, we set

Call–on–call: $V_t^{(1)} - K_2 V_t^{(2)} - K_1 \dfrac{B(t,T_1)}{B(t,T_2)} V_t^{(3)}$

$$S^{(1)} = S^{(2)} = \begin{pmatrix} 1 & 0 \\ 0 & 1 \end{pmatrix} \qquad S^{(3)} = 1$$

Call–on–put: $K_2 V_t^{(2)} - V_t^{(1)} - K_1 \dfrac{B(t,T_1)}{B(t,T_2)} V_t^{(3)}$

$$S^{(1)} = S^{(2)} = \begin{pmatrix} -1 & 0 \\ 0 & -1 \end{pmatrix} \qquad S^{(3)} = -1$$

Put–on–call: $K_1 \dfrac{B(t,T_1)}{B(t,T_2)} V_t^{(3)} - (V_t^{(1)} - K_2 V_t^{(2)})$

$$S^{(1)} = S^{(2)} = \begin{pmatrix} -1 & 0 \\ 0 & 1 \end{pmatrix} \qquad S^{(3)} = -1$$

Put–on–put: $K_1 \dfrac{B(t,T_1)}{B(t,T_2)} V_t^{(3)} + (V_t^{(1)} - K_2 V_t^{(2)})$

$$S^{(1)} = S^{(2)} = \begin{pmatrix} 1 & 0 \\ 0 & -1 \end{pmatrix} \qquad S^{(3)} = 1$$

Applying (4.52) to these inputs results in the compound option formula first derived by Geske (1979).

4.3.4 Options on maxima or minima

Two–colour rainbow options. These options are calls or puts on the maximum or minimum of two assets. For example, the payoff of a call on the maximum of two assets is given by

$$\max(\max(X_1(T_1), X_2(T_1)) - K, 0) \tag{4.70}$$

	$V^{(1)}$	$V^{(2)}$	$V^{(3)}$	$V^{(4)}$	$V^{(5)}$	$V^{(6)}$	$V^{(7)}$
# of assets $N =$	2	2	2	2	1	1	2
Payoff dimension $n =$	2	2	2	2	1	1	2
# of observation times $M =$	1	1	1	1	1	1	1
Exercise dimension $m =$	2	2	2	2	1	1	2
Asset index mapping $\bar{\imath}(j) =$	j	j	j	j	1	2	j
Time index mapping $\bar{k}(k) =$	1	1	1	1	1	1	1
Payoff powers α	$(1\ 0)$	$(0\ 1)$	$(0\ 0)$	$(0\ 0)$	0	0	$(0\ 0)$
Sign matrix S	$S_{j,j} = 1$						
Indicator powers A	$\begin{pmatrix} 1 & -1 \\ 1 & 0 \end{pmatrix}$	$\begin{pmatrix} -1 & 1 \\ 0 & 1 \end{pmatrix}$	$\begin{pmatrix} 1 & -1 \\ 1 & 0 \end{pmatrix}$	$\begin{pmatrix} -1 & 1 \\ 0 & 1 \end{pmatrix}$	1	1	$\begin{pmatrix} 1 & 0 \\ 0 & 1 \end{pmatrix}$
Thresholds a	$(1\ K)$	$(1\ K)$	$(1\ K)$	$(1\ K)$	K	K	$(K\ K)$

Table 4.5 Inputs to calculate the price of a call on the maximum of two assets using the Skipper/Buchen formula. The option value can be expressed as $V^{(1)} + V^{(2)} - K(V^{(3)} + V^{(4)})$ or $V^{(1)} + V^{(2)} - K(V^{(5)} + V^{(6)} - V^{(7)})$.

Using the indicator function notation, this can be written as

$$
\underbrace{X_1(T_1)\mathbb{I}\left\{\frac{X_1(T_1)}{X_2(T_1)} > 1\right\}\mathbb{I}\{X_1(T_1) > K\}}_{V_{T_1}^{(1)}}
$$

$$
\underbrace{+ X_2(T_1)\mathbb{I}\left\{\frac{X_2(T_1)}{X_1(T_1)} \geq 1\right\}\mathbb{I}\{X_2(T_1) > K\}}_{V_{T_1}^{(2)}}
$$

$$
\underbrace{- K\,\mathbb{I}\left\{\frac{X_1(T_1)}{X_2(T_1)} > 1\right\}\mathbb{I}\{X_1(T_1) > K\}}_{V_{T_1}^{(3)}} \underbrace{- K\,\mathbb{I}\left\{\frac{X_2(T_1)}{X_1(T_1)} \geq 1\right\}\mathbb{I}\{X_2(T_1) > K\}}_{V_{T_1}^{(4)}}
$$

Note that the payoff events in $V^{(3)}$ and $V^{(4)}$ are disjoint (i.e. mutually exclusive) and we can write

$$
\left(\left\{\frac{X_1(T_1)}{X_2(T_1)} > 1\right\} \cap \{X_1(T_1) > K\}\right) \cup \left(\left\{\frac{X_2(T_1)}{X_1(T_1)} \geq 1\right\} \cap \{X_2(T_1) > K\}\right)
$$

$$
= \left(\left\{\frac{X_1(T_1)}{X_2(T_1)} > 1\right\} \cup \left\{\frac{X_2(T_1)}{X_1(T_1)} \geq 1\right\}\right) \cap \left(\{X_1(T_1) > K\} \cup \left\{\frac{X_2(T_1)}{X_1(T_1)} \geq 1\right\}\right)
$$

$$
\cap \left(\left\{\frac{X_1(T_1)}{X_2(T_1)} > 1\right\} \cup \{X_2(T_1) > K\}\right) \cap (\{X_1(T_1) > K\} \cup \{X_2(T_1) > K\})
$$

$$
= \Omega \cap \left(\{X_1(T_1) > K\} \cup \{X_2(T_1) > K\} \cup \left\{\frac{X_2(T_1)}{X_1(T_1)} \geq 1\right\}\right)
$$

$$
\cap \left(\{X_1(T_1) > K\} \cup \{X_2(T_1) > K\} \cup \left\{\frac{X_1(T_1)}{X_2(T_1)} > 1\right\}\right)
$$

$$
\cap (\{X_1(T_1) > K\} \cup \{X_2(T_1) > K\})
$$

$$
= (\{X_1(T_1) > K\} \cup \{X_2(T_1) > K\})
$$

and thus

$$
K(V_{T_1}^{(3)} + V_{T_1}^{(4)}) = K\mathbb{I}\{\{X_1(T_1) > K\} \cup \{X_2(T_1) > K\}\}
$$

$$
= \underbrace{K\,\mathbb{I}\{X_1(T_1) > K\}}_{V_{T_1}^{(5)}} + \underbrace{K\,\mathbb{I}\{X_2(T_1) > K\}}_{V_{T_1}^{(6)}}
$$

$$
\underbrace{- K\,\mathbb{I}\{X_1(T_1) > K\}\mathbb{I}\{X_2(T_1) > K\}}_{V_{T_1}^{(7)}}
$$

replacing the evaluation of one bivariate normal cumulative distribution function with the evaluation of two univariate normal cumulative distribution functions, which is somewhat more efficient computationally. The Skipper/Buchen inputs for $V_{T_1}^{(1)}$ to $V_{T_1}^{(7)}$ are given in Table 4.5.

Generalisation. Skipper and Buchen (2003) provide a useful theorem which allows the unified treatment of contingent payoffs which condition on the maximum or minimum of a set of asset prices. Expressed directly in terms of the M–Binary payoff notation in (4.51), it reads

Theorem 4.8 *Let X be as in (4.51) and let K be a positive scalar. Then $X_{\bar{\imath}(p)}(T_{\bar{k}(p)})$ is the maximum (resp. minimum) of all $X_{\bar{\imath}(\cdot)}(T_{\bar{k}(\cdot)})$ and K if and only if*

$$\prod_{l=1}^{m} \mathbb{I}\left\{ S_{l,l} \prod_{j=1}^{n} X_{\bar{\imath}(j)}^{A_{l,j}}(T_{\bar{k}(j)}) > S_{l,l}a_l \right\} = 1 \qquad (4.71)$$

with $S_{l,l} = 1$ for the maximum and $S_{l,l} = -1$ for the minimum for all l and

$$A_{l,j} = \begin{cases} 1 & \text{if } j = p \\ -1 & \text{if } l = j \neq p \\ 0 & \text{otherwise} \end{cases} \qquad a_l = \begin{cases} K & \text{if } l = p \\ 1 & \text{if } l \neq p \end{cases}$$

The payoff dimension n and the exercise dimension m, as well as the index mapping functions $\bar{\imath}(\cdot)$ and $\bar{k}(\cdot)$, are determined by the particular payoff in question. For example, when there are N assets X_i observed at a single time point T_1, and the payoff is given by one unit of a particular asset $X_p(T_1)$ if $X_p(T_1)$ is the maximum of all $X_i(T_1)$ and K, then we have a case of what Skipper and Buchen (2003) call a *quality binary*. Then, to price this payoff using (4.52) and the above theorem, set $M = 1$, $m = N$, $n = N$ and

$$\bar{\imath}(i) = i \qquad \bar{k}(k) = 1 \qquad \alpha_j = \begin{cases} 1 & \text{if } j = p \\ 0 & \text{if } j \neq p \end{cases}$$

Denote the time t price of a quality binary on the p-th asset being the maximum of the cash amount K and n asset prices $X_{\bar{\imath}(\cdot)}(T_{\bar{k}(\cdot)})$ assembled in an n-dimensional vector X by $Q_t^+(X, K, p)$, and the corresponding price of a quality binary on the minimum by $Q_t^-(X, K, p)$. These can then be used as building blocks to assemble the prices of certain *multicolour rainbow options*. Of the payoffs considered by Rubinstein (1991), Skipper and Buchen (2003) show that two types are amenable to representation in terms of quality binaries:[24] Best/worst options and calls/puts on the maximum or minimum of several assets.

Best and worst options. These options pay the maximum ("best") or minimum("worst") of a cash amount K and a set of N asset prices $X_i(T_1)$. The

---

[24] Rubinstein (1991) gives formulas for *two–colour rainbow options*. For this case, as in the previous section, the Skipper/Buchen formula yields the same results. However, some of the payoffs considered by Rubinstein cannot be represented in the multiplicative form (4.51), in particular spread options, portfolio (basket) options and dual–strike options. The sums and/or differences of lognormal random variables making up these payoffs are not lognormal, and there are no closed form pricing formulas for these cases.

time t value of a "best" option is given by

$$\sum_{p=1}^{N} Q_t^+(X, K, p) + KV_t^-$$

where KV_t^- represents the value of the cash amount K paid if

$$\max_{1 \le i \le N} X_i(T_1) < K$$

V_t^- is calculated along the lines of Theorem 4.8, but setting $S_{l,l} = -1$ and

$$A_{l,j} = \begin{cases} 1 & \text{if } l = j \\ 0 & \text{otherwise} \end{cases} \qquad a_l = K \ \forall \ l$$

with $M = 1$, $m = N$, $n = N$ and

$$\bar{\imath}(i) = i \qquad \bar{k}(k) = 1 \qquad \alpha_j = 0 \ \forall \ j$$

Similarly, the time t value of a "worst" option is

$$\sum_{p=1}^{N} Q_t^-(X, K, p) + KV_t^+$$

with V_t^+ defined as V_t^-, except that $S_{l,l} = 1 \ \forall \ l$.

Options on the maximum or minimum. These options were first priced by Johnson (1987), and an application of the Skipper/Buchen methodology yields the same result. For a call on the maximum of N assets, i.e. with payoff

$$\left(\max_{1 \le i \le N} X_i(T_1) - K \right) \mathbb{I} \left\{ \max_{1 \le i \le N} X_i(T_1) > K \right\} \qquad (4.72)$$

the time $t < T_1$ value is

$$\sum_{p=1}^{N} Q_t^+(X, K, p) + KV_t^- - KB(t, T_1)$$

and, similarly, the value of a put on the minimum of the N assets is

$$KB(t, T_1) - \left(\sum_{p=1}^{N} Q_t^-(X, K, p) + KV_t^+ \right)$$

A put on the maximum and a call on the minimum cannot be directly expressed in terms of the quality binaries defined using Theorem 4.8. Rather, the matrix A and the vector a in Theorem 4.8 must be appropriately redefined — this is left as an exercise for the reader.

Discrete lookback options. Theorem 4.8 can also be applied to payoffs conditioning on the maximum or minimum of a set of M prices $X_1(T_k)$ of a single asset[25] X_1 observed at discrete monitoring times T_k, $1 \le k \le M$. Again fol-

---

[25] This can also be extended in a straightforward manner to several assets observed at discrete monitoring times.

lowing Skipper and Buchen (2003), define the *lookback binary* $L^+(X, K, p)$ as an option which pays $X_1(T_p)$ at time T_M if and only if

$$\max_{1 \leq k \leq M} X_1(T_k) = X_1(T_p) \qquad \text{and} \qquad X_1(T_p) > K$$

Conversely, $L^-(X, K, p)$ pays $X_1(T_p)$ if and only if

$$\min_{1 \leq k \leq M} X_1(T_k) = X_1(T_p) \qquad \text{and} \qquad X_1(T_p) < K$$

To price these lookback binaries using (4.52) and Theorem 4.8, set $m = M$, $n = M$ and

$$\bar{\imath}(i) = 1 \qquad \bar{k}(k) = k \qquad \alpha_j = \begin{cases} 1 & \text{if } j = p \\ 0 & \text{if } j \neq p \end{cases}$$

The time $t < T_1$ price of a fixed strike discrete maximum lookback call, with payoff

$$\left(\max_{1 \leq k \leq M} X_1(T_k) - K \right) \underbrace{\mathbb{I}\left\{ \max_{1 \leq k \leq M} X_1(T_k) > K \right\}}_{=1 - \mathbb{I}\left\{ \max_{1 \leq k \leq M} X_1(T_k) \leq K \right\}}$$

is then given by

$$\sum_{p=1}^{M} L_t^+(X, K, p) - K(B(t, T_M) - \hat{V}_t^-(K))$$

where $\hat{V}_t^-(K)$ is calculated in line with Theorem 4.8, but setting $S_{l,l} = -1$ and

$$A_{l,j} = \begin{cases} 1 & \text{if } l = j \\ 0 & \text{otherwise} \end{cases} \qquad a_l = K \,\forall\, l \qquad \alpha_j = 0 \,\forall\, j$$

If some of the monitoring dates lie in the past, there will already be an existing maximum $X_{\max}(t)$ (or minimum $X_{\min}(t)$) over the asset prices observed at those dates. In this case the payoff becomes

$$\left(\max \left(X_{\max}(t), \max_{1 \leq k \leq M} X_1(T_k) \right) - K \right)$$

$$\mathbb{I}\left\{ \max \left(X_{\max}(t), \max_{1 \leq k \leq M} X_1(T_k) \right) > K \right\}$$

and the time $t < T_1$ price is

$$\sum_{p=1}^{M} L_t^+(X, \max(X_{\max}(t), K), p)$$

$$- KB(t, T_M) + \max(X_{\max}(t), K)\hat{V}_t^-(\max(X_{\max}(t), K)) \quad (4.73)$$

There are a total of eight variants of discrete lookback options: One can have either fixed or floating strike, call or put, on a maximum or minimum, though

	Call	Put
	$\sum_{p=1}^M L_t^+(X, \max(X_{\max}(t),K),p) - KB(t,T_{M1})$ $+ \max(X_{\max}(t),K)\hat{V}_t^-(\max(X_{\max}(t),K))$	$\left.\begin{cases} 0 & \text{if } X_{\max}(t) \geq K \\ \sum_{p=1}^M \left(L_t^+(X,K,p) - L_t^+(X,X_{\max}(t),p)\right) \\ \quad + K\hat{V}_t^-(K) - X_{\max}(t)\hat{V}_t^-(X_{\max}(t)) & \text{if } X_{\max}(t) < K \end{cases}\right.$
# of assets $N =$		
Payoff dimension $n =$	1	1
# of observation times $M =$	M	M
Exercise dimension $m =$	M	M
Asset index mapping $\bar{i}(i) =$	M	M
Time index mapping $\bar{k}(k) =$	1	1
	k	k
	$L_t^+(\cdot, \boldsymbol{\xi}, \boldsymbol{p})$	$\hat{V}_t^-(\boldsymbol{\xi})$
Payoff powers α	$\alpha_j = \begin{cases} 1 & \text{if } j = p \\ 0 & \text{if } j \neq p \end{cases}$	$\alpha_j = 0$
Sign matrix S	$S_{j,j} = 1$	$S_{j,j} = -1$
Indicator powers A	$A_{l,j} = \begin{cases} 1 & \text{if } j = p \\ -1 & \text{if } l = j \neq p \\ 0 & \text{otherwise} \end{cases}$	$A_{l,j} = \begin{cases} 1 & \text{if } l = j \\ 0 & \text{otherwise} \end{cases}$
Thresholds a	$a_l = \begin{cases} \xi & \text{if } l = p \\ 1 & \text{if } l \neq p \end{cases}$	ξ

Table 4.6 Inputs to calculate the price of a discretely monitored fixed strike maximum lookback options using the Skipper/Buchen formula. Option values are given by (4.73) for calls and analogously for puts.

	Call	Put
	$0 \qquad$ if $X_{\min}(t) \leq K$ $X_{\min}(t)\hat{V}_t^+(X_{\min}(t)) - K\hat{V}_t^+(K)$ $\left. -\sum_{p=1}^{M}\left(L_t^-(X,K,p) \right. \right.$ $\left. \left. - L_t^-(X,X_{\min}(t),p) \right) \right\}$ if $X_{\min}(t) > K$	$KB(t,T_M) -$ $\sum_{p=1}^{M} L_t^-(X,\min(X_{\min}(t),K),p)$ $- \min(X_{\min}(t),K)\hat{V}_t^+(\min(X_{\min}(t),K))$
# of assets $N =$	1	
Payoff dimension $n =$	M	
# of observation times $M =$	M	
Exercise dimension $m =$	M	
Asset index mapping $\bar{\imath}(i) =$	1	
Time index mapping $\bar{k}(k) =$	k	
	$L_t^-(\cdot,\boldsymbol{\xi},\boldsymbol{p})$	$\hat{V}_t^+(\boldsymbol{\xi})$
Payoff powers α	$\alpha_j = \begin{cases} 1 & \text{if } j = p \\ 0 & \text{if } j \neq p \end{cases}$	$\alpha_j = 0$
Sign matrix S	$S_{j,j} = -1$	$S_{j,j} = 1$
Indicator powers A	$A_{l,j} = \begin{cases} 1 & \text{if } j = p \\ -1 & \text{if } l = j \neq p \\ 0 & \text{otherwise} \end{cases}$	$A_{l,j} = \begin{cases} 1 & \text{if } l = j \\ 0 & \text{otherwise} \end{cases}$
Thresholds a	$a_l = \begin{cases} \xi & \text{if } l = p \\ 1 & \text{if } l \neq p \end{cases}$	ξ

Table 4.7 *Inputs to calculate the price of a discretely monitored fixed strike minimum lookback options using the Skipper/Buchen formula.*

certain combinations do not make much sense — a floating strike minimum put would pay

$$\left(\min_{1 \leq k \leq M} X_1(T_k) - X(T_M) \right) \mathbb{I} \left\{ \min_{1 \leq k \leq M} X_1(T_k) > X(T_M) \right\}$$

but the indicator function would always evaluate to zero. Similarly, a floating strike maximum call would never be exercised. Conversely, floating strike minimum calls and floating strike maximum puts will always be exercised, and it makes sense to implement pricing formulas for these cases. The Skipper/Buchen inputs for the four fixed strike discrete lookbacks are listed in Tables 4.6 and 4.7. The two floating strike discrete lookbacks use the same building blocks L_t^-, L_t^+, $\hat{V}_t^-$ and $\hat{V}_t^+$, with the floating strike minimum call given by

$$X(t) - \sum_{p=1}^{M} L_t^- (X, X_{\min}(t), p) - X_{\min}(t) \hat{V}_t^- (X_{\min}(t)) \qquad (4.74)$$

and the put by

$$\sum_{p=1}^{M} L_t^+ (X, X_{\max}(t), p) + X_{\max}(t) \hat{V}_t^+ (X_{\max}(t)) - X(t) \qquad (4.75)$$

Note that another exotic, the *one–clique option*, can easily be expressed as a discrete lookback. For example, Zhang (1998) gives the payoff of a one–clique call as (translated to our notation)

$$\max(X_1(T_2) - K, X_1(T_1) - K, 0)$$

which is equivalent to

$$\max(\max(X_1(T_1), X_1(T_2)) - K, 0)$$

i.e. the payoff of a fixed strike maximum call with two monitoring dates T_1 and T_2.

4.3.5 Continuous monitoring and the method of images

For discretely monitored barrier and lookback options, the exercise dimension increases with the number of monitoring dates, quickly compounding the computational effort required to price the option. Although in practice there can be no such thing as truly continuous monitoring — any set of asset price observations is necessarily discrete — it therefore makes sense to approximate very frequent monitoring with the continuous case. In a series of papers, Buchen and Konstandatos have shown how to express the prices of a wide range of continuously monitored barrier and lookback options using a common set of building blocks.[26]

---

[26] See Buchen (2001, 2012), Konstandatos (2004, 2008) and Buchen and Konstandatos (2005).

Barrier options. For the eight standard continuously monitored barrier options, we shall see that once the price of one has been derived, the remaining seven follow in a straightforward manner. Following Buchen (2001), we start with a down–and–out call. In the Black/Scholes framework, any derivative financial instrument must satisfy the partial differential equation (4.9), with the appropriate boundary and terminal conditions. Taking σ and r to be constant in the present analysis, we have for the value V_{do} of the down–and–out option

$$\mathcal{L}V_{\text{do}}(x,t) = \frac{\partial V_{\text{do}}}{\partial t} - rV_{\text{do}} + rx\frac{\partial V_{\text{do}}}{\partial x} + \frac{1}{2}\sigma^2 x^2 \frac{\partial^2 V_{\text{do}}}{\partial x^2} = 0 \quad (4.76)$$

$$V_{\text{do}}(b,t) = 0$$

$$V_{\text{do}}(x,T) = f(x)$$

on $x > b$, $t < T$, where $f(x)$ is the option payoff. The key to solving (4.76) by the method of images for the Black/Scholes equation is given by the following[27]

Theorem 4.9 *To solve the problem (4.76), first solve the related full–range problem with the introduction of an indicator function in the initial condition:*

$$\mathcal{L}V_b(x,t) = 0 \quad \text{for } x > 0 \text{ and } t < T \quad (4.77)$$

$$V_b(x,T) = f(x)\mathbb{I}\{x > b\}$$

The solution for $V_{do}(x,t)$ is then given by

$$V_{do}(x,t) = V_b(x,t) - \overset{*}{V}_b(x,t) \quad (4.78)$$

where

$$\overset{*}{V}_b(x,t) = \left(\frac{b}{x}\right)^\alpha V_b\left(\frac{b^2}{x},t\right), \qquad \alpha = 2r/\sigma^2 - 1 \quad (4.79)$$

is the image of $V_b(x,t)$ relative to the operator $\mathcal{L}$ and the barrier b.

It is easily verified that $\overset{*}{V}_b(x,t)$ satisfies the five properties of an image function:

1. The image of the image is the original function, i.e.

$$\left(\overset{*}{V}_b(x,t)\right)^* = \left(\frac{b}{x}\right)^\alpha \left(\frac{b}{b^2/x}\right)^\alpha V_b\left(\frac{b^2}{b^2/x},t\right) = V_b(x,t) \quad (4.80)$$

2. If $V_b(x,t)$ satisfies the Black/Scholes PDE $\mathcal{L}V_b = 0$, then it follows that $\overset{*}{V}_b(x,t)$ also satisfies $\mathcal{L}\overset{*}{V}_b = 0$, as applying the Black/Scholes operator $\mathcal{L}$

---

[27] This is Theorem 3.2.2 of Konstandatos (2004).

to $\overset{*}{V}_b$ as given by (4.79) yields

$$\mathcal{L}\overset{*}{V}_b = \left(\frac{b}{x}\right)^\alpha \partial_2 V_b - r\left(\frac{b}{x}\right)^\alpha V_b + rx\left(-\frac{\alpha b^\alpha}{x^{\alpha+1}}V_b - \left(\frac{b}{x}\right)^{\alpha+2}\partial_1 V_b\right)$$

$$+ \frac{1}{2}\sigma^2 x^2\left((\alpha+1)\frac{\alpha b^\alpha}{x^{\alpha+2}}V_b + \frac{\alpha b^\alpha}{x^{\alpha+1}}\left(\frac{b}{x}\right)^2\partial_1 V_b\right.$$

$$\left. + \frac{b^{\alpha+2}}{x^{\alpha+3}}(\alpha+2)\partial_1 V_b + \left(\frac{b}{x}\right)^{\alpha+4}\partial_{1,1}V_b\right)$$

$$= \left(\frac{b}{x}\right)^\alpha\left(\partial_2 V_b - rV_b + r\frac{b^2}{x}\partial_1 V_b + \frac{1}{2}\sigma^2\left(\frac{b^2}{x}\right)^2\partial_{1,1}V_b\right)$$

$$= \left(\frac{b}{x}\right)^\alpha \mathcal{L}V_b\left(\frac{b^2}{x},t\right) = 0$$

3. When $x = b$,
$$\overset{*}{V}_b(b,t) = \left(\frac{b}{b}\right)^\alpha V_b\left(\frac{b^2}{b},t\right) = V_b(b,t)$$

4. If $x > b$ (respectively $x < b$) is the active domain of $V_b(x,t)$, then $x < b$ (respectively $x > b$) is the active domain of $\overset{*}{V}_b(x,t)$.

5. The imaging operator is linear:
$$(cV_1(x,t) + dV_2(x,t))^* = \left(\frac{b}{x}\right)^\alpha\left(cV_1\left(\frac{b^2}{x},t\right) + dV_2\left(\frac{b^2}{x},t\right)\right)$$
$$= c\overset{*}{V}_1(x,t) + d\overset{*}{V}_2(x,t)$$

The representation (4.78) of the down–and–out option price then follows from the following argument. We need to verify that $V_b(x,t) - \overset{*}{V}_b(x,t)$ solves (4.76):

1. Since by assumption $V_b(x,t)$ solves (4.77), it follows that $\overset{*}{V}_b(x,t)$ also satisfies $\mathcal{L}\overset{*}{V}_b = 0$, and therefore, due to the linearity of the operator $\mathcal{L}$, $\mathcal{L}V_{\text{do}} = 0$.

2. Since $V_b(b,t) = \overset{*}{V}_b(b,t)$, $V_{\text{do}}(b,t) = 0$.

3. For the terminal value $V_{\text{do}}(x,T)$, only the case $x > b$ is relevant, since otherwise the option has been knocked out at the barrier b. For $x > b$, $V_b(x,T) = f(x)$ and $\overset{*}{V}_b(x,T) = 0$, so $V_{\text{do}}(x,T) = f(x)$ as required.

Thus it only remains to obtain $V_b(x,t)$. (4.77) is the Black/Scholes option pricing problem for a European option with time T payoff
$$f(X(T))\mathbb{I}\{X(T) > b\}$$

If $f(X(T))$ is the call option payoff
$$(X(T) - K)\mathbb{I}\{X(T) > K\}$$

then

$$V_b(X(T), T) = \begin{cases} (X(T) - K)\mathbb{I}\{X(T) > K\} & \text{for } b \leq K \\ X(T)\mathbb{I}\{X(T) > b\} - K\mathbb{I}\{X(T) > b\} & \text{for } b > K \end{cases}$$

and the value $V_b(X(t), t)$ for $t < T$ can be written in terms of known Black/Scholes solutions (without having to solve (4.77) explicitly) by a simple static replication argument. If $b \leq K$, the payoff $V_b(X(T), T)$ is identical to that of a European call option, while for $b > K$, V_b can be represented as the difference of an asset–or–nothing option (4.59) (also called an *asset binary*) and K cash–or–nothing options (4.57) (also called a *bond binary*).

Making these results explicit for the case of a down–and–out call option with barrier $b > K$, we have

$$V_b(X(t), t) = X(t)\mathcal{N}(h_1) - Ke^{-r(T-t)}\mathcal{N}(h_2) \tag{4.81}$$

$$h_{1,2} = \frac{\ln\frac{X(t)}{b} + \left(r \pm \frac{1}{2}\sigma^2\right)(T-t)}{\sigma\sqrt{T-t}}$$

and thus

$$\begin{aligned} V_{\text{do}}(X(t), t) &= V_b(X(t), t) - \overset{*}{V}_b(X(t), t) \\ &= X(t)\mathcal{N}(h_1) - Ke^{-r(T-t)}\mathcal{N}(h_2) \\ &\quad - \left(\frac{b}{X(t)}\right)^{\frac{2r}{\sigma^2}-1}\left(\frac{b^2}{X(t)}\mathcal{N}(h_1^*) - Ke^{-r(T-t)}\mathcal{N}(h_2^*)\right) \end{aligned} \tag{4.82}$$

with

$$h_{1,2}^* = \frac{\ln\frac{b}{X(t)} + \left(r \pm \frac{1}{2}\sigma^2\right)(T-t)}{\sigma\sqrt{T-t}}$$

which matches the corresponding formula given by Reiner and Rubinstein (1991).

Given the price of a down–and–out call option, the price of a down–and–in call option can be obtained by in/out parity, i.e.

$$V_{\text{di}}(X(t), t) = V_O(X(t), t) - V_{\text{do}}(X(t), t) \tag{4.83}$$

where V_O is the price of a standard call option.

To obtain the "up" options, we can make use of what Buchen (2001) calls *up/down symmetry*, which gives

$$V_{\text{ui}}(X(t), t) = \overset{*}{V}_{\text{di}}(X(t), t) \tag{4.84}$$

To see that this property holds, it suffices to note that

$$V_{\text{ui}}(b, t) = \overset{*}{V}_{\text{di}}(b, t) = V_O(b, t),$$

that the active domains of the up–and–in and the image of the down–and–in option are both $x < b$, and that the terminal value at expiry (if the barrier wasn't hit) is zero in both cases. In summary, we have the main result of

Buchen (2001):

$$V_{\text{do}} = V_b - \overset{*}{V}_b \tag{4.85}$$

$$V_{\text{di}} = V_O - \left(V_b - \overset{*}{V}_b \right) \tag{4.86}$$

$$V_{\text{ui}} = \overset{*}{V}_O - \left(\overset{*}{V}_b - V_b \right) \tag{4.87}$$

$$V_{\text{uo}} = \left(V_O - \overset{*}{V}_O \right) - \left(V_b - \overset{*}{V}_b \right) \tag{4.88}$$

This result holds analogously for puts, in which case V_O is the price of a standard European put option and

$$V_b(X(T), T) = \begin{cases} (K - X(T))\mathbb{I}\{X(T) < K\} & \text{for } b > K \\ K\mathbb{I}\{X(T) < b\} - X(T)\mathbb{I}\{X(T) < b\} & \text{for } b \leq K \end{cases}$$

can be represented either as a standard put or a difference between K bond binaries and an asset binary. (4.85)–(4.88) then become (note the reversal of "up" and "down")

$$V_{\text{uo}} = V_b - \overset{*}{V}_b \tag{4.89}$$

$$V_{\text{ui}} = V_O - \left(V_b - \overset{*}{V}_b \right) \tag{4.90}$$

$$V_{\text{di}} = \overset{*}{V}_O - \left(\overset{*}{V}_b - V_b \right) \tag{4.91}$$

$$V_{\text{do}} = \left(V_O - \overset{*}{V}_O \right) - \left(V_b - \overset{*}{V}_b \right) \tag{4.92}$$

The derivation of the barrier option prices in terms of V_O, V_b and their images was fairly generic in terms of the payoff function $f(x)$. Consequently, this approach can be applied directly to price options which combine exotic payoff functions $f(x)$ with barrier features, as long as it is possible to evaluate V_O and V_b.

Lookback options. Buchen and Konstandatos (2005) develop the method reviewed in the previous section further to yield a similar modular pricing approach for lookback options. This analogy between lookback and barrier options is not surprising; for example, a down–and–out option conditions on the minimum value of the underlying asset over the life of the option in the sense that the option is knocked out if and only if this minimum is less than the barrier.

Starting point of the analysis is the PDE representation of the option price, which can be stated as the following

Theorem 4.10 *Let X be the price process of a non–dividend–paying asset following Black/Scholes–type dynamics with constant continuously compounded*

interest rate r and volatility σ. Denote by

$$Y(t) = \min_{0 \le \tau \le t} X(\tau)$$

$$Z(T) = \max_{0 \le \tau \le t} X(\tau)$$

the continuously monitored minimum and maximum of the asset price from the inception of the option contract to time t. Then the price U of a European option with time T payoff of

$$U(x, y, T) = f(x, y) \qquad resp. \qquad U(x, z, T) = F(x, z)$$

satisfies

$$\mathcal{L}U = -\frac{\partial U}{\partial t} + rU - rx\frac{\partial U}{\partial x} - \frac{1}{2}\sigma^2 x^2 \frac{\partial^2 U}{\partial x^2} = 0 \tag{4.93}$$

with

$$U(x, y, T) = f(x, y) \qquad and \qquad \partial_2 U(y, y, t) = 0$$

resp.

$$U(x, z, T) = F(x, z) \qquad and \qquad \partial_2 U(z, z, t) = 0$$

This is a commonly cited, but non-trivial result, as the role of the maximum resp. minimum has been reduced from that of a variable to that of a parameter. It is worth outlining the derivation, which appears in Wilmott, Howison and Dewynne (1995). Define

$$I_n(t) = \int_0^t (X(s))^n ds \tag{4.94}$$

and

$$Z_n(t) = (I_n(t))^{\frac{1}{n}} \tag{4.95}$$

Under the Black/Scholes assumptions, sample paths of the process $X(s)$ are almost surely continuous, and as Wilmott et al. note, results from stochastic calculus allow us to assume continuity without loss of generality. Consequently,

$$\lim_{n \to \infty} Z_n(t) = \max_{0 \le \tau \le t} X(\tau) \tag{4.96}$$

$$\lim_{n \to -\infty} Z_n(t) = \min_{0 \le \tau \le t} X(\tau) \tag{4.97}$$

By its definition as the integral (4.94), $I_n(t)$ (and consequently $Z_n(t)$) follows a process of finite variation, and thus

$$dZ_n(t) = \frac{1}{n}(I_n(t))^{\frac{1}{n}-1} dI_n(t) = \frac{1}{n}\frac{X(t)^n}{Z_n(t)^{n-1}} dt \tag{4.98}$$

Approximating $U(x, z, t)$ by $U(x, z_n, t)$ and applying Ito's Lemma, we have

$$dU_n = \frac{\partial U}{\partial t} dt + \frac{\partial U}{\partial X} dX + \frac{\partial U}{\partial Z_n} dZ_n + \frac{1}{2}\frac{\partial^2 U}{\partial X^2}(dX)^2 \tag{4.99}$$

where any second–order terms involving Z_n vanish because Z_n is of finite variation. Setting up a portfolio

$$\Pi = U - \Delta X \qquad \text{with} \qquad \Delta = \frac{\partial U}{\partial X}$$

we get the locally riskless dynamics

$$d\Pi = \frac{\partial U}{\partial t}dt + \frac{\partial U}{\partial Z_n}dZ_n + \frac{1}{2}\frac{\partial^2 U}{\partial X^2}(dX)^2 \qquad (4.100)$$

which on the other hand must equal

$$d\Pi = r\Pi dt = r\left(U - \frac{\partial U}{\partial X}X\right)dt \qquad (4.101)$$

in order to preclude arbitrage. Combining (4.100) and (4.101) and inserting the dynamics of Z_n and X results in the PDE

$$\frac{\partial U}{\partial t} + \frac{\partial U}{\partial Z_n}\frac{1}{n}\frac{X(t)^n}{Z_n(t)^{n-1}} + \frac{\partial U}{\partial X}X + \frac{1}{2}\sigma^2 X^2 \frac{\partial^2 U}{\partial X^2} - rU = 0 \qquad (4.102)$$

Since

$$X(t) \le \max_{0 \le \tau \le t} X(\tau) = \lim_{n \to \infty} Z_n(t),$$

the coefficient of the partial derivative with respect to Z_n tends to zero as n tends to infinity, as well as for n to $-\infty$, since

$$X(t) \ge \min_{0 \le \tau \le t} X(\tau) = \lim_{n \to -\infty} Z_n(t).$$

Thus (4.93) must hold.

The boundary conditions on the partial derivative of the lookback option price with respect to the maximum (resp. minimum) can be justified by noting that when the current price of the underlying asset is equal to the current maximum (resp. minimum), it follows from the properties of Brownian motion that there is zero probability that the current maximum (resp. minimum) is still the maximum (resp. minimum) at expiry of the option. Therefore, in this situation the option price must be insensitive to small changes in the extremum.   □

Buchen and Konstandatos (2005) then note that the minimum–type lookback option pricing problem of Theorem 4.10 "closely resembles the PDE satisfied by a down–and–out barrier option with barrier level $x = y$, the only difference being the boundary condition at $x = y$."[28] This resemblance becomes evident when one considers the partial differential equation, which $V := \partial_2 U$ must satisfy: Since the second argument of U acts only as a parameter in the PDE (4.93), $\mathcal{L}U = 0$ implies $\mathcal{L}V = 0$. The boundary conditions become $V(x,y,T) = \partial_2 f(x,y)$ (resp. $V(x,z,T) = \partial_2 F(x,y)$) and $V(y,y,t) = 0$ (resp. $V(z,z,t) = 0$). For the minimum–type lookback, this problem matches (4.76), the PDE and boundary conditions satisfied by a down–and–out barrier option, and the maximum–type lookback can analogously be matched with an

---

[28] See Buchen and Konstandatos (2005), p. 247.

up–and–out barrier. From Theorem 4.9, it then follows that the terminal condition on V (its "payoff") can equivalently be represented as

$$V_{eq}(x, y, T) = \partial_2 f(x, y)\mathbb{I}\{x > y\} - (\partial_2 f(x, y))^* \mathbb{I}\{x < y\} \tag{4.103}$$

resp.

$$V_{eq}(x, z, T) = \partial_2 F(x, z)\mathbb{I}\{x < z\} - (\partial_2 F(x, z))^* \mathbb{I}\{x > z\} \tag{4.104}$$

where $*$ represents the image operator introduced in the previous section. Integrating with respect to the second argument yields

Theorem 4.11 *The payoff $f(x, y)$ (resp. $F(x, z)$) of a European lookback option can be represented as*

$$U_{eq}(x, y, T) = f(x, y)\mathbb{I}\{x > y\} + g(x, y)\mathbb{I}\{x < y\} \tag{4.105}$$

resp.

$$U_{eq}(x, z, T) = F(x, y)\mathbb{I}\{x < z\} + G(x, y)\mathbb{I}\{x > z\} \tag{4.106}$$

with

$$
\begin{aligned}
g(x, y) &= f(x, x) - \int_x^y (\partial_2 f(x, \xi))^* d\xi \\
G(x, z) &= F(x, x) + \int_z^x (\partial_2 F(x, \xi))^* d\xi
\end{aligned}
$$

Taking a modular approach, Buchen and Konstandatos (2005) apply this result to price two building blocks from which the prices of the various lookback options can then be assembled: an option which pays the observed minimum (i.e. $f(x, y) = y$) and an option which pays the observed maximum (i.e. $F(x, z) = z$). Recalling the definition (4.79) of the image operator, the equivalent payoff for the minimum is

$$
\begin{aligned}
m_{eq} &= y\mathbb{I}\{x > y\} + \left(x - \int_x^y \left(\frac{\xi}{x}\right)^\alpha d\xi\right)\mathbb{I}\{x < y\} \\
&= y\mathbb{I}\{x > y\} + x\mathbb{I}\{x < y\} - \frac{x}{\alpha + 1}\left(\left(\frac{y}{x}\right)^{\alpha+1} - 1\right)\mathbb{I}\{x < y\} \\
&= y\mathbb{I}\{x > y\} + \left(1 + \frac{1}{\alpha}\right)x\mathbb{I}\{x < y\} - \frac{y}{\alpha + 1}\left(\frac{y}{x}\right)^\alpha \mathbb{I}\left\{y < \frac{y^2}{x}\right\}
\end{aligned}
$$

Thus this building block can be statically replicated using three previously encountered binaries:

1. y units of the cash–or–nothing binary $\mathbb{I}\{x > y\}$
2. $\left(1 + \frac{1}{\alpha}\right)$ units of the asset–or–nothing binary $x\mathbb{I}\{x < y\}$
3. $-y(\alpha + 1)^{-1}$ units of the cash–or–nothing binary image

$$\overset{*}{\mathbb{I}}\{x > y\} = \left(\frac{y}{x}\right)^\alpha \mathbb{I}\left\{y < \frac{y^2}{x}\right\}$$

Similarly, the equivalent payoff for the maximum is

$$M_{\text{eq}} = z\mathbb{I}\{x < z\} + \left(x + \int_z^x \left(\frac{\xi}{x}\right)^\alpha d\xi\right)\mathbb{I}\{x > z\}$$

$$= z\mathbb{I}\{x < z\} + \left(1 + \frac{1}{\alpha}\right)x\mathbb{I}\{x > z\} - \frac{z}{\alpha + 1}\left(\frac{z}{x}\right)^\alpha \mathbb{I}\left\{z > \frac{z^2}{x}\right\}$$

and the static replication is the same as for the minimum, except that the inequalities in the indicator functions are reversed.

Proceeding to the continuous counterparts of the discrete lookback options considered in Section 4.3.4, note that we are already in a position to price the two meaningful floating strike lookbacks. The payoff of the floating strike minimum call is

$$\max\left(0, X(T) - \min_{0 \leq \tau \leq T} X(\tau)\right)$$

This will always be exercised, so its time t value can be written as

$$X(t) - m_{\text{eq}}(X(t), Y(t), t)$$

where $m_{\text{eq}}(x, y, t)$ is assembled from the time t prices of the three binaries listed above. Similarly, the floating strike maximum put will always be exercised at expiry and its time t price is

$$M_{\text{eq}}(X(t), Z(t), t) - X(t)$$

To reduce fixed strike lookback options to simpler building blocks, note that if the current maximum $Z(t)$ is replaced by $\max(Z(t), K)$, for some cash amount K, i.e. $M_{\text{eq}}(X(t), \max(Z(t), K), t)$, this will ensure that the terminal payoff of the "maximum binary" M_{eq} will be the greater of K and the terminal maximum $Z(T)$. Similarly, $m_{\text{eq}}(X(t), \min(Y(t), K), t)$ yields an instrument which pays the lesser of K and the terminal minimum $Y(T)$. The four fixed strike lookback options can thus be represented in terms of the existing building blocks as summarised in Table 4.8. The representation is in terms of static replication of the payoffs at expiry; the value of the lookback options at any $t < T$ is thus given by the time t values of the components of the replicating portfolio. For example, the time t value of a fixed strike maximum lookback call with payoff $[Z(T) - K]^+$ is

$$M_{\text{eq}}(X(t), \max(Z(t), K), t) - e^{-(T-t)}K$$

Continuously monitored options with more complex payoffs. The techniques reviewed above for the pricing of continuously monitored barrier and lookback options can also be employed to derive closed–form or near closed–form expressions for the prices of yet more complex path–dependent payoffs. Buchen and Konstandatos (2005) present two examples: partial–price and partial–time lookbacks.

For partial–price lookback options,[29] the value of the extremum in the op-

---

[29] See also Conze and Viswanathan (1991).

Payoff	Static replication
$[Y(T) - K]^+$	$m_{eq}(X(T), Y(T), T) - m_{eq}(X(T), \min(Y(T), K), T)$
$[Z(T) - K]^+$	$M_{eq}(X(T), \max(Z(T), K), T) - K$
$[K - Y(T)]^+$	$K - m_{eq}(X(T), \min(Y(T), K), T)$
$[K - Z(T)]^+$	$M_{eq}(X(T), \max(Z(T), K), T) - M_{eq}(X(T), Z(T), T)$

Table 4.8 *Static replication of fixed strike lookback options using the "lookback binaries"* m_{eq} *and* M_{eq}

tion payoff is scaled by a constant, typically in such a way as to reduce the price of the option. For example, a partial–price floating strike minimum lookback call will have the payoff $[X(T) - \lambda Y(T)]^+$, with typically $\lambda \geq 1$.[30] To apply Theorem 4.11, set $f(x, y) = (x - \lambda y)\mathbb{I}\{x > \lambda y\}$ and (for $\lambda \geq 1$) $f(x, x) = 0$, $\partial_2 f(x, y) = -\lambda\mathbb{I}\{x > \lambda y\}$. The static replication in terms of known building blocks is then given by

$$(x - \lambda y)\mathbb{I}\{x > \lambda y\}\mathbb{I}\{x > y\} + \mathbb{I}\{x < y\} \int_x^y \lambda \left(\frac{\xi}{x}\right)^\alpha \mathbb{I}\left\{\frac{\xi^2}{x} > \lambda\xi\right\} d\xi$$

$$= (x - \lambda y)\mathbb{I}\{x > \lambda y\} + \mathbb{I}\{x < y\} \int_x^y \lambda \left(\frac{\xi}{x}\right)^\alpha \mathbb{I}\{\xi > \lambda x\} d\xi$$

$$= x\mathbb{I}\{x > \lambda y\} - \lambda y\mathbb{I}\{x > \lambda y\} - \frac{\lambda^{\alpha+2}}{\alpha+1} x\mathbb{I}\left\{x < \frac{y}{\lambda}\right\}$$

$$+ \frac{\lambda y}{\alpha+1} \left(\frac{y}{x}\right)^\alpha \mathbb{I}\left\{\lambda y < \frac{y^2}{x}\right\}$$

i.e. a position in

1. one unit of the asset–or–nothing binary $\mathbb{I}\{x > \lambda y\}$

2. $-\lambda y$ units of the cash–or–nothing binary $\mathbb{I}\{x > \lambda y\}$

3. $-\lambda^{\alpha+2}/(\alpha + 1)$ units of the asset–or–nothing binary $\mathbb{I}\{x < y/\lambda\}$

4. $\lambda y/(\alpha + 1)$ units of the cash–or–nothing binary image $\overset{*}{\mathbb{I}}\{x > \lambda y\}$ (where the image is taken at y)[31]

---

[30] Note that in this case the option is *not* always exercised, and all four possible partial–price floating strike lookback options (i.e. call/put, maximum/minimum) potentially have non-trivial payoffs.

[31] Buchen and Konstandatos (2005) use a slightly different notation and take the image at y/λ.

replicates the option. Making the time t option price explicit, we have

$$
X(t)\mathcal{N}\left(\frac{\ln\frac{X(t)}{\lambda Y(t)} + \left(r + \frac{1}{2}\sigma^2\right)(T - t)}{\sigma\sqrt{T - t}}\right)
$$

$$
- \lambda Y(t)e^{-r(T-t)}\mathcal{N}\left(\frac{\ln\frac{X(t)}{\lambda Y(t)} + \left(r - \frac{1}{2}\sigma^2\right)(T - t)}{\sigma\sqrt{T - t}}\right)
$$

$$
- \frac{\lambda^{\alpha+2}}{\alpha + 1}X(t)\mathcal{N}\left(-\frac{\ln\frac{\lambda X(t)}{Y(t)} + \left(r + \frac{1}{2}\sigma^2\right)(T - t)}{\sigma\sqrt{T - t}}\right)
$$

$$
+ \frac{\lambda}{\alpha + 1}Y(t)e^{-r(T-t)}\mathcal{N}\left(\frac{\ln\frac{Y(t)}{\lambda} + \left(r - \frac{1}{2}\sigma^2\right)(T - t)}{\sigma\sqrt{T - t}}\right)\left(\frac{Y(t)}{X(t)}\right)^{\alpha} \quad (4.107)
$$

where, as defined in Theorem 4.9, $\alpha = 2r/\sigma^2 - 1$. This result matches the pricing formula for this option given in equation (22) of Conze and Viswanathan (1991). The other three cases of partial–price floating strike lookbacks can be solved analogously.

A *partial–time lookback call* has a payoff at time $T_2 > T_1$ of $[X(T_2) - Y(T_1)]^+$, i.e. it is a floating strike minimum lookback call where the minimum is "frozen" at time T_1. Viewed from another perspective, it is a forward start call where the strike is given by the observed minimum up to time T_1, rather than the time T_1 asset price: From time T_1, the option is a standard call with strike $Y(T_1)$. Thus one can set $f(x, y) = C(x, y, \tau)$, where this denotes the Black/Scholes price of a call option with strike y and time to maturity $\tau = T_2 - T_1$, given the current asset price x. Applying Theorem 4.11, the time T_1 payoff can be represented as

$$
C(x, y, \tau)\mathbb{I}\{x > y\} + \left(C(x, x, \tau) + \int_x^y (\partial_2 C(x, \xi, \tau))^* d\xi\right)\mathbb{I}\{x < y\} \quad (4.108)
$$

Consider first the indefinite integral for $\tau = 0$:

$$
\begin{aligned}
\int (\partial_2 C(x, \xi, 0))^* d\xi &= \int \overset{*}{\mathbb{I}}\{x > \xi\}d\xi \\
&= \int \left(\frac{\xi}{x}\right)^{\alpha}\mathbb{I}\left\{\frac{\xi^2}{x} > \xi\right\}d\xi \\
&= \frac{1}{\alpha + 1}\left(\xi\left(\frac{\xi}{x}\right)^{\alpha} - x\right)\mathbb{I}\left\{\frac{\xi^2}{x} > \xi\right\} \\
&= \frac{1}{\alpha + 1}\left(\xi\overset{*}{\mathbb{I}}\{x > \xi\} - x\mathbb{I}\{\xi > x\}\right) \quad (4.109)
\end{aligned}
$$

Since C satisfies the Black/Scholes PDE, so does the derivative $\partial_2 C$, because the second argument of C is just a parameter. Applying the image operator and then integrating with respect to the parameter ξ preserves this property, so

$$\int (\partial_2 C(x, \xi, \tau))^* d\xi \tag{4.110}$$

satisfies the Black/Scholes PDE, and the right hand side of (4.109) gives the equivalent representation of the payoff. So the value of (4.110) for any time to maturity $\tau = T_2 - T_1$ can be statically replicated by $\xi(\alpha + 1)^{-1}$ units of the cash–or–nothing binary image $\overset{*}{\mathbb{I}}\{x > \xi\}$ and $-(\alpha + 1)^{-1}$ units of the asset–or–nothing binary $\mathbb{I}\{\xi > x\}$. Using the Buchen/Konstandatos notation (given in equations (4.59) and (4.60)), the value of (4.110) can be written as

$$\frac{\sigma^2}{2r} \left(\xi \overset{*}{B}{}_{\xi}^{+} (x, \tau) - A_{\xi}^{-} (x, \tau) \right)$$

Substituting this into (4.108), one obtains

$$C(x, y, \tau)\mathbb{I}\{x > y\} + \Bigg(C(x, x, \tau)$$

$$+ \frac{\sigma^2}{2r} \left(y \overset{*}{B}{}_{y}^{+} (x, \tau) - A_y^{-} (x, \tau) - x \overset{*}{B}{}_{x}^{+} (x, \tau) + A_x^{-} (x, \tau) \right) \Bigg) \mathbb{I}\{x < y\} \tag{4.111}$$

As per Section 4.3.2, the standard call option prices C can likewise be expressed in terms of cash–or–nothing and asset–or–nothing binaries, e.g.

$$C(x, y, \tau) = A_y^{+} (x, \tau) - y B_y^{+} (x, \tau)$$

Furthermore, note that at the money, i.e. when the strike equals the current value of the underlying asset, the binary building blocks simplify to

$$\begin{aligned} A_x^s(x, \tau) &= x\mathcal{N}(s\bar{h}_1\sqrt{\tau}) \\ B_x^s(x, \tau) &= e^{-r\tau}\mathcal{N}(s\bar{h}_2\sqrt{\tau}) \end{aligned}$$

with

$$\bar{h}_{1,2} = \frac{r \pm \frac{1}{2}\sigma^2}{\sigma}$$

so (4.111) becomes

$$(A_y^{+} (x, \tau) - y B_y^{+} (x, \tau))\mathbb{I}\{x > y\} + \eta(\tau)x\mathbb{I}\{x < y\}$$

$$+ \frac{\sigma^2}{2r} \left(y \overset{*}{B}{}_{y}^{+} (x, \tau) - A_y^{-} (x, \tau) \right) \mathbb{I}\{x < y\} \tag{4.112}$$

with

$$\eta(\tau) = \mathcal{N}(\bar{h}_1\sqrt{\tau}) - e^{-r\tau}\mathcal{N}(\bar{h}_2\sqrt{\tau}) + \frac{\sigma^2}{2r} \left(\mathcal{N}(-\bar{h}_1\sqrt{\tau}) - e^{-r\tau}\mathcal{N}(\bar{h}_2\sqrt{\tau}) \right)$$

For time T_1 payoffs of the type

$$A_{\xi_2}^{s_2}(X(T_1), \tau)\mathbb{I}\{s_1 X(T_1) > s_1 \xi_1\} \tag{4.113}$$

Buchen and Konstandatos (2005) introduce the notation $A_{\xi_1 \xi_2}^{s_1 s_2}$ and call this a *second order asset binary*. (4.113) represents the value $A_{\xi_1 \xi_2}^{s_1 s_2}$ at time T_1 and consequently its time T_2 payoff can be written as

$$X(T_2)\mathbb{I}\{s_2 X(T_2) > s_2\xi_2\}\mathbb{I}\{s_1 X(T_1) > s_1\xi_1\}$$

Thus $A_{\xi_1 \xi_2}^{s_1 s_2}$ has two event dates T_1 and T_2 and its value at time $t < T_1$ can be calculated using the Skipper/Buchen formula (4.52) with the inputs $N = 1$, $n = M = m = 2$, $\bar{\imath}(i) = 1$, $\bar{k}(k) = k$ and

$$\alpha = \begin{pmatrix} 0 \\ 1 \end{pmatrix} \quad S = \begin{pmatrix} s_1 & 0 \\ 0 & s_2 \end{pmatrix} \quad A = \begin{pmatrix} 1 & 0 \\ 0 & 1 \end{pmatrix} \quad a = \begin{pmatrix} \xi_1 \\ \xi_2 \end{pmatrix} \quad (4.114)$$

Similarly, a second order bond binary $B_{\xi_1 \xi_2}^{s_1 s_2}$ has time T_2 payoff

$$\mathbb{I}\{s_2 X(T_2) > s_2\xi_2\}\mathbb{I}\{s_1 X(T_1) > s_1\xi_1\}$$

resulting in the same Skipper/Buchen inputs (4.114) with the exception of $\alpha = (0 \ 0)^\top$. Now all the building blocks are in place to write the time $t < T_1$ value of the equivalent payoff (4.112) of a partial–time lookback call in terms of first– and second–order binaries and their images as

$$A_{yy}^{++}(X(t), \tau_1, \tau_2) - Y(t)B_{yy}^{++}(X(t), \tau_1, \tau_2) + \eta(T_2 - T_1)A_y^-(X(t), \tau_1)$$
$$+ \frac{\sigma^2}{2r}\left(Y(t)\overset{*}{B}_{yy}^{+-}(X(t), \tau_1, \tau_2) - A_{yy}^{--}(X(t), \tau, \tau_2)\right)$$

with $\tau_i := T_i - t$.

The method of images can be taken further to provide similarly modular forms of closed–form and near closed–form expressions for ever more complex payoffs. Konstandatos (2004) pursues this path and provides solutions for lookback and barrier options with partial–time, multiple–window and double–barrier features, at times combining several of these in a single option and impressively demonstrating the power of this approach. For our purposes, it is the modular nature of the Skipper and Buchen (2003) and Buchen and Konstandatos (2005) results that facilitates an object–oriented implementation of the option pricing formulas. This is the topic of the next section.

4.4 Implementation of closed form solutions

As we have seen in the previous sections, the "Quintessential Option Pricing Formula" (4.52) of Skipper and Buchen (2003) covers nearly all cases where the prices of options with a set of discrete event dates are available in closed form in the Black/Scholes framework. It also permits pricing of the building blocks, into which continuously monitored barrier and lookback options can be decomposed via the method of images. Thus this formula represents the ideal centerpiece around which to structure a C++ implementation of option pricing formulas. It is implemented in the class `MBinary`, the data members of which collect the payoff representation (4.51), the underlying assets X_i and the current (deterministic) term structure of interest rates — all of which are required to calculate the option price via (4.52) in `MBinary::price()`.

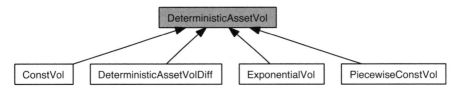

Figure 4.1 *Classes representing deterministic volatility functions*

The term structure of interest rates can be represented by any class derived from the abstract base class `TermStructure`,[32] while the underlying asset(s) X_i are given as a `std::vector` of pointers to `BlackScholesAsset` objects. The class `BlackScholesAsset` serves to represent assets with deterministic proportional volatility. We have already encountered this class briefly in Section 3.4, where it was used to represent the underlying asset modelled in a binomial lattice. In the present context, the key feature of a `BlackScholesAsset` is its deterministic volatility.

4.4.1 Representing the deterministic volatility of assets with Black/Scholes–type dynamics

Though volatility was assumed to be constant in the seminal work of Black and Scholes (1973), we have seen that permitting volatility to depend deterministically on time still results in Black/Scholes–type option pricing formulae. This time dependence can come in many forms, including constant (the original assumption), piecewise constant (the simplest form permitting a fit to a "term structure of volatility") and exponential (occurring, for example, in Gaussian models of the term structure of interest rates — see Chapter 8). The abstract base class `DeterministicAssetVol` defines the interface, which classes implementing deterministic volatility functions must satisfy, and Figure 4.1 shows the class hierarchy of the different representations of deterministic volatility in the current implementation. The volatilities of the underlying assets enter the pricing formula (4.52) via the covariance matrix of logarithmic asset prices given by (4.54). Thus we need to be able to evaluate integrals over sum products of volatilities (recall that each asset price process is driven by a d–dimensional Brownian motion and volatilities are, correspondingly, d–dimensional vectors). This is done via the member function `DeterministicAssetVol::volproduct(t,ttm,v)`,[33] which is defined to return the integral between `t` and `t+ttm` over the sum product between the volatilities given by `this` and `DeterministicAssetVol *v`. The implementa-

---

[32] See Section 2.6.3.

[33] Note that when setting up the covariance matrix during the initialisation of `MBinary`, the member function `BlackScholesAsset::covariance()` is used, which then calls `DeterministicAssetVol::volproduct()`. This is done because the pointer to a `DeterministicAssetVol` instance is a private data member of `BlackScholesAsset`, in order to maintain encapsulation.

tion of `volproduct()` for concrete cases is a non-trivial matter, however, since the calculation of the integral in closed form requires knowledge of the specific functional form of the volatility `this` *and* `v`. Inside the virtual function `volproduct()`, we know the specific type of volatility function for `this`, but not for `v`. In some cases, this can be resolved relatively easily. When `this` represents a constant or piecewise constant volatility, the constant can be taken outside of the integration, e.g.

$$\int_t^{t+\Delta t} \sigma_1(s)\sigma_2(s)ds = \sigma_1(t) \int_t^{t+\Delta t} \sigma_2(s)ds \qquad (4.115)$$

if $\sigma_1(s)$ is constant on $s \in [t, t + \Delta t]$. The integral on the right hand side of (4.115) can then be calculated by calling the member function `v->integral(t, `Δ`t)` on the `DeterministicAssetVol*` `v`. In other cases, this is not so simple.

So the question is how to implement `volproduct()` in such a way that one does not have to modify the implementation for every existing volatility class every time a new volatility class is added, when they seem so inextricably intertwined. This is precisely the type of situation for which Gamma, Helm, Johnson and Vlissides (1995) suggest the use of the *Mediator* design pattern, writing that a Mediator is "an object that encapsulates how a set of objects interact. Mediator promotes loose coupling by keeping objects from referring to each other explicitly, and lets you vary their interaction independently." In our case, `DeterministicVolMediator` collects the interaction between classes derived from `DeterministicAssetVol` in one place, and so it is only here that additional functionality has to be added when a new type of `DeterministicAssetVol` is derived. This does not necessarily reduce the programming effort, but it does localise the required changes in a single class. Thus calling the virtual function `volproduct()` on an instance of `ExponentialVol` results in a call to `DeterministicVolMediator::vol product_ExponentialVol`, identifying the caller as an `ExponentialVol`. The member function of `DeterministicVolMediator` then calculates the required integral based on the type of the second volatility `v`, and it is here that additional cases would have to be added if additional classes are derived from `DeterministicAssetVol` (note that the case identifiers `FLAT` and `EXPONENTIAL` returned by the virtual member function `type()` of the volatility classes are defined in an `enum` in `DeterministicVolMediator` — thus if any additional types are required, again only this class has to be modified).

For convenience, `BlackScholesAsset` also implements the pricing formula for an option to exchange one asset for another (i.e. Theorem 4.7), in the general case as well as for standard calls and puts. In the general case, we make use of the fact that the proportional volatility of the quotient of two assets is simply the difference between the two volatilities — represented by a `DeterministicAssetVolDiff`, which is itself a class derived from `DeterministicAssetVol`.

The interface of `DeterministicAssetVol` provides a number of further

```
class MBinary {
public:
  MBinary(/// Vector of pointers to underlying assets.
          const std::vector<const BlackScholesAsset*>& xunderlying,
          /// (Deterministic) term structure of interest rates.
          const TermStructure&                         xts,
          /// Time line collecting all event dates.
          const Array<double,1>&                       xtimeline,
          /// A 2 x N matrix of indices, where each column represents
          /// the indices of an (asset,time) combination affecting
          /// the payoff.
          const Array<int,2>&                          xindex,
          ///< Payoff powers.
          const Array<double,1>&                       xalpha,
          ///< Exercise indicators.
          const Array<int,2>&                          xS,
          ///< Exercise condition matrix.
          const Array<double,2>&                       xA,
          ///< Strike vector.
          const Array<double,1>&                       xa
          );
  /* ... */
};
```

Listing 4.1: Declaration of the `MBinary` constructor.

functions. However, these will be discussed in the relevant context in Chapters 7 and 8.

4.4.2 Prices and hedge sensitivities of options with discrete event dates

Consider any one of the options with discrete event dates described in Section 4.3, for example the barrier options in Table 4.2 — taken together with a set of event dates T_k and an underlying `BlackScholesAsset`, the entries in this table completely characterise the option. Thus, in general, the constructor of the class `MBinary` requires the inputs in Listing 4.1, with the vectors a (input `xa`), α (input `xalpha`) and matrices S (input `xS`), A (input `xA`) defined as in (4.51), and where the $2 \times n$ matrix of indices represents the index mappings $\bar{\imath}(i)$ and $\bar{k}(k)$ (recall that n is the *payoff dimension* of the payoff representation). Thus for a barrier option on a single underlying asset with 4 monitoring dates, where $\bar{\imath}(i) = 1$ and $\bar{k}(k) = k$, we have (note the shift of the asset indices because C++ indexing starts from zero):

$$\text{xindex} = \begin{pmatrix} 0 & 0 & 0 & 0 \\ 1 & 2 & 3 & 4 \end{pmatrix} \tag{4.116}$$

1	4.475086394
2	0.494481802
3	0.176394120
4	0.089075387
5	0.053365158
6	0.035431662
7	0.025206325
8	0.018846073

Table 4.9 *The first eight eigenvalues of the covariance matrix for observations of an underlying asset at 52 weekly event dates over a year. The volatility of the underlying asset is constant at 42% per annum.*

For an M-Binary paying the maximum of two assets, where $\bar{\imath}(i) = i$ and $\bar{k}(k) = 1$, we have

$$\texttt{xindex} = \begin{pmatrix} 0 & 1 \\ 1 & 1 \end{pmatrix} \tag{4.117}$$

With these inputs, we are then in a position to initialise the variables β, h and C as defined in Section 4.3.1. This is done in `MBinary::initialise()`.

To price the option via (4.52), one calls the member function `price()`. Here, the only remaining task is to evaluate the cumulative distribution function of the multivariate normal distribution. This task is delegated to an instance of the class `MultivariateNormal`, introduced in Section 2.2.3.

For some types of payoffs, for example barrier or lookback options with a large number of event dates, one might question the usefulness of the Skipper/Buchen approach: If the best way to evaluate a very high–dimensional normal distribution is by Monte Carlo methods, might it not be better to implement the option price directly by Monte Carlo simulation, as will be done in Chapter 7? One might say that if the multivariate cumulative distribution function is evaluated by Monte Carlo simulation, the two approaches are essentially equivalent.[34] However, especially when there is a large number of event dates, the covariance matrix numerically will not have full rank, in the sense that not all eigenvalues will be sufficiently different from zero to matter in the calculation. As an example, Table 4.9 gives the first eight eigenvalues of the covariance matrix for a set of 52 weekly event dates over a year. Only the leading eigenvalues are large enough to be relevant, significantly reducing the computational effort — in this example, the first eigenvalue covers more than 80% of the variability, the first two cover about 90%, the first five more than 95%, and the first eight about 97.5%. This is particularly important

---

[34] Formulas such as (4.52) are still commonly considered "closed form" even though the evaluation of a high–dimensional normal cumulative distribution function is everything but, which goes to show that the distinction between "closed form" and numerical solutions is more semantic than practical.

when using quasi-random methods.[35] Furthermore, in cases of nearly continuous monitoring (e.g. of barriers or lookbacks), the corresponding continuously monitored options give a very close approximation of the correct price. The formulae presented in Section 4.3.5 express these options as combinations of building blocks, which are M–Binaries with a very low exercise dimension (e.g. 1 or 2). From our perspective, the main advantage of the Skipper/Buchen formula is that it allows us to adopt an "implement once" approach, i.e. the pricing formula is implemented once and a very large class of exotic options can be priced simply by specifying their payoffs.

This "implement once" strategy also applies to hedge sensitivities (the "Greeks"). If V_t denotes the price of the M–Binary at time t, we are primarily interested in the first and second derivatives of V_t with respect to the underlying asset(s) (the option *delta* and *gamma*) and the first derivative with respect to volatility (the option *vega*).[36]

In order to perform the required differentiation, we express the cumulative distribution function of the multivariate normal in (4.52) as the multiple integral

$$\int_{-\infty}^{s_{11}h_1} \cdots \int_{-\infty}^{s_{mm}h_m} (2\pi)^{-\frac{1}{2}p}(\det_p SCS)^{-\frac{1}{2}} \exp\left\{-\frac{1}{2}z^\top (SCS)^+ z\right\} dz \quad (4.118)$$

where z is an m–dimensional vector. This representation of the multivariate normal density from van Perlo-ten Kleij (2004) allows for a singular covariance matrix SCS, in which case $(SCS)^+$ is the Moore/Penrose inverse[37] of SCS. p is the rank of the covariance matrix and $\det_p SCS$ is the product of the p positive eigenvalues. When $p = m$, (4.118) coincides with the representation of the multivariate normal typically found in the literature.[38]

To facilitate implementation, represent V_t as $P_t I_t$, where I_t is given by (4.118) and

$$P_t = \beta \prod_{j=1}^{n}(X_{\bar{i}(j)}(t))^{\alpha_j}$$

For the option *delta*, we have

$$\frac{\partial V_t}{\partial X_i(t)} = P_t\frac{\partial I_t}{\partial X_i(t)} + I_t\frac{\partial P_t}{\partial X_i(t)} \quad (4.119)$$

The two partial derivatives are implemented in the private member functions

---

[35] For a further discussion of this issue, see Chapter 7.

[36] Other sensitivities, with respect time, interest rate and strike, can be similarly calculated, but are of less practical significance. Time and strike are simply features of the option contract, while any uncertainty in interest rates is typically dominated by the volatility of the underlying asset — when this is not the case, stochastic interest rates can be explicitly modelled; see Chapter 8.

[37] In our context, where SCS is a positive semidefinite matrix, the Moore/Penrose inverse can easily be calculated by singular value decomposition (SVD). If the SVD of SCS is $U\Lambda U^\top$, where Λ is the diagonal matrix of eigenvalues of SCS, then $(SCS)^+ = U\Lambda^+ U^\top$, with Λ^+ a diagonal matrix with $(\Lambda)^+_{ii} = (\Lambda_{ii})^{-1}$.

[38] See e.g. Kotz, Balakrishnan and Johnson (2000).

`dPdX(int i)` and `dIdX(int i)`. For the former,

$$\frac{\partial P_t}{\partial X_i(t)} = \bar{\alpha}_i \frac{P_t}{X_i(t)} \tag{4.120}$$

where $\bar{\alpha}_i$ is the sum of α_j for all $\bar{\imath}(j) = i$. To implement `dIdX(int i)`, note that this can be written as

$$\frac{\partial I_t}{\partial X_i(t)} = \sum_{j=1}^{m} \frac{\partial h_j}{\partial X_i(t)} \hat{\kappa}_j \int_{-\infty}^{\hat{h}_1(j)} \cdots \int_{-\infty}^{\hat{h}_{m-1}(j)} (2\pi)^{-\frac{1}{2}\hat{p}_j} (\det_{\hat{p}_j} \hat{C}^{(j)})^{-\frac{1}{2}}$$

$$\exp\left\{ -\frac{1}{2} (\hat{z}^{(j)})^\top (\hat{C}^{(j)})^+ \hat{z}^{(j)} \right\} d\hat{z}^{(j)} \tag{4.121}$$

To arrive at the "hat" parameters, reduce the dimension of the multidimensional integral (4.118) by one by "freezing" $z_j = h_j$ (since we are applying chain rule and taking the derivative with respect to h_j, which is the upper bound of integration on one of the integrals): Simplifying notation by defining

$$\bar{C} = SCS \qquad \text{and} \qquad \eta = (2\pi)^{-\frac{1}{2}p} (\det_p SCS)^{-\frac{1}{2}}$$

the density becomes

$$\eta \exp\left\{ -\frac{1}{2} \sum_{l=1}^{m} \sum_{k=1}^{m} z_l \bar{C}_{lk}^+ z_k \right\}$$

and after setting $z_j = h_j$,

$$\eta \exp\left\{ -\frac{1}{2} \sum_{l\neq j} \sum_{k\neq j} z_l \bar{C}_{lk}^+ z_k - \frac{1}{2} \sum_{l\neq j} h_j \bar{C}_{jl}^+ z_l - \frac{1}{2} h_j^2 \bar{C}_{jj}^+ \right\} \tag{4.122}$$

Note that[39] for any valid covariance matrix H and n–vectors μ and y

$$\exp\left\{ -\frac{1}{2}(y - H\mu)^\top H^+ (y - H\mu) \right\}$$

$$= \exp\left\{ \mu^\top y - \frac{1}{2}\mu^\top H\mu \right\} \exp\left\{ -\frac{1}{2} y^\top H^+ y \right\} \tag{4.123}$$

Now define a new $(n-1) \times (m-1)$ matrix $\hat{C}^{(j)}$ by setting

$$(\hat{C}^{(j)})_{lk}^+ = \hat{C}_{lk}^+ \qquad \text{for } l, k \neq j \tag{4.124}$$

and dropping the j-th row and column. $\hat{C}^{(j)}$ is then calculated by

$$\hat{C}^{(j)} = \left((\hat{C}^{(j)})^+ \right)^+$$

---

[39] Skipper and Buchen (2003) call this the *Gaussian Shift Theorem*. See Section 3.9 in Buchen (2012).

Furthermore, define the $(m-1)$–vectors $\hat{\mu}^{(j)}$ and $\hat{z}^{(j)}$ by

$$\hat{z}_k^{(j)} = z_k$$

$$\hat{\mu}_k^{(j)} = -\frac{1}{2}h_j\bar{C}_{jk}^+ \quad \text{for } k \neq j$$

Then (4.122) can be written as

$$\eta \exp\left\{-\frac{1}{2}(\hat{z}^{(j)})^\top(\hat{C}^{(j)})^+\hat{z}^{(j)} + (\hat{\mu}^{(j)})^\top\hat{z}^{(j)} - \frac{1}{2}h_j^2\bar{C}_{jj}^+\right\} =$$

$$\eta \exp\left\{-\frac{1}{2}(\hat{z}^{(j)} - \hat{C}^{(j)}\hat{\mu}^{(j)})^\top(\hat{C}^{(j)})^+(\hat{z}^{(j)} - \hat{C}^{(j)}\hat{\mu}^{(j)})\right\}$$

$$\cdot \exp\left\{\frac{1}{2}\left((\hat{\mu}^{(j)})^\top\hat{C}^{(j)}\hat{\mu}^{(j)} - h_j^2\bar{C}_{jj}^+\right)\right\} \quad (4.125)$$

If $\hat{p}_j$ is the rank of $\hat{C}^{(j)}$ and the $(m-1)$–vector $\hat{h}^{(j)}$ is defined by

$$\hat{h}^{(j)} = h^{(-j)} - \hat{C}^{(j)}\hat{\mu}^{(j)} \quad (4.126)$$

where $h^{(-j)}$ is the vector h with the j-th element removed, then we arrive at the representation (4.121), with

$$\hat{\kappa}_j = \sqrt{\frac{\det_{\hat{p}_j}\hat{C}^{(j)}}{(2\pi)^{p-\hat{p}_j}\det_p\bar{C}}} \exp\left\{\frac{1}{2}\left((\hat{\mu}^{(j)})^\top\hat{C}^{(j)}\hat{\mu}^{(j)} - h_j^2\bar{C}_{jj}^+\right)\right\}$$

Similarly, for option *gamma* we have

$$\frac{\partial^2 V_t}{\partial X_i^2(t)} = P_t\frac{\partial^2 I_t}{\partial X_i^2(t)} + 2\frac{\partial P_t}{\partial X_i(t)}\frac{\partial I_t}{\partial X_i(t)} + I_t\frac{\partial^2 P_t}{\partial X_i^2(t)} \quad (4.127)$$

where

$$\frac{\partial^2 P_t}{\partial X_i^2(t)} = \bar{\alpha}_i(X_i(t))^{-2}\left(X_i(t)\frac{\partial P_t}{\partial X_i(t)} - P_t\right) = \bar{\alpha}_i(\bar{\alpha}_i - 1)\frac{P_t}{(X_i(t))^2} \quad (4.128)$$

and

$$\frac{\partial^2 I_t}{\partial X_i^2(t)} = \sum_{j=1}^m \frac{\partial^2 h_j}{\partial X_i^2(t)}\hat{\kappa}_j$$

$$\cdot \underbrace{\int_{-\infty}^{\hat{h}_1^{(j)}}\cdots\int_{-\infty}^{\hat{h}_{m-1}^{(j)}}(2\pi)^{-\frac{1}{2}\hat{p}_j}(\det_{\hat{p}_j}\hat{C}^{(j)})^{-\frac{1}{2}}\exp\left\{-\frac{1}{2}(\hat{z}^{(j)})^\top(\hat{C}^{(j)})^+\hat{z}^{(j)}\right\}d\hat{z}^{(j)}}_{\hat{I}_t^{(j)}}$$

$$+ \sum_{j=1}^m \frac{\partial h_j}{\partial X_i(t)}\left(\frac{\partial\hat{\kappa}_j}{\partial X_i(t)}\hat{I}_t^{(j)} + \hat{\kappa}_j\frac{\partial\hat{I}_t^{(j)}}{\partial X_i(t)}\right) \quad (4.129)$$

The derivative of the multivariate normal integral $\hat{I}_t^{(j)}$ with respect to $X_i(t)$ can be calculated using `dIdX()` on the appropriate `MultivariateNormal` object.

We define option *vega* as the sensitivity with respect to a parallel shift in the volatility function, i.e. substitute $\sigma_{X_i}(s)$ by

$$\hat{\sigma}_{X_i}(s) = \bar{\sigma}_{X_i} \sigma_{X_i}(s) \tag{4.130}$$

where $\bar{\sigma}_{X_i}$ is a scalar constant in s, and evaluate the derivative of V_t with respect to $\bar{\sigma}_{X_i}$ at $\bar{\sigma}_{X_i} = 1$.

4.5 American options

As opposed to European options, American options can not only be exercised at expiry, but also at any prior time. For American call options on assets for which the cost of carry is greater than or equal to the riskless rate of interest $r \geq 0$,[40] we have the Merton (1973b) result that it is never rational to exercise the option early. Consider an option with strike K, expiring at time T. For the option price $C(t)$ at time $t < T$,

$$
\begin{aligned}
C(t) &= e^{-r(T-t)} E\left[\left[S(T) - K\right]^+ | \mathcal{F}_t\right] \\
\Rightarrow \quad C(t) &\geq e^{-r(T-t)} E\left[S(T) - K | \mathcal{F}_t\right] \\
&= S(t) - e^{-r(T-t)} K \\
&> S(t) - K
\end{aligned}
$$

Thus the value of the option prior to expiry $C(t)$ is always greater than the payoff received from exercising the option at time t. Since it is therefore never rational to make use of the right of early exercise, this right is worthless and the price of the American call option equals the price of the European call option, which can be calculated using the Black/Scholes formula.

For American put options in general, and for American call options if the cost of carry is less than the riskless rate of interest, the right to exercise early potentially has value. In this case there is no closed form formula for the option price, and numerical methods must be used, such as for example the binomial lattices of Chapter 3 or the finite difference schemes of Chapter 5. However, these methods are quite computationally intensive. Barone-Adesi and Whaley (1987) and Geske and Johnson (1984) propose approximate closed form solutions, which we will discuss in turn.

4.5.1 The Method of Barone-Adesi and Whaley

The method of Barone-Adesi and Whaley (1987) is based on a quadratic approximation, applied to the partial differential equation for the early exercise premium. Starting point is the observation that the early exercise premium ϵ, defined as the difference between the value of otherwise identical American

---

[40] When defined in terms of continuous compounding, the cost of carry is $r - q$, where r is the riskfree rate of interest, and a $q > 0$ represents a continuous income stream earned by the asset, while a $q < 0$ represents a continuous cost stream incurred by holding the asset (e.g. a storage cost).

and European options, also must satisfy the Black/Scholes PDE (4.9), i.e.

$$\frac{1}{2}\sigma^2 S^2 \frac{\partial^2}{\partial S^2}\epsilon - r\epsilon + (r-q)S\frac{\partial}{\partial S}\epsilon + \frac{\partial}{\partial t}\epsilon = 0 \tag{4.131}$$

Setting $\tau = T - t$, $M = 2r/\sigma^2$ and $N = 2(r-q)/\sigma^2$, (4.131) becomes

$$S^2 \frac{\partial^2}{\partial S^2}\epsilon - M\epsilon + NS\frac{\partial}{\partial S}\epsilon - \frac{M}{r}\frac{\partial}{\partial \tau}\epsilon = 0 \tag{4.132}$$

Defining $\epsilon(S,\kappa) = \kappa(\tau)f(S,\kappa)$ and substituting into (4.132), we have

$$S^2 \frac{\partial^2}{\partial S^2}f + NS\frac{\partial}{\partial S}f - Mf\left(1 + \frac{\frac{\partial}{\partial\tau}\kappa}{r\kappa}\left(1 + \frac{\kappa\frac{\partial}{\partial\kappa}f}{f}\right)\right) = 0 \tag{4.133}$$

which when choosing $\kappa(\tau) = 1 - e^{-r\tau}$ becomes

$$S^2 \frac{\partial^2}{\partial S^2}f + NS\frac{\partial}{\partial S}f - \frac{M}{\kappa}f - \underbrace{(1-\kappa)M\frac{\partial}{\partial\kappa}f}_{(*)} = 0 \tag{4.134}$$

The "quadratic approximation" of Barone-Adesi and Whaley (1987) now consists in arguing that the term $(*)$ is close to zero, i.e. (4.134) can be approximated by dropping $(*)$, resulting in a second–order ordinary differential equation (ODE). This ODE has two linearly independent solutions of the form aS^b. Substituting $f = aS^b$ into the ODE, we have

$$aS^b\left(b^2 + (N-1)b - \frac{M}{\kappa}\right) = 0 \tag{4.135}$$

Thus

$$b_{1,2} = \frac{1}{2}\left(-(N-1) \pm \sqrt{(N-1)^2 + 4\frac{M}{\kappa}}\right) \tag{4.136}$$

and the general solution of the ODE is

$$f(S) = a_1 S^{b_1} + a_2 S^{b_2} \tag{4.137}$$

To determine a_1 and a_2, we must distinguish calls from puts. For calls, we are interested in the behaviour of the early exercise premium, and thus of (4.137), only on the interval $(0, S^*]$, for some critical $S^* < \infty$, above which the option will be exercised early and the value of the option will equal the proceeds of early exercise $S - K$. Conversely, for puts the interval of interest is $[S^*, \infty)$ for some critical S^*, below which the option will be exercised early.

We have that since $M/K > 0$, $b_1 > 0$ and $b_2 < 0$. For the American call option early exercise premium, this means that a_2 must equal zero, because otherwise the early exercise premium would tend to infinity as S tends to zero, contradicting the fact that the call option price must be increasing in S.[41] For puts, the early exercise premium must tend to zero as S tends to infinity and consequently we must have $a_1 = 0$. Expressing the American call option price

---

[41] We cannot have $a_2 < 0$, because the early exercise premium must always be non-negative.

$C(S, \tau)$ in terms of the European call option $c(S, \tau)$ and the early exercise premium, we have

$$C(S, \tau) = c(S, \tau) + \kappa a_1 S^{b_1} \tag{4.138}$$

and analogously for the put options

$$P(S, \tau) = p(S, \tau) + \kappa a_2 S^{b_2} \tag{4.139}$$

At the critical asset prices S^* and $S^\star$, the American option price equals the proceeds from early exercise,

$$C(S^*, \tau) = S^* - K \tag{4.140}$$
$$P(S^\star, \tau) = K - S^\star \tag{4.141}$$

and setting the slopes equal on each side of (4.140) and (4.141) results in the additional equations

$$\frac{\partial}{\partial S} C(S^*, \tau) = 1 \tag{4.142}$$

$$\frac{\partial}{\partial S} P(S^\star, \tau) = -1 \tag{4.143}$$

Combining (4.140) and (4.142) to eliminate κa_1 results in

$$S^* - K = c(S^*, \tau) + (1 - e^{-q\tau} N(h_1(S^*))) \frac{S^*}{b_1} \tag{4.144}$$

with

$$h_1(S^*) = \frac{\ln \frac{S^*}{K} + \left(r - q + \frac{1}{2}\sigma^2\right)\tau}{\sigma\sqrt{\tau}}$$

(4.144) can be solved for S^* using a root search algorithm. Inserting this S^* back into (4.142) and using this to eliminate κa_1 in (4.138), we obtain the approximate American call option formula

$$C(S, \tau) = \begin{cases} c(S, \tau) + A_1 \left(\frac{S}{S^*}\right)^{b_1} & \text{for } S < S^* \\ S - K & \text{for } S \geq S^* \end{cases} \tag{4.145}$$

with

$$A_1 = (1 - e^{-q\tau} N(h_1(S^*))) \frac{S^*}{b_1}$$

Similarly, (4.141) and (4.143) yield a fixed point problem for $S^\star$,

$$K - S^\star = p(S^\star, \tau) + (1 - e^{-q\tau} N(-h_1(S^\star))) \frac{S^\star}{b_2} \tag{4.146}$$

and the resulting approximate American put option formula is

$$P(S, \tau) = \begin{cases} p(S, \tau) + A_2 \left(\frac{S}{S^\star}\right)^{b_2} & \text{for } S > S^\star \\ K - S & \text{for } S \leq S^\star \end{cases} \tag{4.147}$$

with

$$A_2 = (e^{-q\tau} N(-h_1(S^\star)) - 1) \frac{S^\star}{b_2}$$

As starting points for the root searches to solve (4.144) and (4.146), Barone-Adesi and Whaley (1987) suggest that the simplest approach is to take the exercise price K as the initial guess for S^* resp. $S^\star$. Alternatively, they derive the approximate values

$$S^* = K + (S^*(\infty) - K)(1 - e^{d_1}) \tag{4.148}$$

$$d_1 = -((r - q)\tau + 2\sigma\sqrt{\tau})\frac{K}{S^*(\infty) - K}$$

and

$$S^\star = S^\star(\infty) + (K - S^\star(\infty))e^{d_2} \tag{4.149}$$

$$d_2 = ((r - q)\tau - 2\sigma\sqrt{\tau})\frac{K}{K - S^\star(\infty)}$$

where $S^*(\infty)$ and $S^\star(\infty)$ are the critical values when the time to option expiry is infinite, i.e.

$$S^*(\infty) = \frac{K}{1 - 2(-(N-1) + \sqrt{(N-1)^2 + 4M})^{-1}} \tag{4.150}$$

$$S^\star(\infty) = \frac{K}{1 - 2(-(N-1) - \sqrt{(N-1)^2 + 4M})^{-1}} \tag{4.151}$$

4.5.2 Geske and Johnson's approximation using compound options

Geske and Johnson (1984) note that *Bermudan* options, i.e. options which can be exercised at only a discrete set of time points up to expiry, can be evaluated as multiple compound options, which can be priced in closed form (up to evaluation of a multivariate normal CDF of the appropriate dimension). They compute the value of options with respectively one, two, three and four evenly spaced times of potential exercise, and then use Richardson extrapolation to determine the limit value (for a continuum of potential early exercise times, i.e. the American option).

For only one exercise time, we have the European option price; denote this by $P_1(X(T_0), T_0, \overline{T})$. For two times of potential exercise T_1 and T_2, we have the time line

$$\left\{T_0^{(2)} = T_0, T_1^{(2)} = T_0 + \frac{\overline{T} - T_0}{2}, T_2^{(2)} = \overline{T}\right\} \tag{4.152}$$

where $T_0^{(2)} = T_0$ denotes "today" and $\overline{T}$ is the expiry of the option. Once we reach $T_1^{(2)}$, only one future exercise time remains, so the value of holding on to the option at time $T_1^{(2)}$ is given by the European option price $P_1(X(T_1^{(2)}), T_1^{(2)}, \overline{T})$, which will depend on the value $X(T_1^{(2)})$ of the underlying asset X at time $T_1^{(2)}$. If K is the strike of the put option, the value of exercising the option at $T_1^{(2)}$ will be (if the option is in the money)

$$K - X(T_1^{(2)}) \tag{4.153}$$

As the delta of the put option is always greater than -1, the fixed point problem $X(T_1^{(2)})$,

$$K - X(T_1^{(2)}) = P_1(X(T_1^{(2)}), T_1^{(2)}, \overline{T}) \tag{4.154}$$

has a unique solution $\overline{X}_1^{(2)}$ (the *critical stock price*), and the option will be exercised early if and only if $X(T_1^{(2)}) < \overline{X}_1^{(2)}$. Thus the put option with two times of potential exercise has a time $T_1^{(2)}$ payoff of

$$(K - X(T_1^{(2)}))\mathbb{I}_{\{X(T_1^{(2)}) < \overline{X}_1^{(2)}\}} \tag{4.155}$$

and a time $\overline{T}$ payoff of

$$(K - X(T_2^{(2)}))\mathbb{I}_{\{X(T_2^{(2)}) < K\}}\mathbb{I}_{\{X(T_1^{(2)}) \geq \overline{X}_1^{(2)}\}} \tag{4.156}$$

Note that the two payoffs are mutually exclusive. These payoffs can be cast as M–Binaries (4.51) and values using the Skipper/Buchen formula, where the valuation of (4.156) involves a bivariate normal CDF. Denote this value by $P_2(X(T_0), T_0, \overline{T})$.

Next, consider three times of potential exercise, i.e. the time line

$$\left\{ T_0^{(3)} = T_0, T_1^{(3)} = T_0 + \frac{\overline{T} - T_0}{3}, T_2^{(3)} = T_0 + \frac{2(\overline{T} - T_0)}{3}, T_3^{(3)} = \overline{T} \right\} \tag{4.157}$$

Then the critical stock price $\overline{X}_2^{(3)}$ at time $T_2^{(3)}$ solves the fixed point problem

$$K - X(T_2^{(3)}) = P_1(X(T_2^{(3)}), T_2^{(3)}, \overline{T}) \tag{4.158}$$

the critical stock price $\overline{X}_1^{(3)}$ at time $T_1^{(3)}$ solves

$$K - X(T_1^{(3)}) = P_2(X(T_1^{(3)}), T_1^{(3)}, \overline{T}) \tag{4.159}$$

(because there are two potential exercise times remaining after $T_1^{(3)}$). The option then has three mutually exclusive payoffs:

$$(K - X(T_1^{(3)}))\mathbb{I}_{\{X(T_1^{(3)}) < \overline{X}_1^{(3)}\}} \tag{4.160}$$

at time $T_1^{(3)}$, and

$$(K - X(T_2^{(3)}))\mathbb{I}_{\{X(T_2^{(3)}) < \overline{X}_2^{(3)}\}}\mathbb{I}_{\{X(T_1^{(3)}) \geq \overline{X}_1^{(3)}\}} \tag{4.161}$$

at time $T_2^{(3)}$, and

$$(K - X(T_3^{(3)}))\mathbb{I}_{\{X(T_3^{(3)}) < K\}}\mathbb{I}_{\{X(T_2^{(3)}) \geq \overline{X}_2^{(3)}\}}\mathbb{I}_{\{X(T_1^{(3)}) \geq \overline{X}_1^{(3)}\}} \tag{4.162}$$

at time $T_3^{(3)} = \overline{T}$. Valuing these payoffs using the Skipper/Buchen formula, we obtain the price $P_3(X(T_0), T_0, \overline{T})$.

In the general case of n times of potential exercise, we have the time line

$$\left\{ T_0^{(n)} = T_0, \ldots, T_k^{(n)} = T_0 + k\frac{\overline{T} - T_0}{n}, \ldots, T_n^{(n)} = \overline{T} \right\} \tag{4.163}$$

The $n-1$ critical stock prices $\overline{X}_k^{(n)}$, $1 \le k < n$, solve

$$K - X(T_k^{(n)}) = P_{n-k}(X(T_k^{(n)}), T_k^{(n)}, \overline{T}) \qquad (4.164)$$

In terms of the Skipper/Buchen formula (4.52), we can write

$$P_n(X(T_0), T_0, \overline{T}) = \sum_{i=1}^{n} KV_{i,n}^{(1)}(T_0) - V_{i,n}^{(2)}(T_0) \qquad (4.165)$$

with the discounted expected payoffs

$$V_{i,n}^{(j)}(T_0) = B(T_0, \overline{T})E[V_{i,n}^{(j)}(T_i^{(n)})] \qquad j = 1,2 \qquad (4.166)$$

The Skipper/Buchen inputs for $V_{i,n}^{(j)}(T_i^{(n)})$ are (using the notation of (4.51)):

- S is an $i \times i$ diagonal matrix with $S_{\ell,\ell} = 1$ for $\ell < i$, and $S_{i,i} = -1$.

- A is an $i \times i$ identity matrix.

- a is an i-dimensional vector, with $a_\ell = \overline{X}_\ell^{(n)}$ for $\ell < n$, and $a_\ell = K$ for $\ell = n$ (i.e. a will include K only when $i = n$).

- α is an i-dimensional vector identically equal to zero for $j = 1$. When $j = 2$, α is zero in all elements, except $\alpha_i = 1$.

$P_n(X(T_0), T_0, \overline{T})$ is computed in the private member function

```
double Pnformula(double S,
                 double maturity,
                 double strike,
                 const TermStructure& ts,
                 int n,
                 double today = 0.0);
```

of the class `GeskeJohnsonAmericanPut` in `AnalyticalAmericanPut.cpp`. Given $P_n(X(T_0), T_0, \overline{T})$ for $n = 1, 2, \ldots, \overline{n}$, one can apply polynomial Richardson extrapolation as in Geske and Johnson (1984) to approximate the continuous early exercise limit.

Denote by $\Delta t = (\overline{T} - T_0)/\overline{n}$ the smallest step size between potential exercise times, for which the option price was calculated. Denote

$$F(\xi_n \Delta t) = P_n(X(T_0), T_0, \overline{T}) \qquad \text{with } \xi_n = \frac{\overline{n}}{n} \qquad (4.167)$$

Postulate the truncated Taylor series relationship

$$F(\xi_n \Delta t) = F(0) + \sum_{i=1}^{\overline{n}-1} (\xi_n \Delta t)^i b_i + \mathcal{O}((\xi_n \Delta t)^{\overline{n}}) \qquad (4.168)$$

where $F(0)$ represents the continuous early exercise limit that is to be approximated. Dropping the terms of order $\mathcal{O}((\xi_n \Delta t)^{\overline{n}})$, this can be written as

$$
\underbrace{\begin{pmatrix} F(\xi_1 \Delta t) \\ \vdots \\ F(\xi_{\bar{n}-1}\Delta t) \end{pmatrix}^{\top}}_{\mathbf{F}} = F(0) + \underbrace{\begin{pmatrix} b_1 \Delta t \\ b_2 \Delta t^2 \\ \vdots \\ b_{\bar{n}-1}\Delta t^{\bar{n}-1} \end{pmatrix}^{\top}}_{\mathbf{b}} \underbrace{\begin{pmatrix} \xi_1 & \xi_2 & \cdots & \xi_{\bar{n}-1} \\ \xi_1^2 & \xi_2^2 & \cdots & \xi_{\bar{n}-1}^2 \\ \vdots & \vdots & \ddots & \vdots \\ \xi_1^{\bar{n}-1} & \xi_2^{\bar{n}-1} & \cdots & \xi_{\bar{n}-1}^{\bar{n}-1} \end{pmatrix}}_{\Xi}
$$

(4.169)

Solving this system for b yields

$$
\mathbf{b}^{\top} = (\mathbf{F}^{\top} - F(0))\Xi^{-1}
$$

(4.170)

Inserting this back into (4.168) for $n = \bar{n}$ and solving for $F(0)$ yields

$$
F(\Delta t) = F(0) + (\mathbf{F}^{\top} - F(0))\Xi^{-1} \begin{pmatrix} 1 \\ 1 \\ \vdots \\ 1 \end{pmatrix}
$$

$$
\Rightarrow \quad F(0) = \left(F(\Delta t) - \mathbf{F}^{\top}\Xi^{-1} \begin{pmatrix} 1 \\ 1 \\ \vdots \\ 1 \end{pmatrix} \right) \left(1 - (1\ 1\ \cdots\ 1)\Xi^{-1} \begin{pmatrix} 1 \\ 1 \\ \vdots \\ 1 \end{pmatrix} \right)^{-1}
$$

Thus we have the polynomial Richardson extrapolation

$$
F(0) = \sum_{n=1}^{\bar{n}} c_n F(\xi_n \Delta t)
$$

(4.171)

where the coefficients c_n are given by

$$
c_n = \begin{cases} -\{\Xi^{-1}(1\ 1\ \cdots\ 1)^{\top}\}_n c_0^{-1} & \text{for } 1 \le n < \bar{n} \\ c_0^{-1} & \text{for } n = \bar{n} \end{cases}
$$

(4.172)

and $c_0 = 1 - (1\ 1\ \cdots\ 1)\Xi^{-1}(1\ 1\ \cdots\ 1)^{\top}$, denoting by $\{\cdots\}_n$ the n-th element of the vector given by the expression between the brackets. The coefficients c_n, $1 \le n \le \bar{n}$, are returned by the member function `coefficients()` of the class `PolynomialRichardsonExtrapolation`.

Given the general formula $P_n(X(T_0), T_0, \overline{T})$ and the coefficients for the $\bar{n}$-th degree Richardson extrapolation, we have the $\bar{n}$-th degree Geske/Johnson approximation of the American put price implemented in the member function

r	strike	σ	expiry	Geske/Johnson	C++
0.0488	35	0.2	0.0833	0.0062	0.0062
0.0488	35	0.2	0.3333	0.1999	0.2001
0.0488	35	0.2	0.5833	0.4321	0.4321
0.0488	40	0.2	0.0833	0.8528	0.8521
0.0488	40	0.2	0.3333	1.5807	1.5808
0.0488	40	0.2	0.5833	1.9905	1.9906
0.0488	45	0.2	0.0833	4.9985	4.9969
0.0488	45	0.2	0.3333	5.0951	5.0949
0.0488	45	0.2	0.5833	5.2719	5.2718
0.0488	35	0.3	0.0833	0.0774	0.0774
0.0488	35	0.3	0.3333	0.6969	0.6969
0.0488	35	0.3	0.5833	1.2194	1.2191
0.0488	40	0.3	0.0833	1.3100	1.3101
0.0488	40	0.3	0.3333	2.4817	2.4841
0.0488	40	0.3	0.5833	3.1733	3.1718
0.0488	45	0.3	0.0833	5.0599	5.0599
0.0488	45	0.3	0.3333	5.7012	5.7011
0.0488	45	0.3	0.5833	6.2365	6.2365
0.0488	35	0.4	0.0833	0.2466	0.2466
0.0488	35	0.4	0.3333	1.3450	1.3452
0.0488	35	0.4	0.5833	2.1568	2.1543
0.0488	40	0.4	0.0833	1.7679	1.7682
0.0488	40	0.4	0.3333	3.3632	3.3888
0.0488	40	0.4	0.5833	4.3556	4.3556
0.0488	45	0.4	0.0833	5.2855	5.2858
0.0488	45	0.4	0.3333	6.5093	6.5093
0.0488	45	0.4	0.5833	7.3831	7.3813

Table 4.10 *American put option values from Geske and Johnson (1984) versus values from the C++ implementation. The initial stock price is $40.*

```
double GeskeJohnsonAmericanPut::
            option(double maturity,
                   double strike,
                   const TermStructure& ts,
                   int number_of_terms);
```

where $\overline{n}$ = number_of_terms. Table 4.10 replicates the results given in Table I of Geske and Johnson (1984), which uses $\overline{n} = 4$. The slight differences to Geske and Johnson's numbers are due to inaccuracy in the numerical integration.[42] For low volatility and very short times to maturity, the numerical integration of in particular the four–dimensional normal distribution becomes too inaccurate for Richardson extrapolation to yield a reasonable value. In the function `GeskeJohnsonAmericanPut::option()`, this is fixed by discarding any $P_n(X(T_0), T_0, \overline{T}) < P_1(X(T_0), T_0, \overline{T})$ (i.e. any Bermudan option prices lower than the European option price) and conducting the Richardson extrapolation with a smaller number of terms.

[42] When evaluating the CDF, Geske and Johnson (1984) make use of the special structure of the covariance matrix of the multivariate normal distribution, which results in the case where the volatility of the underlying asset is assumed to be constant. The implementation presented here allows for the more general case of time–varying volatility, so this special structure cannot be assumed.

CHAPTER 5

Finite difference methods for partial differential equations

Option pricing problems can typically be represented as a partial differential equation (PDE) subject to boundary conditions, see for example the Black/Scholes PDE in Section 4.2.1 or the option pricing PDE in the presence of stochastic volatility in Section 6.3. The idea behind finite difference methods is to approximate the partial derivatives in the PDE by a difference quotient, e.g.

$$\partial_1 C(t, S(t)) \approx \frac{C(t + \Delta t, S(t)) - C(t, S(t))}{\Delta t} \tag{5.1}$$

$$\partial_2 C(t, S(t)) \approx \frac{C(t, S(t) + \Delta S) - C(t, S(t))}{\Delta S} \tag{5.2}$$

Thus the PDE is discretised in both the time and the space dimension and the resulting solution converges to the continuous (true) solution as the discretisation grid is made finer and finer. Potentially there may be more than one space dimension, for example under stochastic volatility. However, finite difference schemes suffer from the "curse of dimensionality" in the sense that computational effort grows exponentially as the number of space dimensions is increased. For this reason Monte Carlo simulation is typically the method of choice for problems involving more than two space dimensions.

This chapter serves as an introduction to finite difference methods and their object–oriented implementation, and thus focuses on the three most common one–dimensional schemes, explicit, implicit and Crank/Nicolson.

5.1 The object-oriented interface

The interface and usage of classes representing finite difference schemes are identical to those of the general lattice model described in Section 3.3. That is, the member function `apply_payoff(i,f)` is used to apply the payoff function `f` (represented as a `boost::function` object taking a `double` argument and returning a `double`) to the underlying asset in each grid node in period `i`, typically the expiry of the option. As in the binomial model, the payoff is then "rolled back" from the i-th period to period 0 ("today") using the member function `rollback()`, and the time zero option value can be read using the member function `result()`. The implementation of the finite difference methods is such that calculation of the option price also generates output (see below), which makes it easy to determine the option delta and gamma,

```
// underlying asset
BlackScholesAsset stock(&vol,S);
/* closed-form price, with
   mat - time to maturity
   K   - strike
   r   - interest rate */
double CFcall = stock.option(mat,K,r);
cout << "Time line refinement: " << N
     << "\nState space refinement: " << Nj << endl;
FiniteDifference fd(stock,r,mat,N,Nj);
Payoff call(K);
boost::function<double (double)> f;
f = boost::bind(std::mem_fun(&Payoff::operator()),&call,_1);
fd.apply_payoff(N-1,f);
fd.rollback(N-1,0);
cout << "FD call: " << fd.result()
     << "\nDifference to closed form: "
     << CFcall - fd.result() << endl;
cout << "FD call delta: " << fd.delta()
     << "  FD call gamma: " << fd.gamma() << endl;
```

Listing 5.1: Pricing a European call option using explicit finite differences in the Black/Scholes model

```
Payoff put(K,-1);
EarlyExercise amput(put);
boost::function<double (double,double)> g;
g = boost::bind(boost::mem_fn(&EarlyExercise::operator()),
                &amput,_1,_2);
fd.apply_payoff(N-1,f);
fd.rollback(N-1,0,g);
double fdamput = fd.result();
cout << "FD American put: " << fdamput << endl;
cout << "FD American put delta: " << fd.delta()
     << "  \nFD American put gamma: " << fd.gamma() << endl;
```

Listing 5.2: Pricing an American put option using explicit finite differences in the Black/Scholes model — variables not explicitly given are the same as in Listing 5.1.

therefore the finite difference scheme classes provide the additional member functions delta() and gamma() to access these values.

Listing 5.1 and 5.2 give the code necessary to price, respectively, European and American options using the FiniteDifference class.[1] As in the classes implementing binomial lattice models, the member function rollback() is

---

[1] The full working code is given in FDTest.cpp on the website for this book.

overloaded to allow for early exercise and similar path conditions.[2] This results in the path conditions being applied in a naive manner, i.e. at each node in the finite difference grid the state variable x as defined in (5.3) below is converted into the corresponding value S of the underlying asset and the path condition (such as early exercise if the payoff of immediate exercise exceeds the continuation value, i.e. the value of holding on to the option) is evaluated based on this S. There are more sophisticated methods of adapting finite difference schemes to the *free boundary problem* of early exercise, which will not be covered here.[3]

The three finite difference schemes described below are implemented in a class hierarchy, with the simplest scheme (the "explicit" scheme) forming the base class `FiniteDifference`, from which `ImplicitFiniteDifference` and `CrankNicolson` are derived. The public class declarations are given in Listings 5.3, 5.6 and 5.7 — note that the implementations of the schemes only differ in the class constructors and in the member functions `rollback()`. Given that all schemes calculate the solution of the PDE on an entire grid of states j, it is easy to extract the finite difference approximations for the hedge sensitivities delta and gamma. The member functions `delta()` and `gamma()` will report these values for the central (at–the–money) state in the current time slice. The time of the current slice depends on the previous call of `rollback()`, and calling `delta()` or `gamma()` typically only makes sense if we previously "rolled back" through the grid all the way to time zero, in which case the initial–time hedge sensitivities will be reported.

5.2 The explicit finite difference method

Consider equation (4.9), the Black/Scholes PDE,

$$\partial_1 C(t, S(t)) + \frac{1}{2}\partial_{2,2}C(t, S(t))S(t)^2\sigma^2 = r(C(t, S(t)) - \partial_2 C(t, S(t))S(t))$$

Allowing for a constant dividend yield δ, this becomes

$$\partial_1 C(t, S(t)) + \frac{1}{2}\partial_{2,2}C(t, S(t))S(t)^2\sigma^2 = rC(t, S(t)) - (r - \delta)\partial_2 C(t, S(t))S(t)$$

and changing variables to $x = \ln S$ and defining $\nu := r - \delta$ yields a PDE with constant coefficients,

$$-\partial_1 C(t, x(t)) = \frac{1}{2}\sigma^2\partial_{2,2}C(t, x(t)) + \nu\partial_2 C(t, x(t)) - rC(t, x(t)) \qquad (5.3)$$

i.e. there is no dependence of x and t outside of C and its partial derivatives. Approximating the partial derivative with respect to t by a forward difference quotient and the partial derivatives with respect to x by central difference

---

[2] A similar path condition in this sense would be a barrier condition, for example.
[3] See e.g. Sections 4.5 and 4.6 of Seydel (2006).

```
class FiniteDifference {
public:
  FiniteDifference(const BlackScholesAsset& xS,double xr,
                   double xT,int xN,int xNj);
  /// Calculate the value of the underlying asset at a given
  /// grid point.
  inline double underlying(int i, ///< Time index
                           int j  ///< State index
                           ) const;
  inline void underlying(int i,Array<double,1>& u) const;
  inline void apply_payoff(int i,boost::function<double (double)> f);
  virtual void rollback(int from,int to);
  /// Rollback with early exercise (or similar) condition.
  virtual void rollback(int from,int to,
                        boost::function<double (double,double)> f);
  inline double result() const { return gridslice(Nj); };
  double delta() const;
  double gamma() const;
};
```

Listing 5.3: Public interface of the `FiniteDifference` class

quotients, one obtains

$$-\frac{C_{i+1,j} - C_{i,j}}{\Delta t} = \frac{1}{2}\sigma^2 \frac{C_{i+1,j+1} - 2C_{i+1,j} + C_{i+1,j-1}}{\Delta x^2}$$
$$+ \nu \frac{C_{i+1,j+1} - C_{i+1,j-1}}{2\Delta x} - rC_{i+1,j} \quad (5.4)$$

where $C_{i,j} = C(t_i, x_j)$ with $t_{i+1} = t_i + \Delta t$ and $x_{j+1} = x_j + \Delta x$ for all i and j. Following the presentation in Clewlow and Strickland (1998), this can be written as

$$C_{i,j} = p_u C_{i+1,j+1} + p_m C_{i+1,j} + p_d C_{i+1,j-1} \quad (5.5)$$

$$p_u = \Delta t \left(\frac{\sigma^2}{2\Delta x^2} + \frac{\nu}{2\Delta x}\right)$$

$$p_m = 1 - \Delta t \frac{\sigma^2}{\Delta x^2} - r\Delta t$$

$$p_d = \Delta t \left(\frac{\sigma^2}{2\Delta x^2} - \frac{\nu}{2\Delta x}\right)$$

"Rolling back" in this manner from time t_{i+1} to time t_i is implemented in the member function `rollback()` of the class `FiniteDifference` (see Listing 5.4).[4] In the limit, (5.5) is equivalent to taking discounted expectations in a (trinomial) lattice. Also, this would be true at any level of discretisation if one

---

[4] See also the C++ source file `FiniteDifference.cpp` on the website.

```
void FiniteDifference::rollback(int from,int to)
{
  int i,j;
  for (i=from-1;i>=to;i--) {
    for (j=1;j<=2*Nj-1;j++)
      tmp(j) = pu * gridslice(j+1) + pm * gridslice(j)
             + pd * gridslice(j-1);
    gridslice = tmp;
    // boundary conditions
    boundary_condition(gridslice); }
}
```

Listing 5.4: "Rolling back" through the explicit finite difference grid.

approximates $rC(t, x(t))$ in (5.3) by $rC_{i,j}$ instead of $rC_{i+1,j}$ as in (5.4), i.e.

$$-\frac{C_{i+1,j} - C_{i,j}}{\Delta t} = \frac{1}{2}\sigma^2 \frac{C_{i+1,j+1} - 2C_{i+1,j} + C_{i+1,j-1}}{\Delta x^2}$$
$$+ \nu \frac{C_{i+1,j+1} - C_{i+1,j-1}}{2\Delta x} - rC_{i,j} \quad (5.6)$$

which yields

$$C_{i,j} = \frac{1}{1 + r\Delta t}(p_u C_{i+1,j+1} + p_m C_{i+1,j} + p_d C_{i+1,j-1}) \quad (5.7)$$

$$p_u = \frac{1}{2}\Delta t \left(\frac{\sigma^2}{\Delta x^2} + \frac{\nu}{\Delta x}\right)$$

$$p_m = 1 - \Delta t \frac{\sigma^2}{\Delta x^2}$$

$$p_d = \frac{1}{2}\Delta t \left(\frac{\sigma^2}{\Delta x^2} - \frac{\nu}{\Delta x}\right)$$

The implementation in the class `FiniteDifference`, however, uses the representation (5.5), for which p_u, p_m and p_d are initialised in the constructor. Furthermore, the constructor sets up the $(\Delta t, \Delta x)$ grid on which the finite difference scheme operates.

This is where some care must be taken to ensure that the scheme converges. For the purpose of the present argument, set $C_{i,j}$ to zero for any j beyond the grid boundary. Then (5.7) can be written in matrix form as

$$C_{i,\cdot} = AC_{i+1,\cdot} \quad (5.8)$$

```
FiniteDifference::
FiniteDifference(const BlackScholesAsset& xS,
                double xr,double xT,int xN,int xNj)
  : S(xS),r(xr),T(xT),N(xN),Nj(xNj),gridslice(2*xNj+1),
    tmp(2*xNj+1),boundary_condition(linear_extrapolationBC)
{
    // precompute constants
    double sgm = S.volatility(0.0,T);
    dt = T / N;
    nu = r - 0.5 * sgm*sgm;
    /* Adjust nu for dividend yield - note that average dividend yield
       over the life of the option is used. This is OK for European
       options, but inaccurate for path-dependent (including American)
       options when dividends actually are time-varying. */
    nu -= S.dividend_yield(0.0,T);
    dx = sgm * std::sqrt(3.0 * dt);
    edx = std::exp(dx);
    double sgmdx = sgm/dx;
    sgmdx *= sgmdx;
    pu = 0.5 * dt * (sgmdx + nu / dx);
    pm = 1.0 - dt * sgmdx - r * dt;
    pd = 0.5 * dt * (sgmdx - nu / dx);
}
```

Listing 5.5: Constructor of the class `FiniteDifference`.

where

$$
A = \begin{pmatrix}
p_m & p_u & 0 & 0 & \cdots & 0 & 0 \\
p_d & p_m & p_u & 0 & \cdots & 0 & 0 \\
0 & p_d & p_m & p_u & \cdots & 0 & 0 \\
\vdots & \vdots & \ddots & \ddots & \ddots & \vdots & \vdots \\
0 & 0 & \cdots & 0 & p_d & p_m & p_u \\
0 & 0 & \cdots & 0 & 0 & p_d & p_m
\end{pmatrix}
$$

Thus we have

$$C_{0,\cdot} = A^k C_{k,\cdot} \tag{5.9}$$

for rolling back the values in some arbitrary time step k to time zero. Inevitably, the $C_{k,\cdot}$ will depart from the "true," theoretical $C_{k,\cdot}$ by some numerical error $e_{k,\cdot}$. Considering only this error at time k, (5.9) becomes

$$C_{0,\cdot} = A^k(C_{k,\cdot} + e_{k,\cdot}) \tag{5.10}$$

i.e. $C_{0,\cdot}$ will be subject to the error $A^k e_{k,\cdot}$. The question is therefore whether multiplication by A^k dampens or exacerbates the error, and only in the former case can the scheme be considered stable. To answer this question, Seydel (2006) provides two useful lemmas from linear algebra.[5]

Lemma 5.1 *Define $\rho(A)$ as the* spectral radius *of the $M \times M$ matrix A,*

$$\rho(A) := \max_i |\lambda_i^A|$$

where $\lambda_1^A, \ldots, \lambda_M^A$ are the eigenvalues of A. Then

$$
\begin{aligned}
\rho(A) < 1 \quad &\Leftrightarrow \quad A^k z \to 0 \text{ for all } z \text{ and } k \to \infty \\
&\Leftrightarrow \quad \lim_{k\to\infty} \{(A^k)_{ij}\} = 0
\end{aligned}
$$

Therefore the error in (5.10) is dampened if the largest absolute eigenvalue of A is less than 1. In order to derive a sufficient condition on Δx to ensure this, we use

Lemma 5.2 *Let* $G = \begin{pmatrix} \alpha & \beta & & 0 \\ \gamma & \ddots & \ddots & 0 \\ & \ddots & \ddots & \beta \\ 0 & & \gamma & \alpha \end{pmatrix}$ *be an $M \times M$ matrix.*

The eigenvalues λ_j^G and the eigenvectors $v^{(j)}$ of G are

$$\lambda_j^G \;=\; \alpha + 2\beta \sqrt{\frac{\gamma}{\beta}} \cos \frac{j\pi}{M+1} \;,\quad j = 1, \ldots, M$$

$$v^{(j)} \;=\; \left(\sqrt{\frac{\gamma}{\beta}} \sin \frac{j\pi}{M+1}, \left(\sqrt{\frac{\gamma}{\beta}}\right)^2 \sin \frac{2j\pi}{M+1}, \ldots, \left(\sqrt{\frac{\gamma}{\beta}}\right)^M \sin \frac{Mj\pi}{M+1} \right)$$

Thus the error is dampened if

$$\left| p_m + 2p_u \sqrt{\frac{p_d}{p_u}} \cos \frac{j\pi}{M+1} \right| < 1 \qquad \forall\, 1 \le k \le M$$

$$\Leftarrow \quad |p_m + 2\sqrt{p_d p_u}| < 1 \qquad \forall\, 1 \le k \le M$$

Inserting (5.7), this becomes

$$\left| 1 - \Delta t \frac{\sigma^2}{\Delta x^2} \pm \Delta t \sqrt{\left(\frac{\sigma^2}{\Delta x^2}\right)^2 - \frac{\nu^2}{\Delta x^2}} \right| < 1$$

$$\Leftarrow \quad \begin{cases} \sqrt{\left(\frac{\sigma^2}{\Delta x^2}\right)^2 - \frac{\nu^2}{\Delta x^2}} < \frac{\sigma^2}{\Delta x^2} \\[2mm] \Delta t \sqrt{\left(\frac{\sigma^2}{\Delta x^2}\right)^2 - \frac{\nu^2}{\Delta x^2}} < 2 - \Delta t \frac{\sigma^2}{\Delta x^2} \end{cases}$$

---

[5] These lemmas, reproduced as Lemmas 5.1 and 5.2 below, are given as Lemmas 4.2 and 4.3 in Seydel (2006).

Note that unless the interest rate r is extremely high,

$$\frac{\sigma^2}{\Delta x^2} > \frac{\nu}{\Delta x}$$

for any meaningful refinement of the grid, so the first of these two inequalities will hold. For the second inequality, we have

$$\Delta t \sqrt{\left(\frac{\sigma^2}{\Delta x^2}\right)^2 - \frac{\nu^2}{\Delta x^2}} < 2 - \Delta t \frac{\sigma^2}{\Delta x^2}$$

$$\Leftarrow \quad \frac{\sigma^2}{\Delta x^2} \leq \frac{1}{\Delta t} \quad \Leftrightarrow \quad \Delta x \geq \sigma \sqrt{\Delta t}$$

Therefore, refining the state space grid by halving Δx requires refining the timeline by a factor of four, which increases the computational effort substantially. For this reason, one may prefer unconditionally stable methods such as the implicit finite difference and Crank/Nicolson schemes described below. The constructor of `FiniteDifference` follows Clewlow and Strickland (1998) and sets

$$\Delta x = \sigma \sqrt{3 \Delta t}$$

Consider Listing 5.5. The constructor arguments are, respectively, the underlying `BlackScholesAsset`, the (constant) continuously compounded risk–free rate of interest r, the time horizon T, the number of time steps N and the number n of Δx space steps to either side of the initial price of the underlying asset. Thus $\Delta t = T/N$ and M (the number of states in the grid at each time point) is equal to $2n+1$. When rolling back through the grid, these states will be held in the `Array gridslice`.

The constructor of `FiniteDifference` also sets the `boundary_condition` to linear extrapolation. A boundary condition is necessary for calculating the values in the outermost states of the grid when rolling backwards, i.e. (5.7) is not used to calculate these outermost states (see the inner `for` loop in Listing 5.4), because this would require values of C on the right hand side of the equation, which are beyond the edge of the grid. Instead, these are calculated by applying the `boundary_condition` to the `gridslice`, i.e. in the present case of linear extrapolation,

$$C_{i,0} = C_{i,1} + (C_{i,1} - C_{i,2})$$
$$C_{i,2n} = C_{i,2n-1} + (C_{i,2n-1} - C_{i,2n-2})$$

5.3 The implicit finite difference method

This method replaces the finite differences with respect to the state variable x at time $i+1$ in (5.6) by the corresponding finite differences at time i, i.e.

$$-\frac{C_{i+1,j} - C_{i,j}}{\Delta t} = \frac{1}{2}\sigma^2 \frac{C_{i,j+1} - 2C_{i,j} + C_{i,j-1}}{\Delta x^2}$$

$$+ \nu \frac{C_{i,j+1} - C_{i,j-1}}{2\Delta x} - rC_{i,j} \quad (5.11)$$

```
class ImplicitFiniteDifference : public FiniteDifference {
public:
   ImplicitFiniteDifference(const BlackScholesAsset& xS,
                      double xr,double xT,int xN,int xNj);
   virtual void rollback(int from,int to);
   /// Rollback with early exercise (or similar) condition.
   virtual void rollback(int from,int to,
                      boost::function<double (double,double)> f);
};
```

Listing 5.6: Public interface of the `ImplicitFiniteDifference` class

Collecting terms as before, this gives

$$
\begin{aligned}
C_{i+1,j} &= p_u C_{i,j+1} + p_m C_{i,j} + p_d C_{i,j-1} & (5.12)\\
p_u &= -\frac{1}{2}\Delta t\left(\frac{\sigma^2}{\Delta x^2} + \frac{\nu}{\Delta x}\right)\\
p_m &= 1 + \Delta t\frac{\sigma^2}{\Delta x^2} + r\Delta t\\
p_d &= -\frac{1}{2}\Delta t\left(\frac{\sigma^2}{\Delta x^2} - \frac{\nu}{\Delta x}\right)
\end{aligned}
$$

where the coefficients p can no longer be interpreted as probabilities. Working backward through the grid, each equation (5.12) has three unknowns (the $C_{i,\cdot}$ on the right hand side of the equation). However, if one views all states at a particular time i in the grid simultaneously, at the edges of the grid setting

$$
\begin{aligned}
C_{i,1} - C_{i,0} &= C_{i+1,1} - C_{i+1,0}\\
C_{i,2n} - C_{i,2n-1} &= C_{i+1,2n} - C_{i+1,2n-1}
\end{aligned}
\qquad (5.13)
$$

```
class CrankNicolson : public FiniteDifference {
public:
  CrankNicolson(const BlackScholesAsset& xS,
                double xr,double xT,int xN,int xNj);
  virtual void rollback(int from,int to);
  /// Rollback with early exercise (or similar) condition.
  virtual void rollback(int from,int to,
                        boost::function<double (double,double)> f);
};
```

Listing 5.7: Public interface of the `CrankNicolson` class

then one obtains the tridiagonal system of equations

$$
\begin{pmatrix}
-1 & 1 & 0 & 0 & \cdots & 0 & 0 \\
p_d & p_m & p_u & 0 & \cdots & 0 & 0 \\
0 & p_d & p_m & p_u & \cdots & 0 & 0 \\
\vdots & \vdots & \ddots & \ddots & \ddots & \vdots & \vdots \\
0 & 0 & \cdots & 0 & p_d & p_m & p_u \\
0 & 0 & \cdots & 0 & 0 & -1 & 1
\end{pmatrix}
\begin{pmatrix}
C_{i,0} \\
C_{i,1} \\
\vdots \\
\vdots \\
C_{i,2n-1} \\
C_{i,2n}
\end{pmatrix}
=
$$

$$
\begin{pmatrix}
C_{i+1,1} - C_{i+1,0} \\
C_{i+1,1} \\
\vdots \\
\vdots \\
C_{i+1,2n-1} \\
C_{i+1,2n} - C_{i+1,2n-1}
\end{pmatrix}
$$

which must be solved for each time step. The coefficient matrix for this system of equations is set up in the constructor of the class `ImplicitFiniteDifference`, and the system is solved by calling

$$\text{interfaceCLAPACK::SolveTridiagonal()} \tag{5.14}$$

(which wraps the appropriate CLAPACK library function) from the member function `rollback()`.[6]

5.4 The Crank/Nicolson scheme

Crank and Nicolson (1947) obtained a finite difference scheme, which like the implicit scheme is unconditionally stable, but achieves a better order of convergence ($\mathcal{O}(\Delta t^2)$ instead of $\mathcal{O}(\Delta t)$) in the time dimension. Their basic idea is to average the forward and the backward difference methods, in effect

---

[6] See `FiniteDifference.cpp` on the website for the implementation details.

approximating the space and time derivatives at $(i + \frac{1}{2})$. Again following the notation of Clewlow and Strickland (1998), this results in

$$-\frac{C_{i+1,j} - C_{i,j}}{\Delta t} =$$

$$\frac{1}{2}\sigma^2 \frac{(C_{i+1,j+1} - 2C_{i+1,j} + C_{i+1,j-1}) + (C_{i,j+1} - 2C_{i,j} + C_{i,j-1})}{2\Delta x^2}$$

$$+ \nu \frac{(C_{i+1,j+1} - C_{i+1,j-1}) + (C_{i,j+1} - C_{i,j-1})}{4\Delta x} - r\frac{C_{i+1,j} + C_{i,j}}{2} \quad (5.15)$$

Collecting terms as before, this gives

$$p_u C_{i,j+1} + p_m C_{i,j} + p_d C_{i,j-1} = -p_u C_{i+1,j+1} - (p_m - 2)C_{i+1,j} - p_d C_{i+1,j-1}$$

$$p_u = -\frac{1}{4}\Delta t \left(\frac{\sigma^2}{\Delta x^2} + \frac{\nu}{\Delta x} \right)$$

$$p_m = 1 + \Delta t \frac{\sigma^2}{2\Delta x^2} + \frac{r\Delta t}{2}$$

$$p_d = -\frac{1}{4}\Delta t \left(\frac{\sigma^2}{\Delta x^2} - \frac{\nu}{\Delta x} \right)$$

Using the same boundary conditions (5.13) as in the implicit scheme, this yields the tridiagonal system of equations

$$\begin{pmatrix} -1 & 1 & 0 & 0 & \cdots & 0 & 0 \\ p_d & p_m & p_u & 0 & \cdots & 0 & 0 \\ 0 & p_d & p_m & p_u & \cdots & 0 & 0 \\ \vdots & \vdots & \ddots & \ddots & \ddots & \vdots & \vdots \\ 0 & 0 & \cdots & 0 & p_d & p_m & p_u \\ 0 & 0 & \cdots & 0 & 0 & -1 & 1 \end{pmatrix} \begin{pmatrix} C_{i,0} \\ C_{i,1} \\ \vdots \\ \vdots \\ C_{i,2n-1} \\ C_{i,2n} \end{pmatrix} =$$

$$\begin{pmatrix} C_{i+1,1} - C_{i+1,0} \\ -p_u C_{i+1,2} - (p_m - 2)C_{i+1,1} - p_d C_{i+1,0} \\ \vdots \\ \vdots \\ -p_u C_{i+1,2n} - (p_m - 2)C_{i+1,2n-1} - p_d C_{i+1,2n-2} \\ C_{i+1,2n} - C_{i+1,2n-1} \end{pmatrix}$$

As in the implicit scheme, the coefficient matrix for this system of equations is set up in the constructor of the class CrankNicolson, and solved for each time step in its member function rollback().[7]

---

[7] This is implemented in FiniteDifference.cpp, available on the website.

## CHAPTER 6

Implied volatility and volatility smiles

Recall the Black and Scholes (1973) formula for a European call option in its simplest form,

$$C(t) = X(t)\mathcal{N}(h_1) - Ke^{-r(T-t)}\mathcal{N}(h_2) \qquad (6.1)$$

$$h_{1,2} = \frac{\ln\frac{X(t)}{K} - r(T-t) \pm \frac{1}{2}\sigma_X^2(T-t)}{\sigma_X\sqrt{T-t}}$$

on a non-dividend–paying asset $X(t)$ with constant (scalar) volatility σ_X, strike K, time to maturity $T-t$ and constant continuously compounded risk–free interest rate r. K and $T-t$ are specified in the option contract, and $X(t)$ and r can be assumed to be observable in the market.[1] Only the volatility σ_X is not directly observable. In the absence of any other market information, a simple approach would be to estimate it as the annualised standard deviation of past logarithmic returns of the underlying asset. However, if options on X are liquidly traded, the option price $C(t)$ can be observed in the market, allowing us to invert the formula (6.1) to determine the *implied volatility* σ_X. Due to the highly non-linear way in which σ_X enters (6.1), this cannot be done analytically, but because (6.1) is monotonic in σ_X, solving for σ_X is a well–behaved fixed point problem with a unique solution (for any arbitrage–free option price $C(t)$). Thus it can be calculated using the `Rootsearch` class template, as implemented in the member function `implied_volatility()` of `BlackScholesAsset` (Listing 6.1), which is more general than (6.1) in that it allows for a term structure of interest rates and a term structure of dividend yields.[2]

Using model parameters implied by liquid prices of derivative financial instruments is commonly called *model calibration*, as opposed to statistical *model estimation* based on past time series of underlying asset prices. When it comes to using a model to price other, less liquid derivatives relative to the market, the former is typically preferred. Firstly, this is because it ensures (within the assumptions of the model) that the pricing is arbitrage–free. Secondly, to the extent that liquid prices reflect an aggregation of the views of the market par-

---

[1] As long as interest rates are deterministic, they need not be constant — r can be interpreted as an "average" interest rate in the sense of the continuously compounded yield of a zero coupon bond maturing at time T. Even stochastic interest rates can be accommodated to a certain extent, see Chapter 8. Of course in practice there is some ambiguity as to which interest rate is best used for discounting option payoffs, but that discussion is beyond the present scope.

[2] For Black/Scholes option pricing in this case, see Chapter 4.

```
/// Calculate the implied volatility for a given price.
double BlackScholesAsset::
implied_volatility(double price,double mat,double K,double r,
                   int sign) const
{
  /* Create function for Black/Scholes formula with only the
     volatility variable, all other inputs fixed. Volatility
     is represented as the integral over squared volatility
     over the life of the option. */
  boost::function<double (double)> objective_function
    = boost::bind(&BlackScholesAsset::genericBlackScholes,
         this,(double)(xzero*(*dividend)(mat)/(*dividend)(0.0)),
         (double)(K*exp(-mat*r)),_1,sign);
  // instantiate Rootsearch class
  Rootsearch<boost::function<double (double)>,double,double>
    rs(objective_function,price,0.3,0.2,1e-9);
  // Solve for volatility input.
  double vol = rs.solve();
  // Return implied volatility as annualised volatility.
  return sqrt(vol/mat);
}
```

Listing 6.1: Calculating Black/Scholes implied volatility

ticipants about the future, model calibration is forward looking, while model estimation is backward looking.

Given market prices for options on the same underlying asset with different maturities, one will typically obtain different implied volatilities for different maturities. Usually, this can still be reconciled with the Black/Scholes model if one allows for time–varying deterministic volatility, as for example in Theorem 4.7 in Chapter 4.[3] This is called the *term structure of volatility*. Given market prices for options on the same underlying asset with different strikes, typically one will also obtain different implied volatilities. As the implied volatilities of options away from the money are often higher than those of at–the–money options, this is called the *volatility smile*. A term structure of volatility smiles yields the *volatility surface*. However, the volatility smile cannot be reconciled with the assumptions of the Black/Scholes model; the presence of the smile is an unequivocal statement by the market that the model is wrong,[4] specifically its assumption that the underlying asset follows geometric Brownian motion and has a lognormal distribution for each time horizon.

Much research has gone into constructing models that better reflect option prices observed in the market, though to date none has achieved the dominance

---

[3] An exception would be if the prices of short–term options are unusually high relative to long–term options, so as to imply a negative "forward" volatility.

[4] Nevertheless, in many practical situations the Black/Scholes model remains a useful approximation of reality.

and widespread acceptance to rival Black/Scholes. In this chapter, we will consider two approaches: implied distributions and stochastic volatility.

6.1 Calculating implied distributions

Early on, Breeden and Litzenberger (1978) noted that given option prices for a continuum of strikes for a particular maturity, the risk neutral probability distribution for the value of the underlying asset at that maturity can be recovered: Postulate a risk neutral probability density $f(x)$ for the underlying asset $X(T)$ and consider the forward price[5] P of a put option with strike K,

$$P = E[[K - X(T)]^+] = \int_{-\infty}^{K} (K - x)f(x)dx \qquad (6.2)$$

Differentiating with respect to K, one obtains

$$\frac{\partial P}{\partial K} = (K - K)f(K) + \int_{-\infty}^{K} f(x)dx \qquad (6.3)$$

$$\frac{\partial^2 P}{\partial K^2} = f(K) \qquad (6.4)$$

Fitting a model to the implied risk neutral distribution is the most direct way of ensuring consistency with current market prices. In practice, market prices for options are available only for a small number of strikes, so it becomes necessary to interpolate. This interpolation could be implemented at any one of three levels: the option prices, the implied volatilities, or the implied distribution. At first glance, interpolation at the level of implied volatilities is attractive, because this abstracts from all other market and contract variables that enter into the option price and focuses on the incremental (i.e., volatility) information contained in these prices. However, the conditions one needs to impose on the interpolation of implied volatilities in order to maintain the absence of arbitrage are not straightforward.[6] In contrast, any interpolated implied probability density, as long as it is a valid density, is arbitrage–free by construction.

A particularly tractable way of representing the implied risk neutral density for fitting observed option prices is the Gram/Charlier Type A series expansion. Under the appropriate assumptions, a model where the risk neutral densities of the logarithm of the underlying asset value(s) are given by a Gram/Charlier Type A series expansion arguably is as tractable as the Black/Scholes model: As first demonstrated by Corrado (2007), European options can be priced analytically in terms of the full (untruncated) series

---

[5] Note that considering the forward price of the option removes the need to discount the expected payoff. If interest rates are deterministic, this means that $f(x)$ will still correspond to the usual spot risk neutral probability measure. If interest rates are stochastic, $f(x)$ will correspond to the time T *forward measure* (see Chapter 8).

[6] See for example Fengler (2005) for a discussion of this issue. Alternatively, one possible approach to arbitrage–free interpolation of European call option prices is discussed in Section 6.2 below.

expansion; in fact, the Skipper/Buchen formula (4.52) for pricing a large class of exotic options in the Black/Scholes setting also generalises to the case where the joint distribution of multiple assets at multiple time points is given by Gram/Charlier Type A series expansions. The latter case is beyond the present scope, but the reader so inclined may wish to refer to Schlögl (2013).

In the case of a single asset X and a single maturity T, denote by x the standardised logarithmic value of $X(T)$, i.e.

$$x = \frac{\ln X(T) - E[\ln X(T)]}{\sqrt{\mathrm{Var}[\ln X(T)]}} \qquad (6.5)$$

and make the following

Assumption 6.1 *The standardised risk–neutral distribution, viewed at a fixed time t for a fixed maturity T, of the logarithm of the asset price $X(T)$ is given by the Gram/Charlier Type A series expansion*

$$f(x) = \sum_{j=0}^{\infty} c_j He_j(x)\phi(x) \qquad (6.6)$$

$$c_r = \frac{1}{r!} \int_{-\infty}^{\infty} f(x) He_r(x) dx$$

$$\phi(x) = \frac{1}{\sqrt{2\pi}} e^{-\frac{x^2}{2}}$$

i.e. $\phi(x)$ is the standard normal density function.

The Hermite polynomials $He_i(\cdot)$ are defined by the identity

$$(-D)^i \phi(x) = He_i(x)\phi(x) \qquad (6.7)$$

where

$$D = \frac{d}{dx}$$

is the differential operator. Thus $He_0(x) = 1$ and $He_1(x) = x$, and the higher order polynomials can be generated by the recurrence relation[7]

$$He_{n+1}(x) = x He_n(x) - n He_{n-1}(x), \qquad (6.8)$$

Absence of arbitrage requires what Corrado (2007) calls an "explicit martingale restriction" on $E[\ln X(T)]$, i.e.

Proposition 6.2 *Under Assumption 6.1, the risk-neutral expected value μ of the logarithm of the asset price $X(T)$ must satisfy*

$$\mu = \ln \frac{X(t)}{B(t,T)} - \ln \sum_{j=0}^{\infty} c_j \sigma^j - \frac{1}{2}\sigma^2 \qquad (6.9)$$

---

[7] See e.g. Abramowitz and Stegun (1964).

where $X(t)$ is the current asset value, $B(t, T)$ is the value at time t of a zero coupon bond maturing in T, and σ is the standard deviation of $\ln X(T)$.

Setting $\sigma = \sqrt{\mathrm{Var}[\ln X(T)]}$, one then obtains[8]

Proposition 6.3 *Under Assumption 6.1, the price of a call option at time t on the asset $X(T)$ with expiry T is given by*

$$
\begin{aligned}
C &= B(t, T) \int_{-\infty}^{\infty} [e^{\sigma x + \mu} - K]^+ f(x) dx \\
&= X(t)\mathcal{N}(d^*) - B(t, T)K\mathcal{N}(d^* - \sigma) \quad (6.10) \\
&\quad + X(t) \left(\sum_{j=0}^{\infty} c_j \sigma^j \right)^{-1} \phi(d^*) \sum_{j=2}^{\infty} \sum_{i=1}^{j-1} c_j \sigma^i He_{j-1-i}(-d^* + \sigma)
\end{aligned}
$$

with

$$
d^* = \frac{\mu - \ln K + \sigma^2}{\sigma} \quad (6.11)
$$

Given Proposition 6.3, one can use a set of market option prices for different strikes at a fixed maturity to calibrate the risk–neutral distribution of the underlying asset price. When calculating option prices using (6.10), the infinite sum is truncated after a finite number of terms. It is well known that in this case, due to the polynomial terms in the Gram/Charlier Series A expansion, conditions need to be imposed on the expansion coefficients c_j in (6.6) in order to ensure that $f(x)$ represents a valid probability density with $f(x) \geq 0 \ \forall x.$[9] For truncation after the fourth moment, the case most commonly considered in the literature,[10] Jondeau and Rockinger (2001) characterise the set of permissible values of c_3 and c_4. On the basis of their result, one can constrain the optimisation in the calibration (typically involving the minimisation of some sort of squared distance between market and model prices) to always yield a valid density. Unfortunately, it is not clear how one can extend their approach in a practicable manner to truncated Gram/Charlier expansions involving moments of higher order. However, common well–known unconstrained non-linear optimisation algorithms can be adapted to ensure that the calibrated expansion coefficients in fact yield a valid probability density. Here, we proceed to do so by confining Brent's line search within Powell's algorithm[11] to a convex set of permissible coefficients in a truncated Gram/Charlier expansion. It always holds that $c_0 = 1$, and we standardise $f(x)$ to zero expectation and

---

[8] For a formal proof, see Corrado (2007) or Schlögl (2013).
[9] This condition is sufficient for $f(x)$ to be a valid density, since

$$
\int_{-\infty}^{\infty} f(x) = 1
$$

follows from the properties of Hermite polynomials.
[10] See e.g. Corrado and Su (1996), Bahra (1997), Brown and Robinson (2002) and Jurczenko, Maillet and Negrea (2002).
[11] See Section 2.5.3.

unit variance by setting $c_1 = 0$ and $c_2 = 0$. We truncate the series expansion by setting $c_j = 0 \; \forall j > k$ for some choice of $k \geq 4$. Denote by

$$\mathcal{C}_k = \{(c_3, c_4, \ldots, c_k) \in \mathbb{R}^{k-2} | f(x) \geq 0 \; \forall x\} \qquad (6.12)$$

the set of permissible coefficients when the series is truncated at k. Note that if $c_j \neq 0$ for j odd, then we must have $c_{j+1} \neq 0$ in order to obtain a valid density. This is because if the highest exponent, say j, in the polynomial term of $f(x)$ is odd, for $c_j > 0$ there exists a $\eta < 0$ such that $f(x) < 0 \; \forall x < \eta$, and for $c_j < 0$ there exists a $\eta > 0$ such that $f(x) < 0 \; \forall x > \eta$. Thus, for odd k, $\mathcal{C}_k = \mathcal{C}_{k-1}$, and it is therefore sufficient to consider only even k. Furthermore, we have

Lemma 6.4 *(6.12) defines a convex set.*

PROOF: If the coefficient vectors $c^{(1)}$ and $c^{(2)}$ are elements of $\mathcal{C}_k$, i.e.

$$\sum_{i=0}^{k} c_i^{(1)} \mathrm{He}_i(x) \geq 0 \quad \text{and} \quad \sum_{i=0}^{k} c_i^{(2)} \mathrm{He}_i(x) \geq 0 \quad \forall x$$

then it immediately follows that for any $0 \leq \alpha \leq 1$

$$\sum_{i=0}^{k} (\alpha c_i^{(1)} + (1-\alpha) c_i^{(2)}) \mathrm{He}_i(x) \geq 0 \quad \forall x$$

i.e. $\alpha c^{(1)} + (1-\alpha) c^{(2)} \in \mathcal{C}_k$.

$\square$

Let $x_1, \ldots, x_k$ denote the k (possibly complex) roots of the polynomial

$$\sum_{i=0}^{k} c_i \mathrm{He}_i(x)$$

Then $c \notin \mathcal{C}_k$ if for at least one $1 \leq i \leq k$, $x_i \in \mathbb{R}$ and $x_i = x_j$ for an even number of different $j \neq i$. The boundaries of $\mathcal{C}_k$ along any given direction $d \in \mathbb{R}^{k-2}$ can be determined by two one-dimensional searches. Suppose the current position is $c^{(0)} \in \mathcal{C}_k$. Then a move along the direction d can be represented as $c^{(1)} = c^{(0)} + \lambda d$, $\lambda \in \mathbb{R}$, and we search for the smallest $\lambda_{\mathrm{upper}} > 0$ and the largest $\lambda_{\mathrm{lower}} < 0$ such that $c^{(1)} \notin \mathcal{C}_k$, as determined by the above criterion.[12]

We can then modify Brent's line search to find a $\lambda_{\min}$ moving us from $c^{(0)}$ to the minimum of the calibration objective function along the direction d while ensuring that $\lambda_{\mathrm{lower}} < \lambda_{\min} < \lambda_{\mathrm{upper}}$. Once we have this one-dimensional minimisation algorithm along an arbitrary direction d, we can use Powell's algorithm for minimisation in multiple dimensions.

A probability density $f(x)$ given by a truncated Gram/Charlier Type A

---

[12] The polynomial roots can easily be calculated using Laguerre's method — see e.g. Press, Teukolsky, Vetterling and Flannery (2007).

```
/// Constructor.
GramCharlierAsset(
  GramCharlier& xgc, ///< Gram/Charlier expanded density for the
                     //    standardised risk-neutral distribution.
  double xsgm,       ///< Volatility level.
  double ini,        ///< Initial ("time zero") value of the
                     //    underlying asset.
  double xT          ///< Maturity for which the risk-neutral
                     //    distribution is valid.
) : gc(xgc),sgm(xsgm),xzero(ini),T(xT),prices_(3) { };
```

Listing 6.2: The constructor of `GramCharlierAsset`

series expansion, i.e.

$$f(x) = \sum_{j=0}^{m} c_j \text{He}_j(x)\phi(x) \tag{6.13}$$

is represented by the class `GramCharlier`. The defaults constructor sets $m = 0$, $c_0 = 1$, i.e. $f(x)$ is then the standard normal density $\phi(x)$. Alternatively, the constructor

$$\texttt{GramCharlier(const Array<double,1>\& xcoeff);} \tag{6.14}$$

allows the coefficients c_j to be specified directly. Furthermore, the coefficients c_j and a volatility scaling parameter σ are calibrated by the member function

```
double fit_coefficients(
  boost::function<double (double)> objective_function,
  double sgm,int dim,double eps = 1e-12);
```

by minimising an externally supplied `objective_function`, where `sgm` is an initial guess of σ, `dim` is the truncation threshold m in (6.13), and `eps` determines when the Powell minimisation is deemed to have converged.

The class `GramCharlier` is used by the class `GramCharlierAsset` to represent the risk neutral distribution of the underlying asset; its constructor is given in Listing 6.2. Note that an instance of `GramCharlier` is passed to `GramCharlierAsset` by reference; thus when the risk neutral distribution of `GramCharlierAsset` is calibrated to market data, this changes the coefficients in the original instance of `GramCharlier` accordingly. The member function

```
double call(double K,
            double domestic_discount,                    (6.15)
            double foreign_discount = 1.0) const;
```

implements a slightly more general version of the call option formula (6.10), i.e.

$$C = X(t)\tilde{B}(t,T)\mathcal{N}(d^*) - B(t,T)K\mathcal{N}(d^* - \sigma) \tag{6.16}$$

$$+ X(t)\left(\sum_{j=0}^{\infty} c_j\sigma^j\right)^{-1}\phi(d^*)\sum_{j=2}^{\infty}\sum_{i=1}^{j-1} c_j\sigma^i\text{He}_{j-1-i}(-d^* + \sigma)$$

```
/** Best fit calibration to a given set of Black/Scholes implied
    volatilities. */
double GramCharlierAsset::calibrate(const Array<double,1>* xstrikes,
                                    const Array<double,1>* xvols,
                                    double domestic_discount,
                                    double foreign_discount,
                                    int highest_moment)
{
  int i;
  strikes_ = xstrikes;
  int len = strikes_->extent(firstDim);
  if (len>prices_.extent(firstDim)) prices_.resize(len);
  // Forward price of underlying
  double fwd = xzero * foreign_discount/domestic_discount;
  // Convert implied volatilities to call option prices
  for (i=0;i<len;i++)
    prices_(i) = domestic_discount *
      genericBlackScholes(fwd,(*strikes_)(i),
                          (*xvols)(i)*(*xvols)(i)*T,1);
  domestic_ = domestic_discount;
  foreign_  = foreign_discount;
  // Pass objective function to Gram/Charlier calibration routine
  boost::function<double (double)> objective_function =
    boost::bind(std::mem_fun(
      &GramCharlierAsset::calibration_objective_function),this,_1);
  return gc.fit_coefficients(objective_function,
                             (*xvols)(0),highest_moment);
}
```

Listing 6.3: Calibrating the risk neutral distribution of `GramCharlierAsset`

with

$$d^* = \frac{\mu - \ln K + \sigma^2}{\sigma}$$

$$\mu = \ln \frac{X(t)\tilde{B}(t,T)}{B(t,T)} - \ln \sum_{j=0}^{\infty} c_j \sigma^j - \frac{1}{2}\sigma^2$$

where $\tilde{B}(t,T)$ is the "foreign discount factor," i.e. if $X(t)$ is an exchange rate in units of domestic currency per unit of foreign currency, (6.16) is the price of a call option on foreign currency struck in domestic currency, and $\tilde{B}(t,T)$ is the foreign currency discount factor. Alternatively, if $X(t)$ represents an equity index with, say, a dividend yield of y, $\tilde{B}(t,T) = \exp\{-y(T-t)\}$ can be interpreted as a "dividend discount factor." If $X(t)$ represents an asset that does not pay any dividends during the life of the option, simply set $\tilde{B}(t,T) = 1$ (the default in `call()`).

Calibration of the `GramCharlier` risk neutral distribution to market data is implemented in the member function `calibrate()` (Listing 6.3). Market

```
/** Objective function for best fit calibration to a given set of
    Black/Scholes implied volatilities. Note that this function
    updates the volatility parameter sgm of GramCharlierAsset. */
double GramCharlierAsset::calibration_objective_function(double xsgm)
{
  int i;
  double result = 0.0;
  sgm = xsgm;
  for (i=0;i<strikes_->extent(firstDim);i++) {
  double reldiff =
      (prices_(i)-call((*strikes_)(i),domestic_,foreign_))
          / prices_(i);
  result += reldiff*reldiff; }
  return result * 10000.0;
}
```

Listing 6.4: The objective function when calibrating the risk neutral distribution of GramCharlierAsset

data is given in the form of the Black/Scholes implied volatilities xvols for a given set of strikes. highest_moment determines the truncation threshold m. In calibrate(), the Black/Scholes implied volatilities are converted to option prices. The objective of the calibration is to choose the volatility scale parameter σ and the Gram/Charlier coefficients c_j so as to minimise the sum of the squared relative differences between the "market" option prices and the "model" option prices given by the member function call(). This is implemented in the member function calibration_objective_function() (Listing 6.4), which is passed to GramCharlier::fit_coefficients() by calibrate().

As an example, consider the market data for EUR/USD options with one month time to maturity on 24 January 2008 (Table 6.1). Option values are given as "implied volatility quotes," for at–the–money, and 25% and 10% "delta" *risk reversals* (RR) and *butterflies* (BF). The "implied volatility" σ_{RR} of a risk reversal is given by the difference of the Black/Scholes implied volatility of an out–of–the–money call and put, and the "implied volatility" σ_{BF} of a butterfly is given by the average of the Black/Scholes implied volatilities of the out–of–the–money call and put, minus the at–the–money volatility σ_{ATM}. Denoting by σ_C and σ_P the implied volatilities, respectively, of the out–of–the–money call and put, we have at each level of "delta"

$$\sigma_C = \sigma_{\mathrm{BF}} + \frac{1}{2}\sigma_{\mathrm{RR}} + \sigma_{\mathrm{ATM}} \qquad (6.17)$$

$$\sigma_P = \sigma_C - \sigma_{\mathrm{RR}} \qquad (6.18)$$

The "delta" determines the strikes of the out–of–the–money options, which are such that the absolute value of the option delta equals the prescribed

Maturity	ATM	25D RR	25D BF	10D RR	10D BF
1W	10.8750	-0.7500	0.3253	-1.2500	0.5000
2W	10.3750	-0.6250	0.3207	-1.0000	0.5100
3W	10.2500	-0.5000	0.3058	-0.8750	0.6300
1M	9.5750	-0.4500	0.2750	-0.7500	1.1250
2M	9.3250	-0.4500	0.3000	-0.7500	1.2250
3M	9.2250	-0.6000	0.3250	-0.7500	1.3250
4M	9.0505	-0.4950	0.3535	-0.8910	1.4191
6M	8.9500	-0.4750	0.3750	-0.7500	1.4250
9M	8.9500	-0.5250	0.3990	-0.7750	1.6234
1Y	8.7500	-0.5250	0.4000	-0.8000	1.6125
18M	8.6505	-0.5445	0.4091	-0.9801	1.6463
2Y	8.6500	-0.5250	0.4047	-0.8000	1.7189
5Y	7.6000	-0.5500	0.3535	-1.0692	1.7675

Table 6.1 *USD/EUR at–the–money (ATM) implied volatilities, 25%-delta (25D) and 10%–delta (10D) risk reversals (RR) and butterflies (BF) on 24 January 2011. Source: Numerix CrossAsset XL*

level.[13] E.g., the strike of the 25%–delta call option is given by

$$\frac{X(t)\tilde{B}(t,T)}{B(t,T)} \exp\left\{-\sigma_C\sqrt{T-t}\mathcal{N}^{-1}(0.25) + \frac{1}{2}\sigma_C^2(T-t)\right\} \qquad (6.19)$$

and for the 25%–delta put,

$$\frac{X(t)\tilde{B}(t,T)}{B(t,T)} \exp\left\{-\sigma_P\sqrt{T-t}\mathcal{N}^{-1}(0.75) + \frac{1}{2}\sigma_C^2(T-t)\right\} \qquad (6.20)$$

Feeding the market data for the maturity of one month into `GramCharlier-Asset::calibrate()`,[14] we obtain a reasonable fit for $m = 4$, a better fit for $m = 6$, and a perfect fit for $m = 8$, as Table 6.2 shows. Figure 6.1 compares the resulting normalised (i.e. zero mean and unit variance) risk neutral densities with the standard normal density; note the skew and the fatter tails of the implied distribution.

---

[13] Characterising strikes by option delta is more informative than absolute strikes or relative "moneyness," as it is primarily determined (though in a non-linear fashion) by how many standard deviations of the distribution of the underlying asset lie between the forward price of the underlying and the strike. Absolute strikes say nothing about whether an option is in, at or out of the money, while relative "moneyness" is the quotient of the current price of the underlying and the strike, which is independent of volatility.

[14] See `GCempirical.cpp` on the website for this book.

Strike	Implied Volatility	Black/Scholes Price	Fitted Gram/Charlier Price		
			$m = 4$	$m = 6$	$m = 8$
1.44751	0.10075	0.0345393	0.0345391	0.0345243	0.0345393
1.47556	0.09575	0.0162668	0.0162429	0.0162719	0.0162668
1.50405	0.09625	0.0060296	0.0060350	0.0060288	0.0060296
1.41705	0.11075	0.0607606	0.0608989	0.0607624	0.0607606
1.53369	0.10325	0.0020562	0.0020558	0.0020563	0.0020562
Sigma			0.0296962	0.0295336	0.0295042
Skewness			-0.221359	-0.143105	-0.101406
Excess kurtosis			1.65674	1.50254	1.45571
c_5			N/A	0.00475034	0.0106612
c_6			N/A	4.49E-05	0.000152405
c_7			N/A	N/A	0.000682596
c_8			N/A	N/A	0.000114547

Table 6.2 *Fit to one–month maturity USD/EUR option data on 24 January 2011.*

6.2 Constructing an implied volatility surface

Given the Breeden and Litzenberger (1978) relationship (6.4), a sufficient condition for interpolated standard European option[15] prices to be arbitrage free is that the second derivative with respect to the strike exists and is positive. Here, let us consider the method proposed by Kahalé (2004), which satisfies this condition. Assume that for a fixed maturity market prices for call options are given for strikes k_i, $0 < i \leq n$. Interpolate these option prices as a function of the strike k on each interval $[k_i, k_{i+1}]$ by

$$\psi_i(k) = f_i \mathcal{N}(d_1^{(i)}) - k\mathcal{N}(d_2^{(i)}) + a_i k + b_i \qquad (6.21)$$

$$d_1^{(i)} = \frac{\ln \frac{f_i}{k} + \frac{1}{2}\sigma_i^2}{\sigma_i} \qquad d_2^{(i)} = d_1^{(i)} - \sigma_i$$

where $\psi_i(k)$ is taken to be the "forward" price of the option, i.e. if the option expiry is T and the current time t, then the option price $C(k)$ is given by

$$C(k) = B(t,T)\psi_i(k) \qquad \text{for} \quad k \in [k_i, k_{i+1}] \qquad (6.22)$$

with $B(t,T)$ the time t price of a zero coupon bond maturing at time T. Allowing for extrapolation for strikes below k_1 and above k_n, set $k_0 = 0$ and $k_{n+1} = \infty$, though at these strikes we can only consider the limit expression for (6.21), so (6.22) becomes

$$C(k) = B(t,T)\psi_i(k) \qquad \text{for} \quad k \in [k_i, k_{i+1}] - \{0, \infty\} \qquad (6.23)$$

---

[15] By virtue of put/call parity, these can be either put or call options.

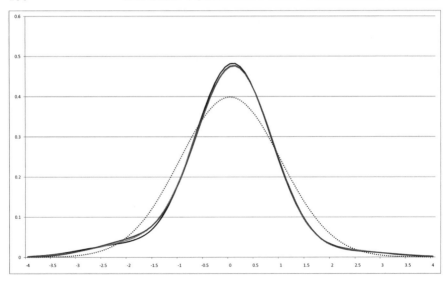

Figure 6.1 *Normalised implied risk–neutral distributions (solid lines) fitted to one–month maturity USD/EUR option data on 24 January 2011, vs. the standard normal distribution (dashed line).*

The coefficients f_i, σ_i, a_i, b_i of (6.21) thus need to be determined for all $0 \leq i \leq n$ ($4n + 4$ variables).

Firstly, we have four limit conditions (the prime denotes differentiation with respect to the strike k),

$$\lim_{k \to 0} C(k) = X(t) \qquad \Rightarrow \qquad f_0 + b_0 = \frac{X(t)}{B(t,T)} \tag{6.24}$$

$$\lim_{k \to 0} C'(k) = -B(t,T) \qquad \Rightarrow \qquad a_0 = 0 \tag{6.25}$$

$$\lim_{k \to \infty} C(k) = 0 \qquad \Rightarrow \qquad b_n = 0 \tag{6.26}$$

$$\lim_{k \to \infty} C'(k) = 0 \qquad \Rightarrow \qquad a_n = 0 \tag{6.27}$$

Secondly, the coefficients must be chosen to fit the observed option prices $\overline{C}_i$ at the strikes k_i, $1 \leq i \leq n$. This yields the $2n$ equations

$$\psi_i(k_i) \quad = \quad \overline{C}_i \qquad \forall\, 1 \leq i \leq n \tag{6.28}$$

$$\psi_i(k_{i+1}) \quad = \quad \overline{C}_{i+1} \qquad \forall\, 0 \leq i < n \tag{6.29}$$

Thirdly, enforce continuous first and second derivatives with respect to the strike by the $2n$ equations,

$$\psi_i'(k_{i+1}) - \psi_{i+1}'(k_{i+1}) \quad = \quad 0 \qquad \forall\, 0 \leq i < n \tag{6.30}$$

$$\psi_i''(k_{i+1}) - \psi_{i+1}''(k_{i+1}) \quad = \quad 0 \qquad \forall\, 0 \leq i < n \tag{6.31}$$

Note that since the second derivative of the ψ_i with respect to the strike is the same as for the Black/Scholes formula, the sufficient condition for the absence of arbitrage, $C''(k) > 0$, holds by construction.

Using the limit conditions (6.24)–(6.27) to eliminate the variables a_0, b_0, a_n, b_n, we are left with a $4n$-dimensional parameter vector

$$\theta = (f_0, \sigma_0, f_1, \sigma_1, a_1, b_1, \ldots, f_{n-1}, \sigma_{n-1}, a_{n-1}, b_{n-1}, f_n, \sigma_n)$$

The values of these parameters can be determined by solving the system of $4n$ equations (6.28)–(6.31), for example using the Newton/Raphson algorithm discussed in Section 2.5.2. In order to implement this, note that

$$\psi_i'(k) = \frac{\partial \psi_i}{\partial k} = -\mathcal{N}(d_2^{(i)}) + a_i \tag{6.32}$$

$$\psi_i''(k) = \frac{\partial^2 \psi_i}{\partial k^2} = \phi(d_2^{(i)})\sigma_i^{-1}k^{-1} \tag{6.33}$$

Furthermore, for the Jacobian we need the partial derivatives of ψ_i, ψ_i', and ψ_i'' with respect to the coefficients f_i, σ_i, a_i and b_i. We have for $1 \leq i \leq n$,

$$\frac{\partial \psi_i}{\partial f_i} = \mathcal{N}(d_1^{(i)}) \qquad \frac{\partial \psi_i'}{\partial f_i} = -\phi(d_2^{(i)})\sigma_i^{-1}f_i^{-1}$$

$$\frac{\partial \psi_i''}{\partial f_i} = -\phi(d_2^{(i)})\sigma_i^{-2}d_2^{(i)}f_i^{-1}k^{-1}$$

and due to the substitution $b_0 = \frac{X(t)}{B(t,T)} - f_0$,

$$\frac{\partial \psi_0}{\partial f_0} = \mathcal{N}(d_1^{(0)}) - 1$$

For the derivatives with respect to σ_i, we have for $0 \leq i \leq n$,

$$\frac{\partial \psi_i}{\partial \sigma_i} = f_i\phi(d_1^{(i)})\left(-\left(\ln\frac{f_i}{k}\right)\sigma_i^{-2} + \frac{1}{2}\right) - k\phi(d_2^{(i)})\left(-\left(\ln\frac{f_i}{k}\right)\sigma_i^{-2} - \frac{1}{2}\right)$$

$$= f_i\phi(d_1^{(i)}) = k\phi(d_2^{(i)})$$

$$\frac{\partial \psi_i'}{\partial \sigma_i} = -\phi(d_2^{(i)})\left(-\left(\ln\frac{f_i}{k}\right)\sigma_i^{-2} - \frac{1}{2}\right)$$

$$\frac{\partial \psi_i''}{\partial \sigma_i} = -\phi(d_2^{(i)})\sigma_i^{-2}k^{-1} + \phi(d_2^{(i)})d_2^{(i)}\left(\left(\ln\frac{f_i}{k}\right)\sigma_i^{-2} + \frac{1}{2}\right)$$

For the derivatives with respect to a_i and b_i, we have for $1 \leq i < n$,

$$\frac{\partial \psi_i}{\partial a_i} = k \qquad \frac{\partial \psi_i'}{\partial a_i} = 1 \qquad \frac{\partial \psi_i''}{\partial a_i} = 0$$

$$\frac{\partial \psi_i}{\partial b_i} = 1 \qquad \frac{\partial \psi_i'}{\partial b_i} = 0 \qquad \frac{\partial \psi_i''}{\partial b_i} = 0$$

6.3 Stochastic volatility

One way to construct a model, which reflects the empirical features of the Black/Scholes implied volatilities observed in the market (i.e. the implied volatility skew and smile), is to allow volatility to evolve stochastically.[16] Following Wilmott (2000), one might specify the dynamics of the underlying asset $S(t)$ as

$$dS(t) = \mu(t)S(t)dt + \sigma(t)S(t)dW, (t) \qquad (6.34)$$

and assume that volatility $\sigma(t)$ follows the stochastic differential equation

$$d\sigma(t) = p(S(t), \sigma(t), t)dt + q(S(t), \sigma(t), t)dW_2(t) \qquad (6.35)$$

with the quadratic variation of the two driving Brownian motions given by

$$< dW_1, dW_2 >_t = \rho dt$$

Given an additional asset V, the time t price of which depends on current volatility $\sigma(t)$, one can proceed analogously to the derivation of the Black/Scholes PDE (4.9) in Section 4.2.1: For some derivative financial instrument $C(t, S(t), \sigma(t))$, construct a portfolio V long one unit of C, short Δ units of S, and short Δ, units of V_1. The change in value of this portfolio is given by

$$
\begin{aligned}
dV = & \left(\partial_1 C(t, S(t), \sigma(t)) + \frac{1}{2}\partial_{2,2}C(t, S(t), \sigma(t))S(t)^2\sigma^2 \right. \\
& + \partial_{2,3}C(t, S(t), \sigma(t))g\sigma(t)q(S(t), \sigma(t), t)S(t) \\
& + \left. \frac{1}{2}\partial_{3,3}C(t, S(t), \sigma(t))q^2(S(t), \sigma(t), t) \right) dt \\
& - \Delta_1 \left(\partial_1 V_1(t, S(t), \sigma(t)) + \frac{1}{2}\partial_{2,2}V_1(t, S(t), \sigma(t))S(t)^2\sigma^2 \right. \\
& + \partial_{2,3}V_1(t, S(t), \sigma(t))g\sigma C t_q(S(t), \sigma(t), t)S(t) \\
& + \left. \frac{1}{2}\partial_{3,3}V_1(t, S(t), \sigma(t))q^2(S(t), \sigma(t), t) \right) dt \\
& + \left(\partial_2 C(t, S(t), \sigma(t)) - \Delta_1\partial_2 V_1(t, S(t), \sigma(t)) - \Delta \right) dS(t) \\
& + \left(\partial_3 C(t, S(t), \sigma(t)) - \Delta_1\partial_3 V_1(t, S(t), \sigma(t)) \right) d\sigma(t) \qquad (6.36)
\end{aligned}
$$

Choosing Δ and Δ_1 to solve

$$\partial_2 C(t, S(t), \partial(t)) - \Delta_1\partial_2 V_1(t, S(t), \sigma(t)) - \Delta = 0$$

$$\partial_3 C(t, S(t), \partial(t)) - \Delta_1\partial_3 V_1(t, S(t), \sigma(t)) - \Delta = 0$$

---

[16] An alternative would be to assume a jump–diffusion process for the underlying stock price. In either case, an option can no longer be replicated by a dynamic hedging strategy in the underlying and a riskless asset alone – to complete the market, further tradeable instruments are required. There is surprisingly little literature on this, with a notable exception being the work of Carr and Schoutens (2007), who extend the stochastic volatility model of Heston (1993) to allow for jumps to default and construct hedging strategies involving credit default swaps and variance swaps as hedge instruments in addition to the underlying and a riskless asset.

eliminates all randomness in (6.36). In this case, the terms involving $dS(t)$ and $d\sigma(t)$ in (6.36) are zero, leaving only the terms involving dt. Additionally assuming a constant risk–free interest rate r, we must have

$$dV = \ldots dt = rV dt = r(C(t, S(t), \sigma()t)) - \Delta S(t) - \Delta_1 V_1(t, S(t), \sigma(t))dt \tag{6.37}$$

Wilmott (2000) notes that one can collect all terms involving C on the left–hand side and all terms involving V_1 on the right–hand side, i.e. (shortening notation by omitting function arguments)

$$\frac{\partial_1 C - \frac{1}{2}\sigma^2 S^2 \partial_{2,2}C + \rho\sigma S_q \partial_{2,3}C + \frac{1}{2}q^2\partial 3, 3C + rS\partial_2 C - rC}{\partial_3 C} = \ldots \tag{6.38}$$

and therefore, since C and V_1 will be derivative financial instruments with different contract specifications, both sides can only be functions of the *independent* variables S, σ and t. This can be expressed as

$$\partial_1 C + \frac{1}{2}\sigma^2 S^2 \partial_{2,2}C + \rho\sigma S_q \partial_{2,3}C + \frac{1}{2}q^2\partial_{3,3}C + rS\partial_2 C - C =$$
$$\Psi(S, \sigma, t)\partial_3 C \tag{6.39}$$

This function Ψ can be expressed in terms of a market price of volatility risk (which could be calibrated to observed option prices). Again following Wilmott (2000), consider a position in $C(t, S(t), \sigma(t))$ which is delta–hedged against changes in the value of underlying asset S only, i.e.

$$V = C(t, S(t), \sigma(t)) - \Delta S(t)$$

The change in this portfolio value is

$$dV = \left(\partial_1 C + \frac{1}{2}\sigma^2 S^2 \partial_{2,2}C + \rho\sigma q S \partial_{2,3}C + \frac{1}{2}q^2\partial_{3,3}C\right)dt$$
$$+ (\partial_2 C - \Delta)dS + \partial_3 C d\sigma \tag{6.40}$$

with $\Delta = \partial_2 C$. Comparing the change in V with what a risk–free investment of the same value would earn results in

$$dV - rV dt = \left(\partial_1 C + \frac{1}{2}\sigma^2 S^2 \partial_{2,2}C + \rho\sigma q S \partial_{2,3}C + \frac{1}{2}q^2\partial_{3,3}C\right.$$
$$\left. - rS\partial_2 C - rC\right)dt + \partial_3 C d\sigma$$
$$= \Psi(S, \sigma, t)\partial_3 C dt + (pdt + qdW_2(t))\partial_3 C \tag{6.41}$$

Defining a function λ by

$$\lambda(S, \sigma, t) = \frac{\Psi(S, \sigma, t) + p(S, \sigma, t)}{q(S, \sigma, t)}$$

we have

$$dV - rV dt = (\lambda(S\sigma, t)dt + q(S, \sigma, t)dW_2(t))\partial_3 C(t, S, \sigma)$$

Thus for every unit of volatility risk, represented by the diffusion term $dW_2(t)$, there are $\lambda(S, \sigma, t)$ units of expected return above the risk-free rate. It is in this that $\lambda(S, \sigma, t)$ can be interpreted as the market price of volatility risk, if the model dynamics (6.34)–(6.35) were specified under the objective (real-world) probability measure.

Within this general framework, due to its tractability, the model of Heston (1993) has found widespread application. As the model is typically calibrated to option prices observed in the market and used primarily to price more exotic instruments relative to the market, it is usually set up directly under the risk neutral measure, rather than under the objective measure. In this case, the dynamics of the underlying asset $S(t)$ are given by

$$dS(t) = r(t)S(t)dt + \sqrt{V(t)}S(t)dW_1(t) \tag{6.42}$$

where $r(t)$ is the (deterministic) continuously compounded short rate of interest and $V(t)$ follows a Cox, Ingersoll and Ross (1985)–type "square root" process

$$dv(t) = k(\theta - v(t))dt + \eta\sqrt{v(t)}dW_2(t) \tag{6.43}$$

k is the speed of mean reversion of $v(t)$ to its long–run mean θ, and η is the "volatility of volatility" parameter.

In the original work of Heston (1993), a semi-analytical pricing formula for standard European options is derived using a Fourier transform technique. However, the characteristic function derived by Heston can be difficult to integrate numerically due to discontinuities. Also using Fourier transforms, Gatheral (2006) derives an equivalent semi-analytical pricing formula in which the characteristic function to be integrated numerically seems to be continuous over the full range of integration, though a rigorous mathematical proof — that it is indeed always continuous — remains outstanding.

Let us focus on the solution presented by Gatheral (2006). Define x as the logarithmic forward moneyness of the option, i.e. at time t with option maturity T and strike K, we have

$$x : = \ln\left(\frac{F_{t,T}}{K}\right)$$

$$F_{t,T} = \frac{S(t)}{B(t,T)}$$

where $B(t,T)$ is the time t price of a zero bond maturing in T. The Heston model satisfies the requisite technical conditions for the existence of a spot risk neutral measure (associated with taking the continuously compounded savings account as the numeraire) as well as the equivalent martingale measure associated with taking the underlying $S(t)$ as the numeraire, and thus forward price of a European call option can be represented as

$$\hat{C}(x,v,\tau) = K(e^x P_1(x,v,\tau) - P_0(x,v,\tau)) \qquad \tau := T - t \tag{6.44}$$

where P_0 is the probability of exercise under the spot risk neutral measure and P_1 is the probability of exercise under the equivalent martingale measure associated with the numeraire $S(t)$. Substituting the risk-neutral dynamics (with market price of risk $\lambda(S, \sigma, t) = 0$) (6.42)–(6.43) into the PDE (6.39), changing variables $\tau := T - t$, and expressing the resulting PDE in terms of the forward price $\hat{C}$ of the option, we have

$$-\frac{\partial \hat{C}}{\partial \tau} + \frac{1}{2} v \partial_1 \hat{C} + \frac{1}{2} \eta^2 v \partial_{2,2} \hat{C} + \rho \eta v \partial_{1,2} \hat{C} + k(\theta - v) \partial_2 \hat{C} = 0 \qquad (6.45)$$

Inserting $\hat{C}$ as given by (6.44) into (6.45) results in partial differential equations for P_0 and P_1, i.e. for $j = 0, 1$

$$-\frac{\partial P_j}{\partial \tau} + \frac{1}{2} v \partial_{1,1} P_j - \left(\frac{1}{2} - j\right) v \partial_1 P_j + \frac{1}{2} \eta^2 v \partial_{2,2} P_j + \rho \eta v \partial_{1,2} P_j$$

$$+ (k\theta - b_j v) \partial_2 P_j = 0 \quad (6.46)$$

with $b_j = k - j\rho\eta$, subject to the terminal condition

$$\lim_{\tau \to 0} P_j(x, v, \tau) = \begin{cases} 1 & \text{if } x > 0 \\ 0 & \text{if } x \leq 0 \end{cases}$$

$$=: \xi(x)$$

Gatheral (2006) solves the PDE for P_j via the Fourier transform of P_j, defined by

$$\tilde{P}(u, v, \tau) = \int_{-\infty}^{\infty} e^{-iux} P(x, v\tau) dx$$

It follows that

$$\tilde{P}(u, v, 0) = \int_{-\infty}^{\infty} e^{-iux} \xi(x) dx = \frac{1}{iu}$$

Substituting the inverse transform

$$P(x, v, \tau) = \int_{-\infty}^{\infty} \frac{1}{2\pi} e^{iux} \tilde{P}(u, v, t) du \qquad (6.47)$$

into (6.46) yields

$$-\frac{\partial \tilde{P}_j}{\partial \tau} - \frac{1}{2} u^2 v \tilde{P}_j - \left(\frac{1}{2} - j\right) iuv \tilde{P}_j$$

$$+ \frac{1}{2} \eta^2 v \partial_{2,2} \tilde{P}_j + \rho \eta iuv \partial_2 \tilde{P}_j + (k\theta - b_j v) \partial_2 \tilde{P}_j = 0 \quad (6.48)$$

Gatheral (2006) defines the variables

$$\alpha = -\frac{u^2}{2} - \frac{iu}{2} + iju$$

$$\beta = k - \rho\eta j - \rho\eta iu$$

$$\gamma = \frac{n^2}{2}$$

so (6.48) can be written as

$$v(\alpha \tilde{P}_j - \beta \partial_2 \tilde{P}_j + \gamma \partial_{2,2} \tilde{P}_j) + a\partial_2 \tilde{P}_j - \frac{\partial \tilde{P}_j}{\partial \tau} = 0 \qquad (6.49)$$

Postulating a solution of the form

$$\tilde{P}_j(u,v,\tau) = \exp\{C(u,t)\theta + D(u,\tau)v\}\tilde{P}_j(u,v,0) = \frac{1}{iu}\exp\{C(u,\tau)\theta + D(u,\tau)v\}$$

the relevant partial derivatives are

$$\frac{\partial P_j}{\partial \tau} = (\theta \partial_2 C + v\partial_2 D)\tilde{P}_j$$

$$\partial_2 \tilde{P}_j = D\tilde{P}_j$$

$$\partial_{2,2}\tilde{P}_j = D^2 \tilde{P}_j$$

The partial differential equation for $\tilde{P}_j$ (6.49) is therefore satisfied if

$$\frac{\partial C}{\partial \tau} = kD$$

$$\frac{\partial D}{\partial \tau} = \alpha - \beta D + \gamma D^2 = \gamma(D - r_+)(D - r_-)$$

where Gatheral (2006) defines

$$r_\pm = \frac{\beta \pm \sqrt{\beta^2 - 4\alpha\gamma}}{2\gamma} =: \frac{\beta \pm d}{\eta^2}$$

and notes that integrating the differential equations for C and D with the terminal conditions $C(u,0) = 0$ and $D(u,0) = 0$ yields

$$D(u,\tau) = r_- \frac{1 - \exp^{-d\tau}}{1 - \exp^{-d\tau}}$$

$$C(u,\tau) = k(r_-\tau - \frac{2}{\eta_2}\ln(\frac{1 - g\exp^{-d\tau}}{1 - g}))$$

with

$$g := \frac{r_-}{r_+}$$

This can then be inserted into the inverse transform equation (6.47) to yield an expression for the P_j as an integral of a real-valued function – the more stable Gatheral (2006) version

$$P_j(x,v,\tau) = \frac{1}{2} + \frac{1}{\pi}\int_0^\infty \Re\left[\frac{\exp\{C_j(u,\tau)\theta + D_j(u,\tau)v + iux\}}{iu}\right]du$$

## CHAPTER 7

Monte Carlo simulation

7.1 Background

In previous chapters (e.g. Chapters 3 and 4) we have seen that option prices can be represented as expected discounted payoffs under some "risk neutral" probability measure. Monte Carlo simulation provides a method to calculate such expected values numerically in the sense that the expectation of a random variable can be approximated by the mean of a sample from its distribution.[1] Furthermore, Monte Carlo simulation can be useful to generate scenarios to evaluate the risk of a particular trading position.

Moving away from the purely probabilistic interpretation, Monte Carlo simulation can also be seen as simply a method of numerical integration. Suppose one wishes to calculate

$$I = \int_a^b f(x)dx \qquad (7.1)$$

Given a method to generate random variates u_i, uniformly distributed on the interval $[a, b]$, this integral can be approximated by

$$\hat{I}_N = (b - a)\frac{1}{N}\sum_{i=1}^N f(u_i) \qquad (7.2)$$

by virtue of the fact that

$$E[f(u)] = \frac{1}{b - a}\int_a^b f(x)dx \qquad (7.3)$$

for a random variable U uniformly distributed on $[a, b]$.

The Monte Carlo estimator (7.2) for I is subject to a random error. This error can be quantified in a probabilistic manner using the *Central Limit Theorem*, which states that the sum of N independent, identically distributed random variables with finite mean μ and finite variance σ^2 is, for large N, approximately normally distributed with mean $N\mu$ and variance $N\sigma^2$.[2] Consequently, $\hat{I}_N$ will be approximately normally distributed with mean $(b-a)E[f(U)] = I$. The variance of the Monte Carlo estimator is typically not known in closed

---

[1] This follows from the Laws of Large Numbers, see e.g. Ash and Doléans-Dade (2000), Sections 4.11.4 and 6.2.5, or any other good textbook on probability theory.

[2] See e.g. Ash and Doléans-Dade (2000), Chapter 7. The Central Limit Theorem also holds for a sum of a large number of independent random vectors — such a sum will have approximately a multivariate normal distribution.

form, but can be estimated from the Monte Carlo sample via

$$\hat{\sigma}_N^2 = \frac{(b-a)^2}{N(N-1)} \sum_{i=1}^{N} (f(u_i) - \hat{I}_N)^2 \tag{7.4}$$

For a given confidence level $1 - \alpha$ (e.g. $1 - \alpha = 95\%$), a confidence interval for I can then be constructed around the Monte Carlo estimate $\hat{I}_N$ by

$$[\hat{I}_N - \mathcal{N}^{-1}(1 - \alpha)\hat{\sigma}_N, \hat{I}_N + \mathcal{N}^{-1}(1 - \alpha)\hat{\sigma}_N] \tag{7.5}$$

where $\mathcal{N}^{-1}(1 - \alpha)$ is the $(1 - \alpha)$–quantile of the standard normal distribution, i.e. the true value I will lie within (7.5) with a probability of $1 - \alpha$.[3] $\hat{\sigma}_N$ is also called the *standard error* of the Monte Carlo estimator. This should always be quoted in conjunction with the Monte Carlo estimate, as a Monte Carlo estimate without error bounds is not very informative.

If, as we are assuming, $f(U)$ has finite variance σ^2,

$$\frac{(b-a)^2}{N-1} \sum_{i=1}^{N} (f(u_i) - \hat{I}_N)^2 \tag{7.6}$$

will converge to $(b-a)^2\sigma^2$ and consequently $\hat{\sigma}_N$ will converge to zero with rate $\mathcal{O}(N^{-\frac{1}{2}})$ as N is increased. The width of the confidence interval (7.5) will converge to zero at the same rate, and it is in this sense that the Monte Carlo estimator can be said to converge to the true value with $\mathcal{O}(N^{-\frac{1}{2}})$. These results generalise to the multidimensional case. It is usually convenient to normalise $[a, b]$ to $[0, 1]$, so in the multidimensional case we have an integration on the unit hypercube and (7.1) becomes

$$I = \int_{[0,1]^d} f(x)dx, \qquad f : [0, 1]^d \to \mathbb{R} \tag{7.7}$$

and the Monte Carlo estimator (7.2) becomes

$$\hat{I}_N = \frac{1}{N} \sum_{i=1}^{N} f(u_i) \tag{7.8}$$

where the u_i are d-dimensional vectors, i.e. the d components of each u_i are independent random variables uniformly distributed on the unit interval.[4] Deterministic one–dimensional numerical integration schemes typically converge with $\mathcal{O}(N^{-2})$ and are thus clearly superior to Monte Carlo simulation for the evaluation of one–dimensional integrals. However, Monte Carlo

---

[3] Strictly speaking, when approximating the variance of the Monte Carlo estimator using (7.4), one should be using a t–distribution with $N - 1$ degrees of freedom instead of the standard normal to determine (7.5). However, in Monte Carlo simulation N is typically sufficiently large (i.e. > 30) for this distinction to be irrelevant.

[4] In the typical option pricing application, f would thus include mapping the u_i to normally distributed random variables representing increments of a diffusion process, mapping the diffusion process to realisations of the underlying financial variables, and mapping the financial variables to a discounted payoff of the option.

1. Initialise the variables RunningSum $= 0$
                                        RunningSumSquared $= 0$
                                        $i = 0$

2. Generate the value of the stochastic process X at each date T_k relevant to evaluate the payoff.
3. Based on the values generated in step 2, calculate the payoff.
4. Add the payoff to RunningSum and the square of the payoff to RunningSumSquared.
5. Increment the counter i.
6. If i is less than the maximum number of iterations, go to step 2.
7. Calculate the simulated mean by dividing RunningSum by the total number of iterations.
8. Calculate the variance of the simulations by dividing RunningSumSquared by the total number of iterations and subtracting the square of the mean.

Algorithm 7.1: Generic algorithm for Monte Carlo simulation

simulation does not suffer from the "curse of dimensionality" that befalls these schemes. Classical numerical integration in d dimensions converges with $\mathcal{O}(N^{-\frac{2}{d}})$, whereas the Monte Carlo estimator converges with $\mathcal{O}(N^{-\frac{1}{2}})$ regardless the dimension. In fact, for the high–dimensional multivariate normal integrals encountered in the explicit evaluation of some exotic options in Chapter 4, Monte Carlo simulation is arguably the most efficient method. This, plus the fact that it is relatively straightforward to implement, makes Monte Carlo simulation the method of choice of pricing many derivative financial instruments.

7.2 The generic Monte Carlo algorithm

In most quantitative finance applications, one is interested in asset prices which follow a stochastic process, say $X(t)$. Monte Carlo simulation of such stochastic dynamics is relatively straightforward in the case where one can directly sample from the conditional distribution of $X(t)$ given $X(s)$, $t > s$. When the asset $X_i(t)$ follows the Black/Scholes–type dynamics (4.50), the logarithmic increment (or *return*) from time s to time t is normally distributed with

$$E\left[\ln\frac{X_i(t)}{X_i(s)}\bigg|\mathcal{F}_s\right] = \ln\frac{X_i(s)}{B(s,t)} - (t-s)d_{X_i} - \frac{1}{2}\int_s^t \sigma_{X_i}^2(u)du \quad (7.9)$$

$$\mathrm{Var}\left[\ln\frac{X_i(t)}{X_i(s)}\bigg|\mathcal{F}_s\right] = \int_s^t \sigma_{X_i}^2(u)du \quad (7.10)$$

```
template <class T>
class MCGatherer {
private:
    T     sum;
    T     sum2;
    size_t c;
public:
    inline MCGatherer() : sum(0.0),sum2(0.0),c(0) { };
    /// Constructor for types T which require a size argument
    /// on construction.
    inline MCGatherer(size_t array_size)
      : sum(array_size),sum2(array_size) { reset(); };
    inline void reset() { sum = 0.0; sum2 = 0.0; c = 0; };
    inline void operator+=(T add)
      { sum += add; sum2 += add*add; c++; };
    inline void operator*=(T f) { sum *= f; sum2 *= f*f; };
    inline T mean() const { return T(sum/double(c)); };
    inline T stddev() const
      { return T(sqrt((sum2/double(c)
                  -mean()*mean())/double(c-1))); };
    inline double max_stddev() const; // Must be specialised!
    inline size_t number_of_simulations() const { return c; };
};
```

Listing 7.1: Class for "gathering" Monte Carlo simulation results

and letting ξ_i denote a random draw from the standard normal distribution, one can simulate $X_i(t)$ by

$$X_i(t) = \frac{X_i(s)}{B(s,t)}$$

$$\cdot \exp\left\{(t-s)d_{X_i} - \frac{1}{2}\int_s^t \sigma_{X_i}^2(u)du + \xi_i\sqrt{\int_s^t \sigma_{X_i}^2(u)du}\right\} \quad (7.11)$$

Note that in the single asset case $X_i(t)$ can be represented in terms of a one–dimensional random variable ξ_i, regardless of the dimension of the driving Brownian motion in (4.50). When simultaneously simulating multiple assets, in general one would need to work out the variance/covariance matrix of the joint, multivariate normal distribution of the returns, and then draw from this distribution. This is necessary even if the driving Brownian motion is one–dimensional: In (4.33), even if $\sigma_X(\cdot)$ and $\sigma_Y(\cdot)$ are scalar–valued functions, the resulting correlation over any time interval $[t, T]$ is generally not one, unless $\sigma_X(\cdot)$ and $\sigma_Y(\cdot)$ are constant on $[t, T]$. By the same token, the variance/covariance matrix of the joint distribution of returns typically will have full rank (rather than rank d).

For most applications, it is sufficient to assume that the vector–valued

```
/** Template for generic Monte Carlo simulation.

    The template parameter class random_number_generator_type
    must implement a member function random(), which returns a
    realisation of the random variable of the type given by
    the template parameter class argtype.
*/
template <class argtype,class rettype,
          class random_number_generator_type>
class MCGeneric {
private:
  NormalDistribution                                             N;
  /// Functor mapping random variable to Monte Carlo payoff.
  boost::function<rettype (argtype)>                             f;
  random_number_generator_type      random_number_generator;
public:
  inline MCGeneric(boost::function<rettype (argtype)> func,
                   random_number_generator_type& rng)
    : f(func),random_number_generator(rng) { };
  void simulate(MCGatherer<rettype>& mcgatherer,
                unsigned long number_of_simulations);
  double simulate(MCGatherer<rettype>& mcgatherer,
                  unsigned long initial_number_of_simulations,
                  double required_accuracy,
                  double confidence_level = 0.95);
};
```

Listing 7.2: Template class declaration for generic Monte Carlo simulation

volatility functions $\sigma_{X_i}(\cdot)$ are piecewise constant. In this case one can simulate the driving Brownian motion component–wise, i.e. for each asset X_i we have

$$X_i(t) = \frac{X_i(s)}{B(s,t)}$$

$$\cdot \exp\left\{(t-s)d_{X_i} - \frac{1}{2}\sigma^2_{X_i}(s)(t-s) + \sum_{j=1}^{d}\xi^{(j)}\sigma^{(j)}_{X_i}(s)\sqrt{t-s}\right\} \quad (7.12)$$

as long as $\sigma_{X_i}(u) \equiv \sigma_{X_i}(s)$ for $u \in [s,t[$. $\xi^{(j)}$ denotes the j-th component of a d–dimensional vector ξ of independent standard normal random variates.

Algorithm 7.1 gives the general procedure for obtaining a Monte Carlo estimate. We implement this in the class templates MCGatherer (Listing 7.1) and MCGeneric (Listings 7.2 and 7.3).

The MCGatherer object serves to collect the Monte Carlo simulation results, i.e. steps 4 and 5 of the algorithm are implemented by the overloaded operator+=() and the member functions mean() and stddev() report back

```
template <class argtype,class rettype,
         class random_number_generator_type>
void MCGeneric<argtype,rettype,random_number_generator_type>::
simulate(MCGatherer<rettype>& mcgatherer,
         unsigned long number_of_simulations)
{
  unsigned long i;
  for (i=0;i<number_of_simulations;i++)
    mcgatherer += f(random_number_generator.random());
}
```

Listing 7.3: The generic Monte Carlo simulation loop

the Monte Carlo estimate and its estimated standard deviation.[5] The template argument of MCGatherer determines the type of the variable representing the payoff. For a single payoff this would typically be double, but the template implementation is sufficiently general that this also could be Array<double,1>, for example, representing multiple payoffs evaluated in the same Monte Carlo simulation.

The generic Monte Carlo algorithm itself is implemented by the template MCGeneric, in the member function

$$\text{void simulate(MCGatherer<rettype>\& mcgatherer,} \atop \text{unsigned long number\_of\_simulations);} \quad (7.13)$$

MCGeneric has three template arguments. The first is class argtype, which is the type of random variable needed to evaluate the payoff — typically, this would be double in the univariate case or Array<double,1> in the multivariate case. The second is class rettype, which is the type of the variable representing the payoff. The third is class random_number_generator_type, which must be a class with a member function

$$\text{argtype random()} \quad (7.14)$$

which returns a draw of the random variable of the required type — in the univariate, normally distributed case, one could use

$$\text{ranlib::Normal<double>} \quad (7.15)$$

from the Blitz++ library.

Each MCGeneric object is initialised at construction with a reference to a random number generator of the required type and a function f mapping the random variable in each draw of the simulation to the (discounted) payoff, where f is represented by the Boost Library functor

$$\text{boost::function<double (argtype)> f;} \quad (7.16)$$

In the main simulation routine in Listing 7.3, Algorithm 7.1 is now very compact:

---

[5] Encapsulating part of the algorithm in the MCGatherer class follows the *strategy* design pattern. See Joshi (2004) for a detailed discussion.

Figure 7.1 *Interaction of objects in the generic Monte Carlo algorithm*

- The initialisations in Step 1 were done by the constructor of the `MCGatherer` object.
- The function call `random_number_generator.random()` generates the random variable in Step 2,
- to which the functor `f` is applied to calculate the payoff in Step 3. Note that `f` encapsulates the entire mapping from random variables to discounted payoff, thus the generic implementation of the Monte Carlo algorithm abstracts from the *stochastic model* (which maps basic random variables to the dynamics of the underlying assets) and any particular *financial instrument* (which maps the prices of underlying assets to a discounted payoff).
- Step 4 is embodied in the overloaded `operator+=()` of the `MCGatherer` object.
- Steps 5 and 6 are managed by the `for` loop.
- Steps 7 and 8 are completed on demand by the member functions `mean()` and `stddev()` of `MCGatherer`.

Thus Steps 2, 3 and 4 are implemented by the statement

$$\texttt{mcgatherer += f(random\_number\_generator.random());} \qquad (7.17)$$

in `MCGeneric::simulate()`, i.e. through the interaction of the random number generator, the functor `f` and the Monte Carlo result accumulator `mcgatherer`, as illustrated in Figure 7.1.

Encapsulating the collection of the simulation results in an `MCGatherer` object also makes it easy to implement a Monte Carlo estimate to a desired accuracy (rather than a fixed number of simulations) in the member function of `MCGeneric`,

```
double simulate(MCGatherer<rettype>& mcgatherer,
                unsigned long initial_number_of_simulations,
                double required_accuracy,
                double confidence_level)
```

as Listing 7.4 demonstrates. After performing an initial set of simulations, the

```
template <class argtype,class rettype,
         class random_number_generator_type>
double MCGeneric<argtype,rettype,random_number_generator_type>::
simulate(MCGatherer<rettype>& mcgatherer,
         unsigned long initial_number_of_simulations,
         double required_accuracy,
         double confidence_level)
{
  unsigned long n = initial_number_of_simulations;
  simulate(mcgatherer,initial_number_of_simulations);
  double d = N.inverse(confidence_level);
  double current_accuracy = d * mcgatherer.max_stddev();
  while (required_accuracy<current_accuracy) {
    double q = current_accuracy/required_accuracy;
    q *= q;
    unsigned long additional_simulations = n * q - n;
    if (!additional_simulations) break;
    simulate(mcgatherer,additional_simulations);
    n += additional_simulations;
    current_accuracy = d * mcgatherer.max_stddev(); }
  return current_accuracy;
}
```

Listing 7.4: Generating a Monte Carlo estimate to a desired accuracy

the Monte Carlo estimator.[6] Additional simulations are then performed until (7.18) is reduced to the desired level. The number of additional simulations required can be predicted by noting that (7.18) will decrease approximately proportionately to the square root of the number of simulations, so we set the size of the next batch of simulations to

$$\left(\frac{\text{current accuracy}}{\text{required accuracy}}\right)^2 n - n \qquad (7.19)$$

As a simple example of using MCGeneric and MCGatherer, consider the Black/ Scholes model implemented in MCGenericTest.cpp,[7] where the stock price under the risk neutral measure at time T is given by

$$S(T) = S(0) \exp\left\{\left(r - \frac{1}{2}\sigma^2\right)T + \sigma W(T)\right\} \qquad (7.20)$$

where $W(T)$ is normally distributed with mean 0 and variance T. We want to

---

[6] In order to cater for multiple payoffs evaluated in the same Monte Carlo simulation (e.g. a rettype of Array<double,1>), $\hat{\sigma}_n$ is accessed via MCGatherer::max_stddev(). The implementation of this template member function is specialised for different cases of rettype: MCGatherer<double>::max_stddev() simply returns MCGatherer::stddev(), which is the standard deviation of the Monte Carlo estimator as defined in (7.4), while MCGatherer<Array<double,1> >::max_stddev() returns the maximum standard deviation of the multiple Monte Carlo estimators collected by this MCGatherer object.

[7] Available on the website for this book.

Paths	MC value	95% CI lower bound	95% CI upper bound	Difference in standard errors
100	14.1060	9.8154	18.3966	-1.4738
400	15.8688	13.5168	18.2209	-1.4558
1600	16.6760	15.5070	17.8450	-1.7933
6400	17.9812	17.3623	18.6002	0.0817
25600	18.0816	17.7796	18.3835	0.7139
102400	18.0850	17.9346	18.2354	1.4710
409600	17.9905	17.9156	18.0653	0.8784
1638400	17.9628	17.9255	18.0002	0.5435

Table 7.1 *Convergence of the Monte Carlo estimator for the price of a European call option to the closed form solution. The closed form price is 17.9505.*

use Monte Carlo simulation to calculate the price of a European call option via the expected discounted value

$$E\left[e^{-rT}\max(0, S(T) - K)\right] \tag{7.21}$$

We can write the discounted payoff inside the expectation as a function f of a standard normal random variable ξ:

$$f(\xi) = e^{-rT}\max\left(0, S(0)\exp\left\{\left(r - \frac{1}{2}\sigma^2\right)T + \sigma\sqrt{T}\xi\right\} - K\right) \tag{7.22}$$

This function can be bound to a `boost::function<double (double)>` functor and passed to an `MCGeneric<double,ranlib::NormalUnit<double> >` object, which is then used to run the Monte Carlo simulations.

Table 7.1 gives the output of `MCGenericTest.cpp` for 100 up to 1638400 simulations.[8] One clearly sees the convergence of the Monte Carlo estimate to the closed form Black/Scholes price. As per (7.5), the 95% confidence interval around the Monte Carlo estimate tightens proportionally to the square root of the number of simulations. The last column gives the difference between the Monte Carlo estimate and the closed form value in terms of the number of standard deviations $\hat{\sigma}_n$, i.e.

$$\frac{\text{MC value} - \text{closed form value}}{\hat{\sigma}_n} \tag{7.23}$$

This is a useful statistic when testing the correctness of a Monte Carlo simulator against a closed form solution. The probability that the absolute value of (7.23) is greater than 2 is about 5%. Thus an absolute value greater than 2 is "suspicious" and indicates a possible bias in the simulation. If one subsequently increases the number of simulations, this value will either drop back

---

[8] The initial stock price was 100, the volatility 30%, the continuously compounded interest rate 5%. The price is for a European call option with strike 100 maturing in 1.5 years.

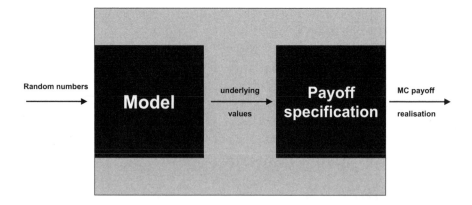

Figure 7.2 *The two–stage mapping of random numbers to asset prices, to the discounted payoff*

below 2 (as in Table 7.1) or continue to increase, reaching ever larger and more improbable values. The latter effectively proves the presence of a bias in the simulator. Whenever possible, an implementation of a Monte Carlo simulator should be tested on a payoff with a known closed form expectation, before proceeding to use the simulator to calculate expectations for which no closed form is available.

7.3 Simulating asset price processes following geometric Brownian motion

In many cases, the payoff that one wishes to evaluate may depend on the value of several assets at several points in time. It is thus convenient to view steps 2 and 3 in Algorithm 7.1 as a two–stage mapping, as illustrated in Figure 7.2: First, a set of random numbers is used to generate the required asset prices at the required points in time, based on the stochastic model. Second, the asset prices thus obtained are mapped to the discounted payoff.

We encapsulate the two steps in the template class MCMapping, the public interface of which is given in Listing 7.5. In turn, the mapping of random numbers to asset prices is delegated to the template parameter class price_process, and the asset prices are then mapped to a payoff by a class derived from the abstract base class MCPayoff.

The logic of MCMapping is best demonstrated in an example. Consider the option to exchange one asset for another in Theorem 4.7. There are two under-

```
template <class price_process,class random_variable>
class MCMapping {
public:
   /// Constructor.
   MCMapping(MCPayoff& xpayoff,
             price_process& xprocess,
             const TermStructure& xts,
             int xnumeraire_index = -1);
   /// Choose the numeraire asset for the equivalent martingale
   /// measure under which the simulation is carried out.
   bool set_numeraire_index(int xnumeraire_index);
   /// The function mapping a realisation of the (often
   /// multidimensional) random variable x to the discounted
   /// payoff.
   double mapping(random_variable x);
   /// The function mapping a realisation of the (often
   /// multidimensional) random variable x to multiple discounted
   /// payoffs.
   Array<double,1> mappingArray(random_variable x);
};
```

Listing 7.5: The public interface of the template class MCMapping

process is two[9] and for simplicity set volatilities to be constant using the class
ConstVol.[10] Thus we could initialise the two assets by

$$
\begin{aligned}
&\texttt{Array<double,1> sgm1(2);}\\
&\texttt{sgm1 = 0.3 0.0;}\\
&\texttt{ConstVol vol1(sgm1);}\\
&\texttt{BlackScholesAsset asset1(\&vol1,100.0);}\\
&\texttt{Array<double,1> sgm2(2);}\\
&\texttt{sgm2 = 0.2 0.1;}\\
&\texttt{ConstVol vol2(sgm2);}\\
&\texttt{BlackScholesAsset asset2(\&vol2,100.0);}
\end{aligned}
\qquad (7.24)
$$

To represent the joint stochastic process of assets with Black/Scholes–type
dynamics and to map random variables representing increments of geometric
Brownian motion to a realisation of the asset price paths, we use the class
GeometricBrownianMotion. This class has a constructor, which takes as its
argument a set of underlying assets, represented by a std::vector<const
BlackScholesAsset*>&; thus we need the following sequence of statements:[11]

---

[9] This is by no means a necessary assumption. However, a dimension of one would lead to
instantaneously perfectly correlated asset dynamics, while a dimension greater than two
is unnecessary when there are only two assets.

[10] In practical applications in the presence of a term structure of volatility one might instead
use PiecewiseConstVol.

[11] The implementation of GeometricBrownianMotion is discussed below.

```
class MCPayoff {
public:
  Array<double,1> timeline;   ///< Timeline collecting all event dates.
  Array<int,2>        index;  ///< A 2 x N matrix of indices, where each
                              ///< column represents the indices of an
                              ///< (asset,time) combination affecting the payoff.
  inline MCPayoff(const Array<double,1>& xtimeline,
                  const Array<int,2>& xindex)
    : timeline(xtimeline),index(xindex) { };
  inline MCPayoff(const Array<double,1>& xtimeline,int number_of_indices)
    : timeline(xtimeline),index(2,number_of_indices) { };
  inline MCPayoff(int number_of_event_dates,int number_of_indices)
    : timeline(number_of_event_dates+1),index(2,number_of_indices) { };
  /// Calculate discounted payoff.
  virtual double operator()(const Array<double,1>& underlying_values,
                            const Array<double,1>& numeraire_values) = 0;
  /// Allow for multidimensional payoff (i.e. portfolio) with meaningful
  /// default (one-dimensional) behaviour.
  virtual Array<double,1>
    payoffArray(const Array<double,1>& underlying_values,
                const Array<double,1>& numeraire_values);
};
```

Listing 7.6: The public interface of the class MCPayoff

$$
\begin{aligned}
&\texttt{std::vector<const BlackScholesAsset*> underlying;} \\
&\texttt{underlying.push\_back(\&asset1);} \\
&\texttt{underlying.push\_back(\&asset2);} \\
&\texttt{GeometricBrownianMotion gbm(underlying);}
\end{aligned}
\tag{7.25}
$$

In order to map the random variables representing increments of Brownian motion to the Monte Carlo payoff(s), we then construct an instance of MCMapping. In addition to an object representing the price process (e.g. Geometric-BrownianMotion), the constructor of MCMapping requires a reference to an instance of a class derived from MCPayoff (Listing 7.6). MCPayoff has a member operator()(), which calculates the discounted payoff based on a set of values of the underlying assets and the numeraire asset at a given set of "event dates." In the barrier option example below, the implementation of a class derived from MCPayoff is discussed in detail — in the present context, let it be sufficient to note that the MBinaryPayoff class introduced in Chapter 4 has MCPayoff as a parent class. Thus payoffs of the type (4.51) can be specified in a manner allowing Monte Carlo and closed–form Black/Scholes pricing on the same payoff object, which is helpful when using the Black/Scholes model as a control variate (see Section 7.6.2, below.) The easiest way to obtain an MCPayoff for an option to exchange one asset for another is to first construct an instance of exotics::Margrabe, which returns the corresponding MCPayoff via the member function get_payoff(), e.g.:

```
double maturity = 2.5;
double strike = 1.1;
FlatTermStructure ts(0.06,0.0,maturity);
exotics::Margrabe margrabe(asset1,asset2,                    (7.26)
                      0.0,maturity,1.0,strike,ts);
boost::shared_ptr<MCPayoff> margrabe_payoff =
                      margrabe.get_payoff();
```

This allows us to construct an instance of MCMapping by

```
MCMapping<GeometricBrownianMotion,Array<double,2> >
mc_mapping(*margrabe_payoff,gbm,ts,numeraire_index);          (7.27)
```

where the optional argument numeraire_index gives the index of the asset, which is to serve as the numeraire, in the set of assets in the Geometric-BrownianMotion object. The default is a numeraire index of -1, which corresponds to taking the riskfree asset as the numeraire (i.e. deterministic discounting) — see Chapter 8 for a discussion of Monte Carlo simulation under a different choice of numeraire.[12]

The second template argument of MCMapping is in this case Array<double,2>: In order to generate a simulated path of the assets, (dimension of the driving Brownian motion)×(number of time steps) independent normal random variables are required. MCMapping then provides the function mapping the random variables to the Monte Carlo payoff, e.g.

```
boost::function<double(Array<double,2>)>
func = boost::bind(&MCMapping<GeometricBrownianMotion,
                Array<double,2> >::mapping,                   (7.28)
                &mc_mapping,_1);
```

To encapsulate the required array of normal random variables, we create the template class RandomArray, e.g. if the dimension of the driving Brownian motion is 2 and there are N time steps,

```
ranlib::NormalUnit<double> normalRNG;
RandomArray<ranlib::NormalUnit<double>,double>               (7.29)
random_container(normalRNG,2,N);
```

i.e. random_container.random() will return a $2 \times N$ Array of independent standard normal random variables. Then we are in a position to construct the MCGeneric object with

```
MCGeneric<Array<double,2>,
            double,
            RandomArray<ranlib::NormalUnit<double>,double>   (7.30)
            > mc(func,random_container);
```

which is used to generate the Monte Carlo estimate for the option price in exactly the same manner as in the single–asset, single–time horizon case in Section 7.2. Thus the core Monte Carlo algorithm remains the same not

---

[12] It is worth noting that one can conduct a Monte Carlo simulation under the "real–world" or "physical" measure in a manner that is consistent with simulation and pricing under a risk neutral measure by selecting as the numeraire an asset, the volatility of which is equal to the market price of risk. For discussion of the numeraire associated with the physical measure, see Bajeux-Besnainou and Portait (1997).

```
class down_and_out_call : public MCPayoff {
public:
  double strike,barrier;
  down_and_out_call(const Array<double,1>& T,
                    int underlying_index,
                    double xstrike,double xbarrier);
  /// Calculate discounted payoff.
  virtual double
    operator()(const Array<double,1>& underlying_values,
               const Array<double,1>& numeraire_values);
};

down_and_out_call::
down_and_out_call(const Array<double,1>& T,
                  int underlying_index,
                  double xstrike,double xbarrier)
: MCPayoff(T,T.extent(firstDim)-1),
          strike(xstrike),barrier(xbarrier)
{
  firstIndex  idx;
  secondIndex jdx;
  index = idx * (jdx+1);
}

double down_and_out_call::
operator()(const Array<double,1>& underlying_values,
           const Array<double,1>& numeraire_values)
{
  int i;
  bool indicator = true;
  double result  = 0.0;
  for (i=0;i<underlying_values.extent(firstDim);i++)
    if (underlying_values(i)<=barrier) indicator = false;
  if (indicator) result = numeraire_values(0)/
    numeraire_values(numeraire_values.extent(firstDim)-1) *
    std::max(0.0,underlying_values(
      underlying_values.extent(firstDim)-1)-strike);
  return result;
}
```

Listing 7.7: An MCPayoff for a down–and–out call option.

only conceptually, but also in implementation.[13] All that has changed is the
random variables being fed into the simulation (a RandomArray vs. straight
ranlib::NormalUnit<double> in the previous example) and the functor mapping the random variables to the discounted payoff.

[13] The code for this example is given in the file MCMargrabe.cpp on the website.

Similarly, we can price a down–and–out barrier option by feeding the appropriate class derived from MCPayoff into MCMapping.[14] Again, one could obtain the required MCPayoff by first constructing an instance of exotics::DiscreteBarrierOut with

```
exotics::DiscreteBarrierOut
  down_and_out_call(asset1,maturity,strike,                      (7.31)
                   barrier,ts,1,-1);
```

and then calling the member function get_payoff(). As before, the MCPayoff then allows us to construct an MCMapping, which provides the functor to feed into MCGeneric.

However, for didactic purposes let us create from scratch a class derived from MCPayoff, as given in Listing 7.7. Let N be the number of (not necessarily equidistant) time steps at which the barrier is monitored between now (time 0) and maturity. For simplicity, we choose the time steps to be equidistant in the present example. We create an instance of this class down_and_out_call_payoff by

```
Array<double,1> T(N+1);
firstIndex idx;
double dt = maturity/N;                                           (7.32)
T = idx*dt;
down_and_out_call_payoff DOpayoff(T,0,strike,barrier);
```

The constructor of down_and_out_call_payoff takes as its first argument the timeline T of "event dates," which is an Array<double,1> of increasing time points starting at 0 (now) and ending at the option maturity. The barrier is monitored at each event date and the payoff occurs at maturity. The second argument is the index of the underlying asset (to allow for a price_process in MCMapping representing more than one underlying asset), followed by the option strike and down–and–out barrier. The constructor initialises the data members strike and barrier, and passes the timeline through to the constructor of the parent class MCPayoff. It also initialises the data member index of MCPayoff, which represents the (asset, time) index combinations which need to be simulated, i.e. in this case asset 0 at each time T_i, $0 < i \leq N$, so

$$\text{index} = \begin{pmatrix} 0 & 0 & \cdots & 0 \\ 1 & 2 & \cdots & N \end{pmatrix}$$

The member operator()() is defined to calculate the discounted payoff from the simulated values of the underlying asset and the numeraire at each date, which MCMapping requests from price_process and passes to an instance of this class down_and_out_call derived from MCPayoff.

The instance DOpayoff of the class down_and_out_call is passed to the constructor of MCMapping, which creates the functor for use in MCGeneric, and the Monte Carlo simulation proceeds as before.

When one wishes to generate a Monte Carlo estimate for the price of an instrument incorporating several MCPayoffs, or value a portfolio (either in total

---

[14] The code for this example is given in the file MCDownandout.cpp on the website.

```
class GeometricBrownianMotion {
public:
  GeometricBrownianMotion(std::vector<const BlackScholesAsset*>&
    xunderlying);
  ~GeometricBrownianMotion();
  /// Query the dimension of the process.
  int dimension() const;
  /// Query the number of factors driving the process.
  inline int factors() const { return dW.extent(firstDim); };
  /// Set process timeline.
  bool set_timeline(const Array<double,1>& timeline);
  /// Get process timeline.
  inline const Array<double,1>& get_timeline() const
    { return *timeline_; };
  inline int number_of_steps() const
    { return T->extent(firstDim)-1; };
  /// Generate a realisation of the process under the martingale
  /// measure associated with deterministic bond prices.
  void operator()(Array<double,2>& underlying_values,
                  const Array<double,2>& x,
                  const TermStructure& ts);
  /// Generate a realisation of the process under the martingale
  /// measure associated with a given numeraire asset.
  void operator()(Array<double,2>& underlying_values,
                  Array<double,1>& numeraire_values,
                  const Array<double,2>& x,
                  const TermStructure& ts,
                  int numeraire_index);
};
```

Listing 7.8: The public interface of `GeometricBrownianMotion`.

or component–wise) of several `MCPayoffs` in the same Monte Carlo simulation, one can use class `MCPayoffList`. In addition to the constructor, the virtual member `operator()()` and function `payoffArray()` declared by the parent class `MCPayoff`, `MCPayoffList` provides the member function

$$\text{void push\_back(boost::shared\_ptr<MCPayoff> xpayoff,} \atop \text{double xcoeff = 1.0);}} \qquad (7.33)$$

to add an `MCPayoff` to the list. It performs various housekeeping operations to ensure that the data members `timeline` and `index` of the parent class `MCPayoff` are updated so that `MCMapping` requests all the necessary (asset, time) pairs from the `price_process`, and that the correct `Arrays` of underlying asset and numeraire values are passed to the member `operator()()` and function `payoffArray()` of the `MCPayoffs` in the list. The member `operator()()` and function `payoffArray()` of `MCPayoffList` call the member `operator()()` of each `MCPayoff` in the list, the difference being that `payoffArray()` returns an `Array<double>` containing the result of `operator()()` for each `MCPayoff`,

```
template <class price_process,class random_variable>
double MCMapping<price_process,random_variable>::
mapping(random_variable x)
{
  int i;
  /* Using the random draw x, generate the values of the price
     process at the dates in the required timeline,
     under the martingale measure associated with the chosen
     numeraire. */
  if (numeraire_index >= 0)
    process(underlying_values,numeraire_values,
            x,ts,numeraire_index);
  else process(underlying_values,x,ts);
  // Map underlying values to the payoff.
  for (i=0;i<mapped_underlying_values.extent(firstDim);i++)
    mapped_underlying_values(i) =
      underlying_values(payoff.index(0,i),payoff.index(1,i));
  // Calculate the discounted payoff.
  return payoff(mapped_underlying_values,numeraire_values);
}
```

Listing 7.9: The member function mapping() of MCMapping.

while operator()() of MCPayoffList returns the sum of the individual re-sults.[15] Note that MCPayoff relates to MCPayoffList via both inheritance and composition, i.e. MCPayoffList is an MCPayoff and contains MCPayoffs.

In keeping with the view of each iteration of Monte Carlo simulation as a mapping of a draw of random variables to a discounted option payoff, the object representing the price process has the task of converting the draw of random variables into a realisation of the price path of the underlying asset(s). For geometric Brownian motion, it is straightforward to generate price paths based on the exact (i.e. lognormal) transition densities, and thus there is no bias in the simulation at any level of discretisation. This is implemented in the class GeometricBrownianMotion, the public interface of which is given in List-ing 7.8. After constructing an instance of GeometricBrownianMotion from a vector of BlackScholesAssets, the timeline of dates, for which realisations of the underlying assets need to be observed, can be set using the member function set_timeline(). When GeometricBrownianMotion is used in con-junction with MCMapping, MCMapping passes the timeline of required observa-tion dates from MCPayoff into GeometricBrownianMotion::set_timeline(). The member function mapping() of MCMapping (see Listing 7.9) maps the draw of random variables to a discounted payoff; in order to generate the re-quired price path of the underlying asset(s), mapping() calls the overloaded operator() of the price process class, e.g. GeometricBrownianMotion. De-

---

[15] See MCCallPut.cpp on the website for example usage.

pending on whether the simulation is being carried out under the usual risk neutral probability measure under deterministic interest rates, or under a martingale measure associated with taking one of the underlying assets as the numeraire, `mapping()` calls (under the usual measure)

```
void operator()(Array<double,2>& underlying_values,
                const Array<double,2>& x,                    (7.34)
                const TermStructure& ts);
```

or (for one of the underlying assets as the numeraire)

```
void operator()(Array<double,2>& underlying_values,
                Array<double,1>& numeraire_values,
                const Array<double,2>& x,                    (7.35)
                const TermStructure& ts,
                int numeraire_index);
```

where `numeraire_index` refers to the index of the numeraire asset in the vector of underlying assets from which the instance of `GeometricBrownianMotion` was constructed. Thus one can change from geometric Brownian motion to alternative stochastic dynamics for the underlying asset(s) by replacing `GeometricBrownianMotion` by a different price process class, which must appropriately implement the member function `set_timeline()` and the two versions (7.34) and (7.35) of the `operator()`. This is done in Chapter 8.

7.4 Discretising stochastic differential equations

If it is not possible (or computationally too expensive) to sample directly from the distribution of the underlying assets, one may instead resort to discretising and simulating the stochastic differential equations (SDE) describing the asset dynamics.

Assume that asset dynamics are given by a vector–valued Itô process

$$dX(t) = a(X(t), t)dt + b(X(t), t)dW(t) \qquad (7.36)$$

i.e. $X(t)$ is an N-dimensional vector of asset prices, $W(t)$ is a d-dimensional standard Brownian motion, $a(\cdot, \cdot)$ is a vector–valued function (dimension N), and $b(\cdot, \cdot)$ is a matrix–valued function (dimension $N \times d$). The drift function a and volatility function b are assumed to satisfy sufficient conditions for (7.36) to have a unique strong solution.[16]

The simplest way to discretise and simulate (7.36) is the *Euler scheme*. Consider a discretisation of the timeline $0 = t_0 < t_1 < \ldots < t_M$. The stochastic dynamics of X can be approximated by

$$\begin{align} X(t_{m+1}) &= X(t_m) + a(X(t_m), t_m)(t_{m+1} - t_m) & (7.37) \\ &\quad + b(X(t_m), t_m)\sqrt{t_{m+1} - t_m}\,\xi_m & (7.38) \end{align}$$

where each ξ_m is a d-dimensional vector of independent standard normal variates.

---

[16] See e.g. Karatzas and Shreve (1991), Section 5.2.

The discretisation (7.37) is only approximate and will typically show *discretisation bias*. That is, a Monte Carlo estimator based on $X(t)$ simulated via (7.37) will converge to a value different from the desired expectation $E[f(X(t))]$ as the number of simulations is increased. The size of this bias depends on the chosen discretisation of the timeline. As $t_{m+1} - t_m$ goes to zero, the discretised dynamics (7.37) will converge to the true dynamics (7.36). There are two types of convergence to be considered, *strong* and *weak*.[17]

Definition 7.1 *1. Denote by $X(T)$ a stochastic process and by $\hat{X}_\delta(T_M)$, $T_M = T$, its discretised approximation with maximum step size*

$$\delta = \max_m t_{m+1} - t_m$$

Then $\hat{X}_\delta(T_M)$ is said to converge strongly *to $X(T)$ at time T if*

$$\lim_{\delta \to 0} E[|X(T) - \hat{X}_\delta(T_M)|] = 0$$

Furthermore, $\hat{X}_\delta$ is said to converge strongly with order $\gamma > 0$ if

$$E[|X(T) - \hat{X}_\delta(T_M)|] \le C\delta^\gamma \quad \forall \delta < \delta_0$$

for some constants C, δ_0.

2. Denote by $\mathcal{C}$ a class of test functions $g : \mathbb{R}^N \to \mathbb{R}$. $\hat{X}_\delta(T_M)$ is said to converge weakly to $X(T)$ at time T if

$$\lim_{\delta \to 0} \left| E[g(X(T))] - E[g(\hat{X}_\delta(T_M))] \right| = 0 \quad \forall g \in \mathcal{C}$$

Furthermore, $\hat{X}_\delta$ is said to converge weakly with order $\beta > 0$ if

$$\left| E[g(X(T))] - E[g(\hat{X}_\delta(T_M))] \right| \le C\delta^\gamma \quad \forall g \in \mathcal{C}, \delta < \delta_0$$

for some constants C, δ_0.

Kloeden and Platen (1992) show that the Euler scheme usually (i.e. under some mild additional technical conditions) converges with weak order $\beta = 1$ and with strong order $\gamma = 0.5$. Thus, when choosing to implement a particular discretisation scheme, it is important to consider whether one requires a good pathwise (strong) approximation or whether a good (weak) convergence of expectations of some function is what is required. Clearly, the latter is the priority when pricing derivative financial instruments, while the former would come into play simulating hedging effectiveness, for example.

Strong order $\gamma = 1$ convergence can be achieved by adding further (second order) terms to (7.37). This is the *Milstein scheme*. In the one–dimensional

---

[17] For a detailed discussion of these convergence concepts, see Kloeden and Platen (1992).

case it is given by

$$
\begin{aligned}
X(T_{m+1}) &= X(t_m) + a(X(t_m), t_m)(t_{m+1} - t_m) \\
&\quad + b(X(t_m), t_m)\sqrt{t_{m+1} - t_m}\xi_m \qquad\qquad (7.39) \\
&\quad + \frac{1}{2}b(X(t_m), t_m)\frac{\partial b(X(t_m), t_m)}{\partial X(t_m)}(t_{m+1} - t_m)(\xi_m^2 - 1)
\end{aligned}
$$

and thus does not require any additional draws of random variates beyond those required by (7.37).

The choice of scheme (Euler, Milstein, or higher order) and refinement of the time line discretisation to reduce discretisation bias can be encapsulated in a class representing a price process, i.e. a class providing the necessary functionality of a valid `price_process` template parameter to the class template `MCMapping`.

7.5 Predictor–Corrector methods

As pricing derivative financial instruments is a much more common requirement than simulating hedging effectiveness, we focus on methods to further improve the order of weak convergence.[18] Predictor–corrector methods can be implemented with relative ease to achieve this. As above, the reader is referred to Kloeden and Platen (1992) for rigorous proofs of the results reproduced here.

Discretisation bias in the Euler scheme stems from the fact that the drift coefficient $a(X(t), t)$ and the diffusion coefficient $b(X(t), t)$ are assumed to be constant on the discretisation time step $[t_m, t_{m+1}]$ and equal to their values at the beginning of the time step, i.e. $a(X(t_m), t_m)$ and $b(X(t_m), t_m)$, thus not taking into account the variation of a and b on $[t_m, t_{m+1}]$ due to the variation of $X(t)$. Predictor–corrector methods aim to achieve a better approximation of the true dynamics of $X(t)$ by first generating a *predictor* $\overline{X}(t_{m+1})$ of $X(t_{m+1})$ (e.g. using an Euler scheme increment) and then *correcting* the initial drift and diffusion coefficients based on $X(t_m)$ at the beginning of the time step by terms based on the predicted value $\overline{X}(t_{m+1})$ at the end of the time step.

A predictor–corrector method with weak convergence of order 1 for (7.36) is given by

$$
\begin{aligned}
&X(t_{m+1}) \qquad\qquad\qquad\qquad\qquad\qquad\qquad\qquad\qquad (7.40) \\
&= X(t_m) + (\alpha \overline{a}_\eta(\overline{X}(t_{m+1}), t_{m+1}) + (1 - \alpha)\overline{a}_\eta(X(t_m), t_m))(t_{m+1} - t_m) \\
&\quad + \sum_{j=1}^{d}(\eta b_j(\overline{X}(t_{m+1}), t_{m+1}) + (1 - \eta)b_j(X(t_m), t_m))\Delta W_j
\end{aligned}
$$

Recall that X is an N-dimensional vector, and so are $\overline{a}_\eta(\cdot, \cdot)$ and $b_j(\cdot, \cdot)$. (7.40)

---

[18] Note that in order to *calculate* hedge ratios (as opposed to simulating hedging effectiveness), weak convergence is again the relevant concept, as a hedge ratio is determined by the sensitivity of the price to an underlying risk factor.

represents a family of schemes parameterised in $\alpha, \eta \in [0,1]$; a typical choice would be $\alpha = 0.5$ and $\eta = 0$. The adjusted drift coefficient $\bar{a}_\eta$ is given by[19]

$$\bar{a}_\eta = a - \eta \sum_{j=1}^{d} \sum_{k=1}^{N} b_{k,j} \frac{\partial b_j}{\partial x_k} \tag{7.41}$$

and the predictor $\overline{X}(t_{m+1})$ is given by

$$\overline{X}(t_{m+1}) = X(t_m) + a(X(t_m), t_m)(t_{m+1} - t_m) + \sum_{j=1}^{d} b_j(X(t_m), t_m) \Delta W_j \tag{7.42}$$

The ΔW_j represent the Brownian motion increments on the time step $[t_m, t_{m+1}]$, and thus are Gaussian random variables with mean zero and variance $\Delta = t_{m+1} - t_m$. However, it is also admissible (and for small time steps possibly more efficient) to choose the ΔW_j as two–point distributed random variables with

$$P(\Delta W_j = \sqrt{\Delta}) = P(\Delta W_j = -\sqrt{\Delta}) = \frac{1}{2} \tag{7.43}$$

In the one–dimensional case (i.e. $d = N = 1$) with $\alpha = 0.5$ and $\eta = 0$, (7.40) becomes

$$\begin{aligned} X(t_{m+1}) &= X(t_m) + \frac{1}{2}\left(a(\overline{X}(t_{m+1}), t_{m+1}) + a(X(t_m), t_m)\right)(t_{m+1} - t_m) \\ &\quad + b(X(t_m), t_m)\Delta W \end{aligned}$$

with predictor

$$\overline{X}(t_{m+1}) = X(t_m) + a(X(t_m), t_m)(t_{m+1} - t_m) + b(X(t_m), t_m)\Delta W \tag{7.44}$$

Depending on the structure of a particular Itô process (7.36), the order of weak convergence achieved by (7.40) can be greater than the guaranteed order of 1. A predictor–corrector method with a guaranteed order 2 weak convergence can also be constructed. In this case, the corrector is given by

$$X(t_{m+1}) = X(t_m) + \frac{1}{2}(a(\overline{X}(t_{m+1}), t_{m+1}) + a(X(t_m), t_m))(t_{m+1} - t_m) + \Psi_m$$

where

$$\begin{aligned} \Psi_m &= \sum_{j=1}^{d} \left(b_j(X(t_m), t_m) + \frac{1}{2} L^0 b_j(X(t_m), t_m)(t_{m+1} - t_m) \right) \Delta W_j \\ &\quad + \frac{1}{2} \sum_{j_1=1}^{d} \sum_{j_2=1}^{d} L^{j_1} b_{j_2}(X(t_m), t_m)(\Delta W_{j_1} \Delta W_{j_2} + V_{j_1,j_2}) \end{aligned}$$

----

[19] This is based on what Kloeden and Platen (1992) call the "fully implicit" Euler scheme. See p. 496ff in this reference.

with the operators L^0 and L^j defined by

$$L^0 = a\frac{\partial}{\partial x} + \frac{1}{2}b^2\frac{\partial^2}{\partial x^2} \qquad L^j = \sum_{k=1}^{N} b_{k,j}\frac{\partial}{\partial x_k}$$

The predictor is given by

$$\begin{aligned}
\overline{X}(t_{m+1}) \;=\; & X(t_m) + a(X(t_m), t_m)(t_{m+1} - t_m) + \Psi_m \\
& + \frac{1}{2}L^0 a(X(t_m), t_m)(t_{m+1} - t_m)^2 \\
& + \frac{1}{2}\sum_{j=1}^{d} L^j a(X(t_m), t_m)\Delta W_j(t_{m+1} - t_m)
\end{aligned}$$

The ΔW_j can be chosen as independent Gaussian random variables with mean zero and variance $\Delta = t_{m+1} - t_m$, or alternatively as independent three–point distributed random variables with

$$P(\Delta W = \sqrt{3\Delta}) = P(\Delta W = -\sqrt{3\Delta}) = \frac{1}{6}, \quad P(\Delta W = 0) = \frac{2}{3}$$

The V_{j_1,j_2} are independent two–point distributed random variables satisfying

$$P(V_{j_1,j_2} = \Delta) = P(V_{j_1,j_2} = -\Delta) = \frac{1}{2}$$

for $j_2 = 1, \ldots, j_1 - 1$ and

$$V_{j_1,j_1} = -\Delta, \qquad V_{j_1,j_2} = -V_{j_2,j_1}$$

for $j_2 = j_1 + 1, \ldots, d$ and $j_1 = 1, \ldots, d$.

As with the choice of discretisation scheme in the previous section, the implementation of predictor–corrector methods can be encapsulated in a class representing a price process.

7.6 Variance reduction techniques

We have seen that the accuracy of a Monte Carlo estimate — as measured by the width of, say, a 95% confidence interval around the estimate — improves in proportion to the square root of the number of simulations. Variance reduction methods aim to reduce $\hat{\sigma}_N$ (the standard deviation of the Monte Carlo estimator given by (7.4)) by reducing the standard deviation σ of $f(U)$ — halving σ will improve the accuracy of the Monte Carlo estimate by as much as would a fourfold increase in the number of simulations.

7.6.1 Antithetic sampling

The variance reduction method which is arguably the easiest to implement is based on antithetic variates. For symmetric distributions, antithetic values are generated by mirroring on the mean, i.e. if u is a random variate uniformly

```
template <class argtype,class rettype,
         class random_number_generator_type>
void MCGeneric<argtype,rettype,random_number_generator_type>::
simulate(MCGatherer<rettype>& mcgatherer,
         unsigned long number_of_simulations)
{
  unsigned long i;
  if (!antithetic)
    for (i=0;i<number_of_simulations;i++)
      mcgatherer += f(random_number_generator.random());
  else {
  for (i=0;i<number_of_simulations;i++) {
    argtype rnd = random_number_generator.random();
    rettype res = f(rnd);
    res = 0.5 * (res + f(antithetic(rnd)));
    mcgatherer += res; }}
}
```

Listing 7.10: The generic Monte Carlo simulation loop with antithetic sampling

```
/// Antithetic variables mirrored around zero.
template <class T>
T normal_antithetic(T arg)
{
  return T(-arg);
}

/// Antithetic for uniform [0,1].
template <class T>
T uniform_antithetic(T arg)
{
  return T(1.0 - arg);
}
```

Listing 7.11: Templates for the typical antithetic functors

distributed on $[a, b]$, its antithetic variate is given by

$$u' = a + b - u$$

In the typical case that u is uniformly distributed on $[0, 1]$, this reduces to $u' = 1 - u$. Where standard normal random variates z are required, one can set $z = N^{-1}(u)$ and $z' = N^{-1}(1 - u)$, or equivalently (and more efficiently) $z' = -z$. Given a random variate and its antithetic value, each draw in the Monte Carlo sample is calculated by

$$\bar{f}_{u_i} = \frac{1}{2}(f(u_i) + f(u'_i)) \tag{7.45}$$

```
// -------------- create underlying asset ---------------
Array<double,1> sgm1(1);
sgm1 = sgm;
ConstVol vol(sgm1);
BlackScholesAsset stock(&vol,S,0.03);
std::vector<const BlackScholesAsset*> underlying;
underlying.push_back(&stock);
// -------------- closed form option price ---------------
cout << "S: " << S << "\nK: " << K << "\nr: " << r
     << "\nT: " << mat << "\nvolatility: " << sgm << endl;
double CFcall = stock.option(mat,K,r);
cout << "Closed form call: " << CFcall << endl;
double CFprice = CFcall;
// flat term structure for discounting
FlatTermStructure ts(r,0.0,mat+10.0);
```

Listing 7.12: Pricing a European call option in a model driven by geometric Brownian motion (i.e. Black/Scholes) — setup and closed form price. Full code available in `AntitheticExample.cpp` on the website.

The Monte Carlo estimator $\hat{I}_N$ in (7.2) then becomes

$$\hat{I}_N = (b-a)\frac{1}{N}\sum_{i=1}^{N}\bar{f}(u_i)$$

and its variance

$$\hat{\sigma}_N = \frac{(b-a)^2}{N(N-1)}\sum_{i=1}^{N}(\bar{f}(u_i) - \hat{I}_N)^2$$

By (7.45) this requires twice as many evaluations of the function $f(\cdot)$ as in plain Monte Carlo, so antithetic variates lead to a worthwhile variance reduction if

$$\mathrm{Var}[\bar{f}(u)] < \frac{1}{2}\mathrm{Var}[f(u)]$$

which is equivalent to

$$\mathrm{Cov}[f(u), f(u')] < 0$$

If $f(\cdot)$ is monotonic in u, this condition is satisfied, but when $f(\cdot)$ is symmetric around the mean of u, antithetic variates would lead to no variance reduction at all. For most models and payoff functions encountered in finance, monotonicity of $f(\cdot)$ will hold.[20]

In the C++ implementation, at the level of the generic Monte Carlo algorithm class template `MCGeneric`, $f(u)$ is calculated for each draw of u and the result collected in an `MCGatherer`. For antithetic sampling, $f(u)$ is replaced by

---

[20] A notable exception would be the "quadratic Gaussian" models such as the one discussed by Beaglehole and Tenney (1991), which involve a quadratic form of the driving state variable.

```
unsigned long n = minpaths;
// instantiate random number generator
ranlib::NormalUnit<double> normalRNG;
RandomArray<ranlib::NormalUnit<double>,double>
random_container2(normalRNG,1,1); // 1 factor, 1 time step
// instantiate stochastic process
GeometricBrownianMotion gbm(underlying);
// 95% quantile for confidence interval
boost::math::normal normal;
double d = boost::math::quantile(normal,0.95);
// boost functor to convert random variates to their antithetics
// (instantiated from template)
boost::function<Array<double,2> (Array<double,2>)>
  antithetic = normal_antithetic<Array<double,2> >;
// instantiate MCGatherer objects to collect simulation results
MCGatherer<double> mcgatherer;
MCGatherer<double> mcgatherer_antithetic;
// instantiate MCPayoff object
MCEuropeanCall mc_call(0,mat,0,call_strike);
// instantiate MCMapping and bind to functor
MCMapping<GeometricBrownianMotion,Array<double,2> >
  mc_mapping2(mc_call,gbm,ts,numeraire_index);
boost::function<double (Array<double,2>)> func2 = boost::bind(
  &MCMapping<GeometricBrownianMotion,Array<double,2> >::mapping,
    &mc_mapping2,_1);
// instantiate generic Monte Carlo algorithm object
MCGeneric<Array<double,2>,double,
        RandomArray<ranlib::NormalUnit<double>,double> >
  mc2(func2,random_container2);
MCGeneric<Array<double,2>,double,
        RandomArray<ranlib::NormalUnit<double>,double> >
  mc2_antithetic(func2,random_container2,antithetic);
```

Listing 7.13: Pricing a European call option by Monte Carlo simulation in a model driven by geometric Brownian motion (i.e. Black/Scholes), with and without antithetic sampling — instantiation of the required Monte Carlo classes. Full code available in **AntitheticExample.cpp** on the website.

$\bar{f}(u)$ as per (7.45), and thus **MCGeneric** is the appropriate place to implement this. We add the functor

$$\texttt{boost::function<argtype (argtype)> antithetic;} \qquad (7.46)$$

as a private data member of **MCGeneric**. This functor serves to convert a realisation of the random variable of type **argtype** into its antithetic value. Listing 7.3 then becomes Listing 7.10, which implements (7.45) if **antithetic** is present.

The public interface of **MCGeneric** is amended by adding

```
cout << "European call option\nPaths,Closed form value,MC value,
        95% CI lower bound,95% CI upper bound,
        Difference in standard errors,";
cout << "Antithetic MC value,95% CI lower bound,
        95% CI upper bound,Difference in standard errors,
        CI width,Antithetic CI width\n";
// run Monte Carlo for different numbers of simulations
while (mcgatherer.number_of_simulations()<maxpaths) {
  mc2.simulate(mcgatherer,n);
  // half as many paths for antithetic
  mc2_antithetic.simulate(mcgatherer_antithetic,n/2);
  cout << mcgatherer.number_of_simulations() << ',' << CFprice
       << ',' << mcgatherer.mean() << ','
       << mcgatherer.mean()-d*mcgatherer.stddev() << ','
       << mcgatherer.mean()+d*mcgatherer.stddev() << ','
       << (mcgatherer.mean()-CFprice)/mcgatherer.stddev() << ',';
  cout << mcgatherer_antithetic.mean() << ','
       << mcgatherer_antithetic.mean()
           -d*mcgatherer_antithetic.stddev() << ','
       << mcgatherer_antithetic.mean()
           +d*mcgatherer_antithetic.stddev() << ','
       << (mcgatherer_antithetic.mean()-CFprice)/
           mcgatherer_antithetic.stddev() << ',';
  cout << 2.0*d*mcgatherer.stddev() << ','
       << 2.0*d*mcgatherer_antithetic.stddev() << ',' << endl;
  n = mcgatherer.number_of_simulations(); }
```

Listing 7.14: Pricing a European call option by Monte Carlo simulation in a model driven by geometric Brownian motion (i.e. Black/Scholes), with and without antithetic sampling — running Monte Carlo for different numbers of simulations. Full code available in **AntitheticExample.cpp** on the website.

$$\text{void set\_antithetic(boost::function<argtype (argtype)>} \atop \text{antifunc)} \tag{7.47}$$

and the overloaded constructor

$$\begin{aligned} &\text{MCGeneric(boost::function<rettype (argtype)> func,} \\ &\qquad \text{random\_number\_generator\_type\& rng,} \\ &\qquad \text{boost::function<argtype (argtype)>} \\ &\qquad \text{antifunc)} \end{aligned} \tag{7.48}$$

to turn on antithetic sampling (it can be turned off again by passing an empty functor to set_antithetic()).

Templates for the typical antithetic functors are given in **QFRandom.hpp** (see Listing 7.11). Thus if we require a functor to convert an array of standard normal random variates into their antithetic values, we can instantiate it by

$$\begin{aligned} &\text{boost::function<Array<double,2> (Array<double,2>)>} \\ &\text{antithetic = normal\_antithetic<Array<double,2> >} \end{aligned} \tag{7.49}$$

Listings 7.12 to 7.14 show how to price a European call option by Monte Carlo

Paths	Without antithetic sampling			With antithetic sampling			Ratio
	MC value	95% conf. interval		MC value	95% conf. interval		
500	12.8025	10.9738	14.6312	16.3369	14.6648	18.009	1.09
1000	13.1877	11.8794	14.4959	15.4681	14.3071	16.6290	1.13
2000	14.4451	13.4658	15.4245	15.1199	14.3093	15.9305	1.21
4000	14.9054	14.1933	15.6175	15.0886	14.5199	15.6573	1.25
8000	15.0729	14.5746	15.5712	15.0503	14.6517	15.4488	1.25
16000	15.1899	14.8420	15.5378	15.2321	14.9503	15.5138	1.23
32000	15.3473	15.1005	15.5940	15.4917	15.2893	15.6940	1.22
64000	15.3164	15.1429	15.4899	15.3567	15.2145	15.4989	1.22
128000	15.1874	15.0652	15.3095	15.2779	15.1774	15.3784	1.22
256000	15.2299	15.1435	15.3163	15.2825	15.2115	15.3536	1.22
512000	15.2225	15.1615	15.2836	15.2181	15.1681	15.2681	1.22
1024000	15.2235	15.1803	15.2666	15.1735	15.1383	15.2088	1.22
2048000	15.2049	15.1744	15.2354	15.1742	15.1493	15.1991	1.22
4096000	15.1979	15.1764	15.2195	15.1680	15.1504	15.1856	1.22
8192000	15.1997	15.1845	15.2150	15.1840	15.1715	15.1965	1.22

Table 7.2 *Monte Carlo simulation of a European call option, with and without antithetic sampling. Antithetic values are generated using half as many paths as the "plain" values. Initial stock price 100, strike 100, time to maturity 1.5 years, interest rate 5% continuously compounded, volatility 30%. Black/Scholes price 15.1867. The last column shows the ratio between the width of the 95% confidence intervals with and without antithetic sampling.*

simulation in a model driven by geometric Brownian motion (i.e. Black/Scholes), with and without antithetic sampling. Listing 7.12 instantiates the class representing the underlying asset and calculates the closed–form Black/Scholes price. Listing 7.13 assembles the classes required to perform the Monte Carlo simulation: the random number generator, the stochastic process, the payoff, the instance of MCMapping mapping the random number to realisations of the underlying asset price and to the option payoff, the antithetic functor, the generic Monte Carlo algorithm and instances of MCGatherer to hold the simulation results.

This example program was used to generate Table 7.2, which compares the standard deviations of the resulting Monte Carlo estimators. The antithetic estimates are based on half as many simulations as the "plain" estimates (since they require about twice as much computational effort per simulation). Antithetic sampling in this case improves the accuracy of the Monte Carlo estimate by about 22%.

7.6.2 Control variates

As before, given a (possibly multidimensional) random variable U, we wish to calculate $E[f(U)]$ by Monte Carlo simulation. Suppose that for another function g, $E[g(U)]$ can be evaluated analytically. If $f(U)$ and $g(U)$ are closely correlated, then $g(U)$ represents an effective control variate. Naively, if $f(U)$

```
template <class target_price_process,
          class controlvariate_price_process,
          class random_variable>
double MCControlVariateMapping<target_price_process,
                              controlvariate_price_process,
                              random_variable>
::mapping(random_variable x)
{
  return target.mapping(x) - controlvariate.mapping(x)
    + controlvariate_values_sum;
}

template <class target_price_process,
          class controlvariate_price_process,
          class random_variable>
double MCControlVariateMapping<target_price_process,
                              controlvariate_price_process,
                              random_variable>
::mappingArray(random_variable x)
{
  return target.mappingArray(x) - controlvariate.mappingArray(x)
    + controlvariate_values;
}
```

Listing 7.15: Code implementing `mapping()` (for a scalar payoff) and `mappingArray()` for the template class `MCControlVariateMapping`

and $g(U)$ are of similar variance and setting $a = 0$, $b = 1$, we can replace the Monte Carlo estimator (7.2) by

$$\hat{I}_N = \left(\frac{1}{N} \sum_{i=1}^{N} (f(u_i) - g(u_i)) \right) + E[g(U)] \qquad (7.50)$$

and the variance of the difference of the two closely correlated random variables $f(U)$ and $g(U)$ is much lower than the variance of $f(U)$ alone:

$$\text{Var}[f(U) - g(U)] = \text{Var}[f(U)] + \text{Var}[g(U)] - 2\rho\sqrt{\text{Var}[g(U)]\text{Var}[f(U)]} \qquad (7.51)$$

where ρ is the correlation between $f(U)$ and $g(U)$.

This approach to control variate variance reduction is implemented in the class template `MCControlVariateMapping`, which can replace `MCMapping` in the implementation of "plain" Monte Carlo as it provides the public member functions `mapping()` and `mappingArray()` used to "map" a random variate draw to a discounted payoff. As noted previously, the "plain" mapping (and thus $f(\cdot)$) encapsulates the model, the discounting and the payoff. A second "plain" mapping represents $g(\cdot)$ and encapsulates the control variate model, discounting and payoff. `MCControlVariateMapping` brings the two plain mappings together, encapsulating $f(\cdot)$, $g(\cdot)$ and the control variate calculation

(7.50). Thus the control variate $g(\cdot)$ may differ from the target $f(\cdot)$ in terms of its "model" (i.e. how the input random variates are mapped to realisations of the underlying financial variables), its discounting or its payoff, the only requirement being that $E[g(U)]$ is known exactly and $f(U)$ and $g(U)$ are closely correlated.

Therefore the constructor of `MCControlVariateMapping` is

```
MCControlVariateMapping(
    MCMapping<target_price_process,
    random_variable>& xtarget,
    MCMapping<controlvariate_price_process,random_variable>&
        xcontrolvariate,
    const Array<double,1>& xcontrolvariate_values)
```

where `target_price_process` and `controlvariate_price_process` may differ, but must be driven by the same `random_variable`. The constructor argument `xcontrolvariate_values` contains the exact value of $E[g(U)]$, which may in general be a vector.

The code implementing `mapping()` (for a scalar payoff) and `mappingArray()` (for a vector–valued payoff) is given in Listing 7.15, and the full implementation of `MCControlVariateMapping` can be found in the file `MCEngine.hpp` on the website.

The textbook example for using control variates in MonteCarlo simulation in finance is the case of the arithmetic average (or Asian) option. This option does not have a closed–form pricing formula in the Black/Scholes model, as the arithmetic average of lognormal random variables is not lognormal. However, the geometric average is lognormal and closely correlated to the arithmetic average, so an option on the geometric average can be priced in closed form and provides an effective control variate for the arithmetic average option.

The code using `MCControlVariateMapping` to price a call on the arithmetic average of the underlying stock price, i.e. with payoff

$$\max\left(0, \frac{1}{n}\sum_{j=1}^{n} S(t_j) - K\right)$$

for an underlying stock price $S(\cdot)$ observed at a discrete set of dates t_j, is given in the file `CVTest.cpp` on the website. The closed–form solution for the geometric average option is calculated using the "Quintessential Option Pricing Formula" from Section 4.3. The template–based approach makes it easy to implement the Monte Carlo simulation in its "plain" version (Case 1), with antithetic sampling (Case 2), with control variates (Case 3), and with both antithetic sampling and control variates (Case 4). Table 7.3 compares all four cases. For a given number of simulations, antithetic sampling and control variate variance reduction each increase the computational effort by about a factor of two, so in order to make the results comparable, in each row of the table Cases 2 and 3 involve half as many simulations as Case 1, and Case 4 involves one quarter as many simulations. Use of the control variate

	Case 1: plain MC		Case 2: antithetic MC	
Paths	MC value	CI width	MC value	CI width
200	20.4445	9.8290	14.4162	5.0281
800	15.9943	3.6264	16.1170	2.7406
3200	16.4933	1.8148	16.0282	1.4640
12800	16.2209	0.8862	16.1802	0.7528
51200	16.1333	0.4409	16.0277	0.3741
204800	15.9870	0.2189	16.0116	0.1866
819200	16.0712	0.1103	16.0925	0.0939
3276800	16.0996	0.0553	16.0591	0.0467

	Case 3: MC with control variate		Case 4: antithetic MC with control va	
Paths	MC value	CI width	MC value	CI width
200	16.6326	1.7580	15.9488	0.9979
800	16.2901	1.0580	16.0943	0.8569
3200	15.9721	0.4697	16.1298	0.4842
12800	16.0441	0.2420	16.1502	0.2278
51200	16.1128	0.1248	16.1698	0.1161
204800	16.0941	0.0605	16.1092	0.0558
819200	16.1015	0.0304	16.0819	0.0270
3276800	16.1004	0.0152	16.0899	0.0136

	Ratios between CI widths			
Paths	Case 1/2	Case 1/3	Case 1/4	Case 3/4
200	1.9548	5.5911	9.8500	1.7617
800	1.3232	3.4277	4.2320	1.2346
3200	1.2396	3.8637	3.7478	0.9700
12800	1.1773	3.6625	3.8903	1.0622
51200	1.1788	3.5320	3.7992	1.0757
204800	1.1730	3.6164	3.9217	1.0844
819200	1.1744	3.6240	4.0896	1.1285
3276800	1.1825	3.6289	4.0691	1.1213

Table 7.3 *Pricing a call option with strike 102 maturing in 2.25 years on the arithmetic average of a stock price observed at 11 equally spaced time points. The model is Black/Scholes with initial stock price 100, 63.64% volatility and an interest rate of 5%. Pricing is done by 'plain" Monte Carlo (Case 1), with antithetic sampling (Case 2), with control variates (Case 3), and with both antithetic sampling and control variates (Case 4). "CI width" refers to the width of a 95% confidence interval around the Monte Carlo estimate.*

improves accuracy by a factor of about 3.6 without antithetic sampling and 4 with antithetic sampling, which for a given accuracy translates to a reduction in computing time by a factor of about 13 and 16, respectively.

7.6.3 Optimal control variate weights

One notes that (7.50) can be generalised to

$$\hat{I}_N = \left(\frac{1}{N} \sum_{i=1}^{N} (f(u_i) - \beta g(u_i)) \right) + \beta E[g(U)] \tag{7.52}$$

An optimal choice of β minimises $\mathrm{Var}[f(U) - \beta g(U)]$, in other words, β is the slope of a linear regression of $f(u)$ on $g(u)$, and thus given by

$$\beta = \frac{\mathrm{Cov}[f(U), g(U)]}{\mathrm{Var}[g(U)]} \tag{7.53}$$

Using β chosen optimally in this way to evaluate (7.52), the variance of the original Monte Carlo estimator (7.2) is reduced by a factor of

$$1 - (\mathrm{Corr}[f(U), g(U)])^2 \tag{7.54}$$

where $\mathrm{Corr}[f(U), g(U)]$ is the correlation between $f(U)$ and $g(U)$. In other words, in order to achieve the same accuracy as (7.52) using the "plain" estimator (7.2), $1/(1 - (\mathrm{Corr}[f(U), g(u)])^2)$ as many simulations are required.

Typically, the covariance between $f(U)$ and $g(U)$, $\mathrm{Cov}[f(U), g(U)]$, will not be known analytically, so we have to calculate β using sample estimates for variance and covariance, setting

$$\bar{f} = \frac{1}{N} \sum_{i=1}^{N} f(u_i)$$

$$\bar{g} = \frac{1}{N} \sum_{i=1}^{N} g(u_i)$$

$$\hat{\beta}_N = \frac{\sum_{i=1}^{N} (f(u_i) - \bar{f})(g(u_i) - \bar{g})}{\sum_{i=1}^{N} (g(u_i) - \bar{g})^2} \tag{7.55}$$

Using $\hat{\beta}_N$ in (7.52) introduces some bias if $\hat{I}_N$ and $\hat{\beta}_N$ are calculated on the same set of simulations. This bias is avoided if $\hat{\beta}_N$ is calculated on an independent (possibly smaller) set of simulations, separately from $\hat{I}_N$.[21]

(7.52) can be generalised further to allow for multiple control variates, i.e.

$$\hat{I}_N = \left(\frac{1}{N} \sum_{i=1}^{N} \left(f(u_i) - \sum_{j=1}^{J} \beta_j g_j(u_i) \right) \right) + \sum_{j=1}^{J} \beta_j E[g_j(U)] \tag{7.56}$$

---

[21] However, experience appears to indicate that this bias tends to be small. See Glasserman (2004), Section 4.1.3, for a discussion of this issue.

```
double MCGatherer<Array<double,1> >::CVestimate(int target,int
CV,double CV_expectation) const
{
   double b = CVweight(target,CV);
   return mean(target) - b*(mean(CV) - CV_expectation);
}


double MCGatherer<Array<double,1> >::CVweight(int target,int CV)
const
{
   double b = 0.0;
   if (CVon) {
   b = (covar(target,CV)/number_of_simulations() -
mean(target)*mean(CV)) / (variance(CV)*number_of_simulations()); }
   return b;
}
```

Listing 7.16: The member functions CVestimate() and CVweight() for Monte Carlo estimates using one control variate

with J control variates $g_j(\cdot)$. The optimal weights β_j can be estimated by multivariate linear regression: Denote by S_g the $J \times J$ matrix with jk entry

$$\frac{1}{N-1} \left(\sum_{i=1}^{N} g_j(u_i)g_k(u_i) - N\bar{g}_j\bar{g}_k \right)$$

and by S_{fg} the J-dimensional vector with j-th entry

$$\frac{1}{N-1} \left(\sum_{i=1}^{N} g_j(u_i)f(u_i) - N\bar{g}_j\bar{f} \right)$$

Then $\hat{\beta}_N$ is the J-dimensional vector which solves linear system of equations

$$S_g\hat{\beta}_N = S_{fg} \tag{7.57}$$

One should note that the use of multiple control variates exacerbates small–sample issues (see e.g. Section 4.1.3 of Glasserman (2004)), but in modern computer implementations one would typically choose N in the thousands, while J would be a small integer (typically less than ten), so the impact of these issues can be expected to be negligible.

The appropriate place to implement control variates with optimal weights is a specialisation of the MCGatherer class template, i.e.[22]

$$\text{template<> class MCGatherer<Array<double,1> >} \tag{7.58}$$

is used to collect simulation results for a set of payoffs, where some payoffs may be used as control variates for one "target" payoff. Use of control variates can be turned on or off with the member function

[22] See also the files MCGatherer.hpp and MCGatherer.cpp on the website.

```
double MCGatherer<Array<double,1> >::
CVestimate(int target,
           const Array<int,1>& CV,
           const Array<double,1>& CV_expectation) const
{
  int i;
  int n = CV.extent(blitz::firstDim);
  Array<double,2> b(CVweight(target,CV));
  double result = mean(target);
  for (i=0;i<n;i++)
    result -= b(i,0) * (mean(CV(i)) - CV_expectation(i));
  return result;
}

Array<double,2> MCGatherer<Array<double,1> >::
CVweight(int target,const Array<int,1>& CV) const
{
  int i,j;
  int n = CV.extent(blitz::firstDim);
  Array<double,2> covCV(n,n),cov(n,1),b(n,1);
  b = 0.0;
  if (CVon) {
    for (i=0;i<n;i++) {
      double meani = mean(CV(i));
      for (j=0;j<i;j++) {
        covCV(i,j) = covCV(j,i) =
          covar(CV(i),CV(j))/number_of_simulations() -
            meani*mean(CV(j)); }
      covCV(i,i) = covar(CV(i),CV(i))/number_of_simulations() -
                    meani*meani;
      cov(i,0)   = covar(target,CV(i))/number_of_simulations() -
                    mean(CV(i))*mean(target); }
    interfaceCLAPACK::SolveLinear(covCV,b,cov); }
  return b;
}
```

Listing 7.17: The member functions CVestimate() and CVweight() for Monte Carlo estimates using several control variates

$$\text{void set\_control\_variate(bool cv)} \qquad (7.59)$$

If control variates have been turned on, the accumulation operator,

$$\text{void operator+=(const Array<double,1>\& add)} \qquad (7.60)$$

records the information necessary to calculate S_g and S_{fg}.

Note that even with control variates simulation turned on, the member functions

```
Array<double,1> mean() const
Array<double,1> stddev() const
Array<double,1> variance() const
```
$$(7.61)$$

Paths	MC value	CI width	CV MC value	CV CI width	Acc. gain
100	0.098103	1.63E-02	0.101249	5.75E-04	28.29
400	0.100593	8.19E-03	0.101322	2.86E-04	28.62
1600	0.101214	4.30E-03	0.101166	1.81E-04	23.79
6400	0.101297	2.17E-03	0.101146	8.48E-05	25.63
25600	0.100603	1.09E-03	0.101104	4.90E-05	22.20
102400	0.101176	5.44E-04	0.101110	2.39E-05	22.77
409600	0.101147	2.71E-04	0.101109	1.18E-05	23.08
1638400	0.101118	1.36E-04	0.101103	5.91E-06	23.01

Table 7.4 *Pricing an option with payoff (7.66), strike 1.47876, power $\alpha = 0.1$ maturing in 1.5 years. Closed form value of the option is 0.101104. The model is Black/Scholes with initial stock price 100, 42.43% volatility and an interest rate of 5%. Pricing is done by "plain" Monte Carlo and with five standard call options with strikes 37.5, 50, 62.5, 75 and 125 as control variates. "CI width" refers to the width of a 95% confidence interval around the Monte Carlo estimate.*

return the vector of "plain" Monte Carlo estimates, Monte Carlo estimate standard deviations and variances, respectively. The template specialisation also provides the member functions

```
double mean(int i) const
double stddev(int i) const                         (7.62)
double variance(int i) const
```

to access the mean, standard deviation and variance of the Monte Carlo estimate for a particular element in the vector of Monte Carlo estimates collected by MCGatherer<Array<double,1> >.

Monte Carlo estimates using one control variate are accessed using the member function

```
double CVestimate(int target,int CV,
                  double CV_expectation) const       (7.63)
```

where target is the index of the target payoff $f(\cdot)$ and CV is the index of the control variate payoff $g(\cdot)$ in the vector of Monte Carlo estimates, and CV_expectation is the exact value of $E[g(U)]$. CVestimate() calls the member function CVweight(target,CV) to calculate β using (7.55) and then calculates the Monte Carlo estimate using (7.52), see Listing 7.16.

The member function

```
double CVestimate_stddev(int target,int CV) const   (7.64)
```

uses (7.54) to calculate the standard deviation of the Monte Carlo estimate. The file CVtest.cpp on the website demonstrates this version of the control variate simulation for the case of an option on the arithmetic average, with an option on the geometric average as the control variate.

For multiple control variates, CVestimate() is overloaded to

```
double CVestimate(int target,
                  const Array <int,1>& CV,
                  const Array<double,1>&
                  CV_expectation) const
```

$$(7.65)$$

where CV now is an Array of indices of the control variate payoffs $g_j(\cdot)$ in the vector of Monte Carlo estimates, and CV_expectation is the vector of exact values $E[g_j(U)]$. This version of CVestimate() implements (7.56), where the optional control variate weights are calculated in an overloaded version of CVweight(target,CV) by solving (7.57), see Listing 7.17.

To get a reliable confidence interval indicating the accuracy of a Monte Carlo estimate with multiple control variates, it is recommended to run a number of simulations to estimate the optional control variate weights, and then hold those weights fixed in MCGatherer<Array<double,1> > by calling the member function

```
void fix_weights(int target,
                 const Array<int,1>& CV,
                 const Array <double,1>& CV_expectation)
```

This resets the variables collecting the simulation results to zero, discarding the simulations used to calculate the weights. This means that the Monte Carlo estimates using the fixed weights will be calculated based on a set of simulations that are independent of those used to determine the weights, thus removing this source of potential bias. Furthermore, fix_weights() sets a flag weights_fixed, which causes the accumulation operator+= to calculate each simulated value of the target payoff by

$$f(u_i) - \sum_{j=1}^{J} \beta_j \left(g_j(u_i) - E[g_i(U)] \right)$$

which is now possible because the weights β_j are fixed.

Consequently, mean(target) will now return the Monte Carlo estimate for $E[f(U)]$ using the control variates (as opposed to the "plain" Monte Carlo estimate when weights_fixed is not set) and stddev(target) will return the corresponding standard deviation.

The file CVTestPower.cpp on the website demonstrates this two–staged approach for variance reduction with multiple control variates in the case of a power option with payoff

$$\max(0, S(T)^\alpha - K) \qquad (7.66)$$

where five standard options with different strikes are used as control variates. Table 7.4 illustrates the resulting variance reduction: By using five control variates, the accuracy of the Monte Carlo estimate is improved by a factor of about 23. Since this requires six times as many payoff evaluations as plain Monte Carlo, computational efficiency is improved by a factor of about 3.8. It is also worth noting that the regression (i.e. solving (7.57)) to determine the optimal weights is equivalent to approximating the power option payoff by a linear combination (i.e. portfolio) of the payoffs of the standard options used as

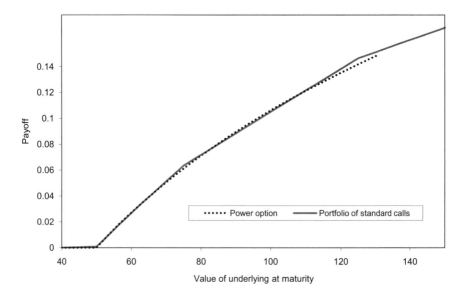

Figure 7.3 *Power option payoff approximated by payoff of a portfolio of five standard call options, based on the optimal control variate weights*

control variates: Figure 7.3 plots the payoff of the power option and the payoff of a portfolio of the standard options, where the portfolio weights are given by the (estimated) optimal control variate weights $\hat{\beta}_N$. Clearly, whenever a target payoff can be closely approximated by a portfolio of control variate payoffs, the portfolio and the target will be closely correlated, and the portfolio will be an effective control variate. Conversely, as Figure 7.3 illustrates, the regression (7.57) will converge for the portfolio payoff to optimally (in a probability–weighted least squares sense) approximate the target payoff.

7.7 Pricing instruments with early exercise features

At each point in time at which an option permits early exercise, the value of the option is given by the greater of the early exercise payoff and the *continuation value* (i.e. the value of not exercising early and instead holding on to the option). When using binomial lattice models (Chapter 3) or finite difference schemes (Chapter 5), one can price the option by backward induction, and thus the continuation value is known whenever the early decision needs to be evaluated. When simulating a Monte Carlo path, the continuation value is *not*

Also, as we will see below, once we are able to determine (an approximation for) the continuation value, we are in a position to also price path dependent options with early exercise features (including barrier options, Asian options and lookback options) in a generic manner that is straightforward to implement.

7.7.1 The regression–based algorithm

The most common approach to approximate the continuation value is to estimate it by regression on a set of basis functions of state variables on a Monte Carlo path. Methods of this type have been proposed by a number of authors, including Carriére (1996), Tsitsiklis and van Roy (1999) and Longstaff and Schwartz (2001). Given a Monte Carlo path, for each time point where early exercise is permitted we can look ahead and see what the value of holding on to the option would be – at the last early exercise time prior to maturity this is simply the discounted payoff at maturity on that path. However, at the time the option holder must make a decision whether or not to exercise early, only information on the path up to that point in time is available, summarised by a set of state variables.[23] Thus we need a function which predicts the continuation value based on the realisation of the state variables. It is obtained by choosing a set of basis functions of the state variables, simulating a set of Monte Carlo paths (the *training paths*), and on these paths performing a linear regression of the realised discounted value of holding on to the option regressed on the values of the basis functions at the time of potential early exercise. Once the coefficients of this regression have been determined and we thus have a function which maps the values of the state variables at the time of potential early exercise to a (predicted) continuation value, this continuation value can be compared to the early exercise payoff in order to decide whether to exercise or not. This early exercise strategy is then used on a fresh (i.e. independent) set of Monte Carlo paths to yield an estimate for the option value.

As the early exercise strategy only uses information that would be available to the option holder at the time of exercise, it can be no better than the best possible (i.e. optimal) strategy based on available information. In other words, the approximation of the continuation value by regression on state variables results in a suboptimal exercise strategy, and the Monte Carlo estimator — in the absence of any other bias — will converge to a lower bound on the value of the option. The quality of the lower bound depends on how well the regression on the chosen set of basis functions approximates the conditional expectation of the discounted value of not exercising the option early — thus a better set of basis functions yields a higher option value. One should note that Longstaff and Schwartz (2001) use the same set of paths to conduct the regression and

---

[23] Depending on the model and the (potentially path–dependent) payoff, the set of state variables may be quite large — in some cases it might include the entire history up to that point in time.

the calculation of the Monte Carlo estimator for the option value. In this case one can no longer make a definitive statement about the direction of the bias, though in practice it appears that the estimator is typically still biased low.

When reliable error bounds are required, it is better to ensure that the regression–based method described here converges to a lower bound: The correct value of an option with the possibility of early exercise can be written in terms of expectations under a risk–neutral measure

$$\sup_{\tau} E[h_{\tau}(X_{\tau})] \tag{7.67}$$

where X_{τ} is the set of state variable at time τ, $h_{\tau}(X_{\tau})$ is the discounted payoff resulting from exercise at time τ, and τ is a stopping time. If the decision to exercise at time $\hat{\tau}$ is based on a continuation value function estimated by regression on an independent (i.e. separate) set of training paths, $\hat{\tau}$ is a stopping time and the Monte Carlo estimator converges to $E[h_{\hat{\tau}}(X_{\hat{\tau}})]$ with

$$E[h_{\hat{\tau}}(X_{\hat{\tau}})] \leq \sup_{\tau} E[h_{\tau}(X_{\tau})] \tag{7.68}$$

One could then make use of a dual formulation (established by Haugh and Kogan (2004) and Rogers (2002)) of the pricing problem to obtain an upper bound, given by

$$\sup_{\tau} E[h_{\tau}(X_{\tau})] \leq \inf_{M} E[\max_{k=1,\ldots,m}(h_{t_k}(X_{t_k}) - M_{t_k})] \tag{7.69}$$

where the infimum is taken over martingales M under the risk–neutral measure, with initial value 0. The t_k, $k = 1, \ldots, m$, are the possible exercise times.

For $M_t \equiv 0 \; \forall \; t$, the expectation on the right–hand side of (7.69) corresponds to early exercise with perfect foresight, which is clearly an upper bound for the value of the option. However, this is not a very good upper bound, and the quality of the upper bound hinges on a good choice of the martingale M. One attractive, computationally tractable choice of M has been proposed by Belomestny, Bender and Schoenmakers (2009).

Many practical applications restrict themselves to the lower–bound, regression–based method, and this is the approach that we will take here. Consider a set of financial variables[24] $X_i, i = 1, \ldots, N$, observed at times $T_k, k = 0, \ldots, K$. Furthermore, denote by $T_{\bar{k}(j)}, 0 \leq \bar{k}(j) \leq K$ the j-th time of potential exercise of an option (i.e. the option holder does not necessarily have the right of early exercise at all times T_k). For each early exercise time $T_{\bar{k}(j)}, 0 \leq \bar{k}(j) < K$, we will approximate the continuation value by a linear combination of n basis

---

[24] These financial variables may or may not be assets as in Chapter 4.

functions,

$$V(T_{\bar{k}(j)}) \;\; = \;\; \sum_{\ell=1}^{n} c_{\ell j} \psi_{lj} \tag{7.70}$$

$$\psi_{\ell j} \;\; := \;\; \psi_\ell \left(T_{\bar{k}(j)}, \{X_i(T_k)\}_{\substack{i=1,\ldots,N, \\ k=0,\ldots,\bar{k}(j)}} \right), \quad \ell = 1,\ldots,n$$

which is to say that the $\psi_{\ell j}$ can, in general, be functions of time and the entire history of observed financial variables until time $T_{\bar{k}(j)}$. Following Longstaff and Schwartz (2001), the algorithm to estimate $c_{\ell j}$ for each $T_{\bar{k}(j)}$ proceeds like this:

Step 1: Generate `npaths` simulated Monte Carlo paths of the financial variables X_i observed at times T_k. (These paths will be used to estimate the $c_{\ell j}$ and then discarded.)

Step 2: Initialise a vector y of length `npaths` with the option payoff at time T_K on each path, discounted to the *last* time point $T_{\bar{k}(J)} < T_K$ of potential early exercise. Set the "current" time $t = T_{\bar{k}(J)}$. The initial value of the index j is J (i.e. we will be proceeding by backward induction).

Step 3: Calculate a vector e of length `npaths`, containing the payoff on each path if the option is exercised at time t.

Step 4: Determine the $c_{\ell j}$ for all $\ell = 1,\ldots,n$ and the current j by least squares fit, i.e. choose the $c_{\ell,j}$ to minimise

$$\sum_{h=1}^{\text{npaths}} (y_h - V^{(h)}(T_{\bar{k}(j)}))^2 \tag{7.71}$$

where $V^{(h)}(T_{\bar{k}(j)})$ is given by (7.70) calculated using the realisations of $X_i(T_k)$ on the h-th path.

Step 5: Using the thus determined $c_{\ell j}$, set each y_h to $\max(e_h, V^{(h)}(T_{\bar{k}(j)}))$ discounted to time $T_{\bar{k}(j-1)}$.

Step 6: If $T_{\bar{k}(j)}$ is the earliest time of potential early exercise, stop. Otherwise, set $j = j - 1$ and $t = T_{\bar{k}(j)}$, and go to Step 3.

Longstaff and Schwartz (2001) suggest that in the objective function (7.71), one may want to omit those paths for which e_h at the "current" time t is zero.

7.7.2 Implementation

As noted in the previous section, in order to guarantee that the Monte Carlo price for an option with early exercise features can be interpreted as a lower bound on the true price (as opposed to having uncertainty about the direction of the bias), it is recommended to conduct the procedure in two independent steps. First, the multivariate regression approximating the early exercise boundary is carried out on a set of simulated asset price paths (call these the *training paths*). The Monte Carlo price is then calculated by applying

```
Array<double,2> paths(8,4);
paths = 1.00, 1.09, 1.08, 1.34,
        1.00, 1.16, 1.26, 1.54,
        1.00, 1.22, 1.07, 1.03,
        1.00, 0.93, 0.97, 0.92,
        1.00, 1.11, 1.56, 1.52,
        1.00, 0.76, 0.77, 0.90,
        1.00, 0.92, 0.84, 1.01,
        1.00, 0.88, 1.22, 1.34;
Array<double,1> T(4);
T = 0.0, 1.0, 2.0, 3.0;
double K = 1.10;
Payoff put(K,-1);
boost::function<double (double)> f;
f = boost::bind(std::mem_fun(&Payoff::operator()),&put,_1);
FlatTermStructure ts(0.06,0.0,10.0);
LongstaffSchwartzExerciseBoundary1D ls(T,ts,paths,f,2);
MCGatherer<double> MCestimate;
Array<double,1> path(4);
for (i=0;i<8;i++) {
  path = paths(i,Range::all());
  MCestimate += ls.apply(path); }
std::cout << MCestimate.mean() << std::endl;
```

Listing 7.18: Longstaff/Schwartz toy example using LongstaffSchwartz-
ExerciseBoundary1D

the (suboptimal, because approximate) early exercise strategy to an independent set of paths. This separation suggests an approach to implementation: The first step can be implemented by constructing an object representing the exercise boundary from the training paths, and the second step involves creating an MCMapping from this exercise boundary, which can then be fed into the generic Monte Carlo algorithm MCGeneric.

Let us consider first the C++ class representing the exercise boundary. In the previous section, the continuation value at any time t could potentially depend on several financial variables observed at a number of time points up to and including time t, and the most general implementation of the exercise boundary needs to reflect this.[25] The class RegressionExerciseBoundary provides this functionality, but for clarity of exposition consider first the simpler case where again following Longstaff and Schwartz (2001) the continuation value is approximated by a polynomial function of the value of a single underlying

---

[25] Note that this may be needed regardless of whether the dynamics of the financial variables are Markovian, because the option payoff itself may be path dependent — for example if one had an Asian option that can be exercised early.

financial variable at time $T_{\bar{k}(j)}$. In this case (7.70) becomes, for some fixed i,

$$V(T_{\bar{k}(j)}) = \sum_{\ell=1}^{n} c_{\ell j}(X_i(T_{\bar{k}(j)}))^{\ell-1} \qquad (7.72)$$

This is implemented in the class `LongstaffSchwartzExerciseBoundary1D`.[26] The regression to estimate the coefficients $c_{\ell j}$ is carried out in the constructor, which takes as arguments the timeline of time points of potential exercise, a (number of paths)×(number of time points) `Array` of "training paths," an initial `TermStructure` (for discounting), a `boost::function<double (double)>` to calculate the (undiscounted) payoff at time of exercise as a function of the underlying financial variable, and the degree $(n-1)$ of the polynomial (7.72). The constructor also has a Boolean optional argument, which if set to `true` causes the regression at time point $T_{\bar{k}(j)}$ to be applied only to those paths for which the early exercise payoff at $T_{\bar{k}(j)}$ is positive (the default is `true`).

Once the early exercise boundary is constructed, the resulting early exercise strategy (to exercise the option if and only if the early exercise payoff is greater than the approximate continuation value given by (7.72)) can be applied to each Monte Carlo path to obtain the discounted payoff on that path. This is implemented in the member function `apply()` of `LongstaffSchwartzExerciseBoundary1D`.[27] Using `LongstaffSchwartzExerciseBoundary1D`, one can now reproduce the toy example of Longstaff and Schwartz (2001) of an American put option valued on eight paths with four time points by the code in Listing 7.18.[28]

The class `LongstaffSchwartzExerciseBoundary` generalises `LongstaffSchwartzExerciseBoundary1D` to allow for multiple state variables (but still only observed at the time of potential early exercise) and arbitrary basis functions (rather than polynomials) in the regression (7.70), as well as discounting by an arbitrary numeraire (rather than deterministic discounting using an initial `TermStructure`). Thus the arguments of the constructor are

- an `Array<double,1>` containing the time points of potential exercise (the timeline)

- a (number of paths)×(number of time points)×(number of underlying financial variables) `Array<double,3>` of training paths

- a (number of paths)×(number of time points) `Array<double,2>` containing the realisations of numeraire values on each training path

- a `boost::function<double (double,const Array<double,1>&)>` functor to calculate the (undiscounted) payoff at time of exercise t as a function

---

[26] See `LongstaffSchwartz.hpp` and `LongstaffSchwartz.cpp` on the website.

[27] Note that `apply()` assumes (but does not check) that the timeline corresponding to the Monte Carlo path (represented by `Array<double,1>`) is the same as the timeline of times of potential early exercise for which the training paths were supplied to the constructor.

[28] This is an excerpt from `LS_test.cpp`, available in full on the website. Note that Longstaff/Schwartz reuse the "training paths" to obtain a Monte Carlo estimate for the price. As stated previously, if one wants to ensure that the Monte Carlo estimate converges to a lower bound for the price, one should not reuse paths in this way.

```
Array<double,2> paths(8,4);
paths = 1.00, 1.09, 1.08, 1.34,
        1.00, 1.16, 1.26, 1.54,
        1.00, 1.22, 1.07, 1.03,
        1.00, 0.93, 0.97, 0.92,
        1.00, 1.11, 1.56, 1.52,
        1.00, 0.76, 0.77, 0.90,
        1.00, 0.92, 0.84, 1.01,
        1.00, 0.88, 1.22, 1.34;
Array<double,1> T(4);
T = 0.0, 1.0, 2.0, 3.0;
double K = 1.10;
Payoff put(K,-1);
boost::function<double (double)> f;
f = boost::bind(std::mem_fun(&Payoff::operator()),&put,_1);
FlatTermStructure ts(0.06,0.0,10.0);
Array<double,3> genpaths(8,4,1);
genpaths(Range::all(),Range::all(),0) = paths;
Array<double,2> numeraire_values(8,4);
for (i=0;i<4;i++) numeraire_values(Range::all(),i) = 1.0/ts(T(i));
boost::function<double (double,const Array<double,1>&)>
  payoff = boost::bind(LSArrayAdapter,_1,_2,f,0);
vector<boost::function<double (double,const Array<double,1>&)> >
  basisfunctions;
int degree = 2;
Array<int,1> p(1);
for (i=0;i<=degree;i++) {
  p(0) = i;
  add_polynomial_basis_function(basisfunctions,p); }
LongstaffSchwartzExerciseBoundary
genls(T,genpaths,numeraire_values,payoff,basisfunctions);
Array<double,2> genpath(4,1);
Array<double,1> num(4);
for (i=0;i<8;i++) {
  genpath = genpaths(i,Range::all(),Range::all());
  num = numeraire_values(i,Range::all());
  MCestimate += genls.apply(genpath,num); }
std::cout << MCestimate.mean() << std::endl;
```

Listing 7.19: Longstaff/Schwartz toy example using LongstaffSchwartz-
ExerciseBoundary

of t and the underlying financial variables at time t (contained in the Array
finvars)

• a std::vector of

```
boost::function<double(double t,
                       const Array<double,1>& finvars)
```

functors representing the basis functions ψ_ℓ in (7.70), where in this case

$$\psi_{\ell j} := \psi_\ell(T_{\bar{k}(j)}, \{X_i(T_{\bar{k}(j)})\}_{i=1,\ldots,N}), \quad \ell = 1, \ldots, n$$

is a function of time and the underlying financial variables observed at times of potential early exercise $T_{\bar{k}(j)}$ *only*.

- an optional Boolean flag as in `LongstaffSchwartzExerciseBoundary1D`

Note that the basic structure of the regression algorithm in the constructor and the application of the early exercise strategy to Monte Carlo paths through the member function `apply()` remains the same,[29] where `apply()` now takes as arguments

- a (number of time points)×(number of underlying financial variables) `Array<double,2>` representing a Monte Carlo path

- a (number of time points) `Array<double,1>` of numeraire values on that path

Listing 7.19 reproduces the toy example from Longstaff and Schwartz (2001) using `LongstaffSchwartzExerciseBoundary`.[30]

The class `RegressionExerciseBoundary` implements the regression (7.70) in its full generality. Compared to `LongstaffSchwartzExerciseBoundary`, the only arguments of the constructor that change are

- the functor to calculate the (undiscounted) payoff at time of exercise t: This is now a

```
boost::function<double (const Array<double,1>& t_history,
                        const Array<double,2>& finvars)>,
```

where `finvars` is a (number of underlying financial variables)×(number of observation time points up to t) `Array` of values of the underlying financial variables on a given path, observed at the time points in the (number of observation time points up to t) `Array t_history`

- the `std::vector` of functors representing the basis functions $\psi_\ell(\cdot, \cdot)$ in (7.70), which are now of the type

```
boost::function<double (const Array<double,1>& t_history,
                        const Array<double,2>& finvars)>,
```

analogously to the payoff functor.

Similarly, the arguments of the member function `apply()` do not change. In `LS_Test.cpp` on the website, the toy example is also reproduced using

---

[29] See `LongstaffSchwartz.cpp` on the website.

[30] This is an excerpt from `LS_Test.cpp`, available in full on the website. The utility functions `add_polynomial_basis_function()` (to create the appropriate functor representing a polynomial) and `LSArrayAdapter()` (to convert a `boost::function<double (double)>` into the corresponding `boost::function<double (double,const Array<double,1>&)>`) are implemented in `LongstaffSchwartz.cpp`.

RegressionExerciseBoundary.[31] One should note that RegressionExerciseBoundary assumes that the history of observations of the underlying financial variables relevant for the evaluation of payoffs and the basis functions
approximating the continuation value is limited to time points of potential
early exercise. This assumption may be too restrictive when dealing with
Bermudan options (with infrequent early exercise opportunities) on path–
dependent payoffs.

Moving beyond the toy example, one can proceed to implement the actual Monte Carlo simulation of American options. All that changes is that the
"paths" of the toy example are replaced by a (typically much larger) set of randomly generated paths of the underlying price process and the numeraire. To
assist with this, we implement the class template MCTrainingPaths[32], which
creates a paths × timepoints × (number of state variables) Array of state variable realisations and the corresponding paths × timepoints Array of numeraire
values, as required by the constructors of LongstaffSchwartzExerciseBoundary and RegressionExerciseBoundary.

Listing 7.20 illustrates the construction of a LongstaffSchwartzExercise-
Boundary for an American put option. The template parameters of MCTrainingPaths are a class representing the underlying stochastic process (in this
case GeometricBrownianMotion) and a class for generating the random numbers to drive the stochastic process (in this case the pseudo–random

```
RandomArray<ranlib::NormalUnit<double>,double>
```

— an alternative would be the quasi–random SobolArrayNormal). The class
representing the stochastic process must implement

- a member function dimension(), which returns the number of state variables (i.e. the dimension) of the process,

- a member function set_timeline(const Array<double,1>& timeline),
  which sets the time points for which state variable (and numeraire) realisations will be generated,

- an

```
operator()(Array<double,2>& underlying_values,
           const random_variable& x,
           const TermStructure& ts)
```

   and an

---

[31] The utility functions add_polynomial_basis_function() (to create the appropriate functor representing a polynomial) and REBAdapter() (to convert a
boost::function<double(double)>   into   the   corresponding   boost::function
<double(const Array<double,1>&,const Array<double,2>&)>   are   implemented   in
LongstaffSchwartz.cpp.

[32] See MCAmerican.hpp on the website.

```
BlackScholesAsset stock(&vol,S);
std::vector<const BlackScholesAsset*> underlying;
underlying.push_back(&stock);
FlatTermStructure ts(r,0.0,mat+10.0);
exotics::StandardOption Mput(stock,T(0),mat,K,ts,-1);
GeometricBrownianMotion gbm(underlying);
gbm.set_timeline(T);
ranlib::NormalUnit<double> normalRNG;
RandomArray<ranlib::NormalUnit<double>,double>
  random_container(normalRNG,gbm.factors(),gbm.number_of_steps());
MCTrainingPaths<GeometricBrownianMotion,
              RandomArray<ranlib::NormalUnit<double>,double> >
  training_paths(gbm,T,train,random_container,ts,numeraire_index);
Payoff put(K,-1);
boost::function<double (double)> f;
f = boost::bind(std::mem_fun(&Payoff::operator()),&put,_1);
boost::function<double (double,const Array<double,1>&)> payoff =
  boost::bind(LSArrayAdapter,_1,_2,f,0);
std::vector<boost::function<double (double,
                          const Array<double,1>&)> >
  basisfunctions;
Array<int,1> p(1);
for (i=0;i<=degree;i++) {
  p(0) = i;
  add_polynomial_basis_function(basisfunctions,p); }
boost::function<double (double,double)> put_option;
put_option =
  boost::bind(&exotics::StandardOption::price,&Mput,_1,_2);
boost::function<double (double,const Array<double,1>&)>
put_option_basis_function =
   boost::bind(LSArrayAdapterT,_1,_2,put_option,0);
if (include_put)
  basisfunctions.push_back(put_option_basis_function);
LongstaffSchwartzExerciseBoundary
  boundary(T,training_paths.state_variables(),
         training_paths.numeraires(),payoff,basisfunctions);
```

Listing 7.20: Construction of a LongstaffSchwartzExerciseBoundary for an American put option

```
operator()(Array<double,2>& underlying_values,
           const random_variable& x,
           const TermStructure& ts,
           int numeraire_index)
```

which generate the realisations of the stochastic process at the dates in the required timeline using a draw x of the random variable (the type of which must be consistent with the chosen class for generating the random

```
LSExerciseStrategy<LongstaffSchwartzExerciseBoundary>
  exercise_strategy(boundary);
MCMapping<GeometricBrownianMotion,Array<double,2> >
  mc_mapping(exercise_strategy,gbm,ts,numeraire_index);
boost::function<double (Array<double,2>)> func =
  boost::bind(&MCMapping<GeometricBrownianMotion,
                         Array<double,2> >::mapping,
              &mc_mapping,_1);
MCGeneric<Array<double,2>,double,
          RandomArray<ranlib::NormalUnit<double>,double> >
  mc(func,random_container);
MCGatherer<double> mcgatherer;
size_t n = minpaths;
boost::math::normal normal;
double d = boost::math::quantile(normal,0.95);
cout << "Paths,MC value,95% CI lower bound,95% CI upper bound"
     << endl;
while (mcgatherer.number_of_simulations()<maxpaths) {
  mc.simulate(mcgatherer,n);
  cout << mcgatherer.number_of_simulations() << ','
       << mcgatherer.mean() << ','
       << mcgatherer.mean()-d*mcgatherer.stddev() << ','
       << mcgatherer.mean()+d*mcgatherer.stddev() << endl;
  n = mcgatherer.number_of_simulations(); }
```

Listing 7.21: Pricing an American put option using LongstaffSchwartzExerciseBoundary and the generic Monte Carlo algorithm implementation

numbers — typically, x will be an Array<double,2>). The first version of the operator() generates the realisations under the martingale measure associated with taking the savings account as the numeraire (corresponding to deterministic discounting in the present example), while the second version uses the martingale measure associated with a chosen numeraire — in the case of GeometricBrownianMotion, the variable numeraire_index allows us to choose one of the assets as the numeraire.

The constructor of MCTrainingPaths then takes as arguments

- an instance of the class representing the underlying stochastic process
- the set of time points for which realisations of the stochastic process are to be generated
- the number of paths to be generated
- an instance of the random variable generator
- the initial TermStructure
- the numeraire_index

The thus constructed instance of MCTrainingPaths is then used to instantiate a LongstaffSchwartzExerciseBoundary via

Paths	Polynomial only			Incl. European put		
	MC price	95% conf. bounds		MC price	95% conf. bounds	
10240	12.357	12.1589	12.5551	12.4288	12.2224	12.6352
20480	12.3821	12.2416	12.5225	12.4645	12.3181	12.6109
40960	12.472	12.3725	12.5715	12.538	12.4343	12.6417
81920	12.4689	12.3986	12.5392	12.5196	12.4464	12.5929
163840	12.4582	12.4086	12.5079	12.5054	12.4537	12.5571
327680	12.4808	12.4457	12.516	12.5312	12.4945	12.5678
655360	12.5039	12.4791	12.5288	12.5525	12.5266	12.5784
1310720	12.4976	12.4801	12.5152	12.5442	12.5259	12.5625

Table 7.5 *Monte Carlo estimates for the price of an American put using* LongstaffSchwartzExerciseBoundary*, polynomial basis functions up to degree 3, with and without the European put with the same maturity included as one of the basis functions. Further inputs: initial stock price 100, strike 102, continuously compounded interest rate 5%, time to maturity 1.5 years, volatility 30%, no dividends. Time line refinement: 10. Number of "training paths" to estimate exercise boundary: 100,000.*

```
LongstaffSchwartzExerciseBoundary
    boundary(T,training_paths.state_variables(),
            training_paths.numeraires(),
            payoff,basisfunctions);
```
(7.73)

In order to apply the generic Monte Carlo algorithm implementation, we need to create a class derived from `MCPayoff`, which calculates the payoff on each path using the member function `apply()` of the exercise boundary. This is implemented as the

```
template<class exercise_boundary>
class LSExerciseStrategy : public MCPayoff
```
(7.74)

Instantiating

```
LSExerciseStrategy<LongstaffSchwartzExerciseBoundary>
    exercise_strategy(boundary);
```
(7.75)

supplies the `MCPayoff` that allows us to instantiate first an `MCMapping` and then an `MCGeneric` object to execute the Monte Carlo simulation as before — see Listing 7.21.[33] Note that if the flag `include_put` evaluates to true in Listing 7.20, the price (at each time of potential early exercise) of the European put expiring at the same time as the American put is included as one of the basis functions. One would expect that this would improve the regression fit of the exercise boundary, since if one ignores the value of subsequent intermediate exercise, the price of the European put can serve as a proxy for

---

[33] The full working code containing Listings 7.20 and 7.21 is given in LSMC.cpp on the website. For further illustration, LSMCpath.cpp on the website replaces `LongstaffSchwartzExerciseBoundary` with the more general `RegressionExerciseBoundary`.

No dividends					
	Polynomial only			Incl. European option	
Paths	MC price	95% conf. bounds		MC price	95% conf. bounds	
102400	23.7961	23.6361	23.9561	23.785	23.6273	23.9426
204800	23.8334	23.7203	23.9465	23.843	23.7314	23.9546
409600	23.8937	23.8135	23.9739	23.8906	23.8115	23.9697
819200	23.8621	23.8054	23.9189	23.8585	23.8025	23.9144
1638400	23.8456	23.8054	23.8857	23.8395	23.7999	23.8791
3276800	23.829	23.8006	23.8574	23.8295	23.8015	23.8575
6553600	23.8445	23.8244	23.8645	23.8404	23.8206	23.8602
13107200	23.859	23.8448	23.8732	23.8596	23.8456	23.8736

With dividends					
	Polynomial only			Incl. European option	
Paths	MC price	95% conf. bounds		MC price	95% conf. bounds	
102400	17.8089	17.696	17.9218	17.9196	17.8091	18.0301
204800	17.818	17.7382	17.8977	17.9389	17.8607	18.017
409600	17.8567	17.8002	17.9132	17.9679	17.9125	18.0232
819200	17.8272	17.7872	17.8672	17.9509	17.9117	17.9901
1638400	17.8288	17.8005	17.8571	17.949	17.9213	17.9767
3276800	17.8182	17.7982	17.8382	17.9456	17.926	17.9652
6553600	17.8266	17.8125	17.8408	17.953	17.9392	17.9669
13107200	17.8348	17.8248	17.8448	17.9609	17.9511	17.9707

Table 7.6 *Monte Carlo estimates for the price of an American Margrabe option using* RegressionExerciseBoundary, *polynomial basis functions up to degree 2, with and without the European Margrabe option with the same maturity included as one of the basis functions. Further inputs: initial stock prices 100 and 105, strike factor 1.02 (i.e. strike = 1.02 · 100/105), continuously compounded interest rate 5%, time to maturity 2.5 years, volatilities 30% and 45%, instantaneous correlation between asset dynamics 0.5, continuously compounded dividend yields 7% and 2%. Time line refinement: 10. Number of "training paths" to estimate exercise boundary: 100,000.*

the "continuation value" of the American put. Numerical experiments bear this out: For a sufficiently large number of simulations, Table 7.5 shows the Monte Carlo price for the American put to be significantly higher when the European put is included as one of the basis functions.

In general it makes sense to include the price of the option without early exercise (when this is available in closed form) as one of the basis functions for the regression of the continuation value. LSMCMargrabe.cpp on the website implements this for the option to exchange one (risky) asset for another — all that changes relative to case of a plain American put is the payoff specification and a different set of basis functions, including the closed–form price of a European Margrabe option. Table 7.6 shows some illustrative numerical results. One should note that in the case of zero dividends on both assets, a result along the lines of Merton (1973b) holds, that it is never optimal to exercise the option early. The fact that in this case the lower bound on the option price obtained using Monte Carlo is less than the European option price reflects the sub-optimality of the exercise strategy obtained using the regression estimate of the exercise boundary. Setting the continuously compounded dividend yield of the asset obtained upon exercise to 7% and the dividend yield of the asset surrendered upon exercise to 2% results in a significant early exercise premium, and a significant improvement of the Monte Carlo price when using the European option price as one of the basis functions, as Table 7.6 shows.

7.8 Quasi-random Monte Carlo

Recall the formulation of the integration problem (7.7), which Monte Carlo methods approximate by

$$\int_{[0,1]^d} f(x)dx \approx \frac{1}{N} \sum_{i=1}^{N} f(u_i), \qquad u_i \in [0,1]^d \qquad (7.76)$$

For pseudo–random Monte Carlo,[34] the u_i represent independent draws from a uniform distribution on the unit hypercube, and we have seen that the pseudo–random Monte Carlo estimate converges with $\mathcal{O}(N^{-\frac{1}{2}})$. Quasi–random Monte Carlo seeks to improve on this convergence by a sequence of u_i that is deterministically chosen to fill the unit hypercube "more evenly" (in a sense to be made precise) than a random or pseudo–random sequence. This can result in an acceleration of convergence to nearly $\mathcal{O}(N^{-1})$ — herein lies the difference to the variance reduction techniques considered in Section 7.6, which only affected the implicit constant in $\mathcal{O}(N^{-\frac{1}{2}})$ convergence.

It is important to note that quasi–random Monte Carlo explicitly depends on the problem dimension. In pseudo–random Monte Carlo a single univariate random number generator was sufficient to generate all the u_i in (7.76), regardless of the dimension d, as all the components of the random vectors u_i are independent random variables, and the pseudo–random generators mimic independent random draws. In quasi–random Monte Carlo, on the other hand, two consecutive elements of the quasi–random sequence are not independent. Consequently, one must distinguish between the set of random variates making up a single random vector u_i required to produce a realisation of the simulation $f(u_i)$, and consecutive realisations of the random vector u_i drawn N times to calculate Monte Carlo estimate in (7.76): the dimension d of the

---

[34] The term "pseudo–random" as opposed to "random" acknowledges the fact that random number generators on a computer are still deterministic algorithms, but they produce sequences of numbers which are statistically indistinguishable from random sequences.

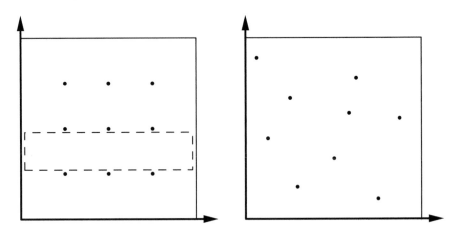

Figure 7.4 A grid of points in the unit   Figure 7.5 Better coverage of the unit
square: the dashed box is completely   square
empty of points

u_i is the required dimension of the quasi–random sequence, while N is the
length of the sequence.

For example, suppose one needs to generate 1,000 paths of geometric Brownian motion observed at 100 time steps, so N is 1,000 and d is 100. In pseudo–
random Monte Carlo, one can generate 100,000 (i.e. $N \times d$) independent univariate uniform random variates, which are transformed into independent, normally distributed Brownian motion increments and then assembled (in batches
of 100) to produce 1,000 paths. In quasi–random Monte Carlo, one has to generate a sequence of 1,000 quasi–random variates of dimension 100, i.e. one for
each path — using 100,000 quasi–random variates of dimension 1 would lead
to an incorrect result. As the rate of convergence of quasi–random methods
tends to deteriorate in higher dimensions, it is recommended to keep the dimension d as low as possible, for example by using the predictor/corrector
method whenever necessary and possible.

7.8.1 Generating quasi–random sequences

There is a large body of literature on quasi–random (also known as low–
discrepancy) sequences, and the underlying theory can only be roughly sketched
here. For a more in–depth treatment, Glasserman (2004) is a good starting
point, who cites Niederreiter (1992) for the rigorous theoretical background.

In just one dimension, generating a sequence that covers the unit interval
as uniformly as possible seems trivial — given N points, simply create a grid
of points $1/(N + 1)$ apart. Even in two dimensions, things are no longer that
simple. Consider Figure 7.4: the nine points are spaced in an even grid in the

unit square, but on each axis there is still empty space of length $1/4$ between points, and viewed in two dimensions there are large rectangles devoid of any points, like the one indicated in the figure. If the points are shifted as in Figure 7.5, these empty spaces are reduced, both in terms of two–dimensional rectangles and one–dimensional projections. In higher dimensions, a grid of points becomes even more problematic. Furthermore, in a grid the total number of points N has to be specified in advance and refining the grid by adding points becomes computationally very costly. Suppose we have points in a grid spaced $1/2^{k+1}$ apart in each of the d dimensions, i.e. a grid of 2^{kd} points. Refining the grid by adding points in the gaps along each dimension requires $2^{(k+1)d} - 2^{kd}$ additional points — this is exponential in d and grows very quickly in k. For appropriately chosen quasi–random sequences, on the other hand, much fewer additional points are needed to maintain uniformity on refinement.

Deviation from uniformity in quasi–random sequences is measured by discrepancy — thus the alternative nomenclature low–discrepancy sequences. These are defined relative to a collection $\mathcal{A}$ of (Lebesgue measurable) subsets of $[0,1]^d$, with the discrepancy of a sequence $\{u_1, \dots, u_N\}$, $u_i \in [0,1]^d$ given by

$$D(u_1, \dots, u_N; \mathcal{A}) = \sup_{A \in \mathcal{A}} \left| \frac{\#\{u_i \in A\}}{n} - \mathrm{vol}(A) \right| \qquad (7.77)$$

where $\#\{u_i \in A\}$ denotes the number of u_i contained in $A \in \mathcal{A}$ and $\mathrm{vol}(A)$ is the volume (measure) of A. As Glasserman (2004) notes, the discrepancy is the supremum over errors in integrating the indicator function of A using the points $u_1, \dots, u_N$. Ordinary (also called extreme) discrepancy is obtained when $\mathcal{A}$ is the collection of all rectangles in $[0,1]^d$ of the form

$$\prod_{j=1}^{d} [s_j, t_j), \qquad 0 \leq s_j < t_j \leq 1 \qquad (7.78)$$

When $\mathcal{A}$ is the collection of all rectangles

$$\prod_{j=1}^{d} [0, s_j), \qquad 0 < s_j \leq 1 \qquad (7.79)$$

one speaks of star discrepancy $D^*(u_1, \dots, u_N; \mathcal{A})$, which is a slightly weaker concept. Glasserman (2004) cites Niederreiter (1992) as stating that "it is widely believed" that in dimensions $d \geq 2$ there is a lower bound to star discrepancy in that any possible sequence $u_1, \dots, u_N$ will satisfy

$$D^*(u_1, \dots, u_N) \geq c_d \frac{(\ln N)^{d-1}}{N} \qquad (7.80)$$

and the first n elements of any sequence $u_1, u_2, \dots$ will satisfy

$$D^*(u_1, \dots, u_n) \geq c_d' \frac{(\ln n)^d}{n} \qquad (7.81)$$

for constants c_d, c_d' depending only on the dimension d. Since there are known sequences which achieve the second lower bound, only such sequences are typically termed "low–discrepancy." In the sense that discrepancy is the supremum over errors in integrating indicator functions, (7.81) represents an order of convergence. As Glasserman notes, the logarithmic term can be absorbed into any power of n, so this order becomes $\mathcal{O}(1/n^{1-\epsilon})$, for all $\epsilon > 0$.

There are a number of explicit constructions of such low–discrepancy sequences, but for finance applications there is a predominant focus on Sobol' (1967) sequences. The reason for this is found in the concept of (t, m, d)–nets and (t, d)–sequences, as defined by Niederreiter (1987), which are a useful way of describing the qualities of a low–discrepancy sequence. Consider subsets of $[0, 1]^d$ of the form

$$\prod_{i=1}^{d} \left[\frac{a_i}{b^{j_i}}, \frac{a_i + 1}{b^{j_i}} \right),\qquad\qquad (7.82)$$

for an integer $b \geq 2$ and with $j_i \in \{0, 1, \ldots\}$ and $a_i \in \{0, 1, \ldots, b^{j_i} - 1\}$.

This is called a b–ary box, also known as an elementary interval in base b. For integers $0 \leq t \leq m$, a (t, m, d)–net in base b is then defined as a set of b^m points in $[0, 1)^d$, such that exactly b^t points fall in each b–ary box of volume b^{t-m}. Therefore the fraction of points b^t/b^m that lie in each box equals the volume of that box. A (t, d)–sequence in base b is a sequence of points $u_1, u_2, \ldots$ in $[0, 1)^d$ such that for all $m > t$ each segment $\{u_i : jb^m < i \leq (j + 1)b^m\}$, $j = 0, 1, \ldots$, is a (t, m, d)–net in base b.

The parameter t describes the uniformity of the net/sequence in the sense that t gives size of the b–ary boxes composing $[0, 1)^d$, which will contain exactly the right number of points — there is no statement made about uniformity within those boxes. A smaller t means smaller boxes and thus, loosely speaking, uniformity at a finer partition of $[0, 1)^d$. Note that (7.82) implies that the rectangles making up the partition can be any shape, from squares to narrow strips. Furthermore, (t, m, d)–nets and (t, d)–sequences in base b describe the uniformity of sets of b^m points, thus larger b means that a larger number of points is required for the uniformity properties to hold. Consequently, one would prefer a smaller base b, all other things being equal.

Taking into consideration both t and b, as well as computational efficiency in constructing the low–discrepancy sequence, leads one to favour Sobol' (1967) sequences over those of, for example, Faure (1982). Sobol' sequences are (t, d)–sequences in base 2 for all d, where t depends on d. Although Faure sequences improve on the latter, i.e. $t = 0$ for all d, they require a base at least as large as d, which becomes unworkable in the large dimensions often required in finance applications. Additionally, a base 2 has the computational advantage that Sobol' sequences can be generated using fast, bit–level operations.

7.8.2 Sobol' sequence construction

Denote by $u_i \in [0,1)^d$ the i–th point in a d–dimensional Sobol' sequence. The j-th component u_{ij} of the d–dimensional vector u_i is given by

$$u_{ij} = i_1 v_{1,j} \oplus i_2, v_{2,j} \oplus \cdots \tag{7.83}$$

where $\oplus$ is the bit–by–bit exclusive–or operator and i_k is the k–th binary digit of i, i.e.

$$i = \sum_k i_k 2^{k-1}$$

is the binary representation of i. The v_{kj} are the so–called *direction numbers* given by

$$v_{kj} \quad := \quad \frac{m_{kj}}{2^k} \tag{7.84}$$

$$m_{kj} \quad := \quad 2a_{1,j}m_{k-1,j} \oplus 2^2 a_{2,j}m_{k-2,j} \oplus \cdots \oplus 2^{s_j-1}a_{s_j-1,j}m_{k-s_j+1,j}$$
$$\oplus \, 2^{s_j}m_{k-s_j,j} \oplus m_{k-s_j,j} \tag{7.85}$$

where the a_{hj} are either 0 or 1 and satisfy the condition that

$$x^{s_j} + a_{1,j}x^{s_j-1} + a_{2,j}x^{s_j-2} + \cdots + a_{s_j-1,j}x + 1 \tag{7.86}$$

is a primitive polynomial, i.e. is irreducible (cannot be factored) and the smallest p for which the polynomial divides $x^p + 1$ is $p = 2^{s_j} - 1$. In order to obtain Sobol' sequence, the initial values $m_{1,j}, m_{2,j}, \ldots, m_{s_j,j}$ may be chosen arbitrarily provided that each $m_{k,j}, 1 \le k \le s_j$, is odd and less than 2^k.

The quality of a Sobol' sequence depends a lot on the choice of direction numbers used to generate it. Joe and Kuo (2008) note that the error bounds for Sobol' sequences given by Sobol' (1967) indicate that one should use primitive polynomials in increasing order of their degrees, and leaving aside the special case for the first dimension of the Sobol' points where all the $m_{k,1}$ are 1, Joe and Kuo (2008) assign one primitive polynomial for each dimension starting from dimension 2. Thus they obtain a Sobol' sequence of dimension 1111 using all primitive polynomials up to degree 13, and dimension 21201 using all primitive polynomials up to degree 18. Subsequently, they seek to obtain "good" values for the first s_j values of m_{kj} required to initialise the recursion (7.85), in particular for Sobol' sequences in high dimensions. "Good" may be manifested by a small t (indicating higher uniformity) of a (t, d)–sequence, but in high dimensions the t of a Sobol' sequence cannot be determined practicably. Instead, Joe and Kuo (2008) proceed iteratively, where given the direction numbers for a Sobol' sequence in dimension $d - 1$, they select the numbers for dimension d to minimise the weighted maximum t–value of the two–dimensional projections of the first 2^m points in the sequence on the dimension (j, d), $1 \le j \le d$, with higher dimensions j weighted lower by an arbitrary factor 0.9999^{j-1} and m varying between 1 and 31. A further adjustment is made to the search criterion to account for the fact that b^{t-m} is the volume of the b–ary boxes in the definition of a (t, m, d)–net, and thus accounting for the tradeoff between small t (fewer point required per box) and large $m - t$ (finer partition involving smaller boxes). Since the number of possible sets of direction number grows exponentially in the dimension d, optimisation of the search criterion by a full search soon becomes infeasible, and Joe and Kuo (2008) deal with this by only considering a number of randomly generated sets of direction numbers when optimising based on the search criterion in dimensions higher than 19 (for $d \leq 19$, they conduct optimisation based on full enumeration of the sets of direction numbers).

Thus there are a lot of compromises made by Joe and Kuo in their search for "optimal" direction numbers, the most significant being the restriction of the search criterion to the t–values of the two–dimensional projections. However, these direction numbers appear to be the best currently available. With regard to their focus on two–dimensional projections, they note that "it is often the low–dimensional projections that are important in certain applications." In many (possibly most) problems in quantitative finance, it appears that the integrand depends mostly on a small number of variables, or can be well approximated by a sum of functions with each depending on only a small number of variables at at time. The optimisation of Joe and Kuo favours the initial dimensions (by proceeding iteratively from dimension $d - 1$ to d and by assigning a lower weight 0.9999^{j-1} to higher dimensions j in the search criterion), but nevertheless it appears that the quality of the Sobol' sequences generated with their direction numbers does not deteriorate substantially in higher dimensions, as opposed to Sobol' sequences that had previously been in widespread use. Thus the approach championed by authors such as Jäckel (2002), to use the leading dimensions for the "most important" components of the problem, for example by simulating diffusions by Brownian Bridge techniques, no longer leads to such a significant improvement in convergence of (high–dimensional) quasi-random Monte Carlo prices — to the point that the improvement may no longer justify the computational effort.[35]

7.8.3 Implementation of quasi–random Monte Carlo

The quasi–random sequence generator class `Sobol` in `QFQuasiRandom.hpp` is a wrapper for the C++ code implementing (7.83)–(7.85), modified and slightly optimised from the code provided by Joe and Kuo (2008) on their website.[36] The constructor

$$\text{Sobol(int dimension, size\_t maximum\_number\_of\_points,} \quad (7.87)$$
$$\text{const char* direction\_number\_file)}$$

takes as arguments the dimension d, the maximum number of points required in the sequence, and the name of a text file containing the initial direction number. The last argument defaults to `"new-joe-kuo-6.21201.dat"`, also provided by Joe and Kuo and containing direction numbers for up to 21201 dimensions. The member function

```
Array<double,1>& random()
```

---

[35] The author is grateful to Karl Gellert for pointing this out.
[36] See http://www.maths.unsw.edu.au/~fkuo/sobol/

```
/* set up timeline - N is the number of time steps,
   mat is the maturity of the option */
Array<double,1> T(N+1);
firstIndex idx;
double dt = mat/N;
T = idx*dt;
// set up first asset - volatility level is sgm on both factors
Array<double,1> sgm1(2);
sgm1 = sgm;
ConstVol vol(sgm1);
BlackScholesAsset stock(&vol,S,0.03);
// vector containing underlying assets
std::vector<const BlackScholesAsset*> underlying;
underlying.push_back(&stock);
// term structure of interest rates
FlatTermStructure ts(r,0.0,mat+10.0);
// set up second asset
Array<double,1> sgm2(2);
sgm2 = 0.1, 0.4;
ConstVol vol2(sgm2);
BlackScholesAsset stock2(&vol2,S,0.04);
underlying.push_back(&stock2);
/* closed form price via formula implemented as member function
   of BlackScholesAsset - K/S is strike factor */
double margrabe_price = stock.Margrabe(stock2,mat,K/S);
cout << "Closed form Margrabe option: " << margrabe_price;
```

Listing 7.22: Closed–form solution for a Margrabe option to benchmark quasi–random Monte Carlo simulation

returns the next point in the sequence, and void reset() sets the state of the sequence generator back to the beginning.

Each draw in a typical Monte Carlo simulation in quantitative finance involves a number (say n) of independent, possibly multidimensional (say j–dimensional) random variables. Thus the dimension of the required Sobol' sequence is $j \times n$ and it may be convenient to rearrange the $(j \times n)$–dimensional Sobol points into a $j \times n$ two–dimensional Array<double,2>. For "raw" Sobol' points this is done by the class SobolArray, with constructor

SobolArray(int rows,int cols,
           size_t maximum_number_of_points,   (7.88)
           const char* direction_number_file)

where the member function Array<double,2>& random() returns the next element in the sequence. Instances of the class SobolArrayNormal are constructed in the same way, with the difference being that random() returns a reference to an Array<double,2> in which all numbers have been transformed by applying the inverse of the cumulative distribution function of the standard normal distribution, thus after multiplication by the square root of the time

```
// set up MBinaries
Array<int,2> xS(1,1);
xS = 1;
Array<int,2> mindex(2,2);
mindex = 0, 1,
         N, N;
Array<double,1> malpha(2);
malpha = 1.0, 0.0;
Array<double,2> mA(1,2);
mA = 1.0, -1.0;
Array<double,1> ma(1);
ma = K/S;
MBinary M1(underlying,ts,T,mindex,malpha,xS,mA,ma);
double M1price = M1.price();
Array<double,1> malpha2(2);
malpha2 = 0.0, 1.0;
MBinary M2(underlying,ts,T,mindex,malpha2,xS,mA,ma);
cout << "MBinary price of Margrabe option: "
     << M1price-K/S*M2.price() << endl;
```

Listing 7.23: Setting up M–binaries for a Margrabe option to benchmark quasi–random Monte Carlo simulation

step length $\sqrt{\Delta t}$ this $(j \times n)$ `Array` can be used to generate n increments of a j–dimensional Brownian motion.

Listings 7.22–7.24 demonstrate how to use quasi–random Monte Carlo to price a Margrabe option (i.e. option to exchange one asset for another, see Theorem 4.7) in a two–factor Black/Scholes model with N time steps, comparing the result to the closed–form solution.[37] The multiple time steps are for demonstration purposes only, as in this case of a European option a single time step would be sufficient. Note that the quasi–random Monte Carlo setup is essentially the same as the pseudo–random setup,[38] the only difference being that an instance of

$$\text{RandomArray<ranlib::NormalUnit<double>,double>} \qquad (7.89)$$

is replaced by an instance of `SobolArrayNormal`, and on a quasi–random simulation the standard deviation of the Monte Carlo estimate (as supplied by `MCGatherer::stddev()`) does not provide a meaningful error estimate.

7.8.4 Randomised quasi–random Monte Carlo

It is difficult to obtain a meaningful error estimate through quasi–random methods alone, but a particularly simple way of getting around this problem

---

[37] The full working code for this example is given in the file QRMargrabeExampleReduced.cpp on the website.

[38] Both quasi–random and pseudo–random Monte Carlo for this option are implemented in the file QRMargrabeExample.cpp on the website.

```
//---- Margrabe option price by quasi-random Monte Carlo ----
// quasi-random number generator
SobolArrayNormal sobol(2,N,maxpaths);
// asset price process
GeometricBrownianMotion gbm(underlying);
// create MCPayoffList from MBinaries
boost::shared_ptr<MBinaryPayoff> M1payoff = M1.get_payoff();
boost::shared_ptr<MBinaryPayoff> M2payoff = M2.get_payoff();
MCPayoffList mcpayofflist;
mcpayofflist.push_back(M1payoff);
mcpayofflist.push_back(M2payoff,-K/S);
cout << "Strike coefficient: " << K/S << endl;
/* MCMapping to map random numbers to asset price realisations
   to discounted payoffs */
MCMapping<GeometricBrownianMotion,Array<double,2> >
  mc_mapping(mcpayofflist,gbm,ts,numeraire_index);
// mapping functor
boost::function<double (Array<double,2>)> func =
  boost::bind(&MCMapping<GeometricBrownianMotion,
              Array<double,2> >::mapping,&mc_mapping,_1);
// generic Monte Carlo algorithm object
MCGeneric<Array<double,2>,double,SobolArrayNormal>
  mc_QR(func,sobol);
MCGatherer<double> mcgathererQR;

/* run quasi-random Monte Carlo simulation and compare to
   closed form value */
unsigned long n = minpaths;
cout << "Minimum number of paths: " << minpaths
     << "\nMaximum number of paths: " << maxpaths << endl;
cout << "Margrabe option\nPaths,Closed form value,";
cout << "QR MC value\n";
double CFprice = margrabe.price();
while (n<=maxpaths) {
  cout << mcgathererQR.number_of_simulations() << ','
       << CFprice << ',';
  // simulate
  mc_QR.simulate(mcgathererQR,n);
  // output Monte Carlo result
  cout << mcgathererQR.mean() << ',' << endl;
  n = mcgathererQR.number_of_simulations(); }
```

Listing 7.24: Pricing a Margrabe option by quasi–random Monte Carlo simulation

```
template <class price_process,
         class random_variable,
         class QR_generator>
class RandomShiftQMCMapping {
private:
  MCMapping<price_process,random_variable> mc_mapping;
  QR_generator&                              QRgen;
public:
  /// Constructor.
  inline RandomShiftQMCMapping(QR_generator& xQRgen,
                               MCPayoff& xpayoff,
                               price_process& xprocess,
                               const TermStructure& xts,
                               int xnumeraire_index = -1)
    : mc_mapping(xpayoff,xprocess,xts,xnumeraire_index),QRgen(xQRgen) { };
  /** Choose the numeraire asset for the equivalent martingale measure
      under which the simulation is carried out. */
  inline bool set_numeraire_index(int xnumeraire_index)
    { return mc_mapping.set_numeraire_index(xnumeraire_index); };
  /** The function mapping a realisation of the shift to the (often
      multidimensional) quasi-random variable x to a discounted payoff. */
  double mapping(random_variable x);
  /** The function mapping a realisation of the shift to the (often
      multidimensional) quasi-random variable x to multiple discounted
      payoffs. */
  Array<double,1> mappingArray(random_variable x);
};
```

Listing 7.25: Class template for randomised quasi–random Monte Carlo simulation

is to randomise (in a pseudo–random sense) over a set of quasi–random simulation runs.[39] A simple way a quasi–random sequence can be randomised is by shifting each point in the sequence by the same random amount v, uniform on $[0,1]^d$. Coordinate values greater than one are "wrapped" using modulo–1 division, i.e. if $u_i^{(j)}$ is the j–th coordinate of the i–th element in the original sequence, its randomised counterpart $\hat{u}_i^{(j)}$ is given by

$$\hat{u}_i^{(j)} = (u_i^{(j)} + v^{(j)}) \mod 1 \tag{7.90}$$

where $v^{(j)}$ is the j–th coordinate of v. Thus we can generate arbitrarily many randomised versions of the original sequence by repeated draws of v.

From an implementation perspective, we now have quasi–random Monte Carlo (simulation using the sequence $\hat{u}$) nested inside pseudo–random Monte Carlo (repeated draws of v), and a meaningful standard deviation of the Monte Carlo estimate (which is now an average over the different quasi–random Monte Carlo estimates for each $\hat{u}$) can be obtained at the level of the outer loop.

---

[39] Glasserman (2004) also notes some results in the literature which indicate that randomising quasi–random methods may actually lead to better accuracy.

```
template <class price_process,
          class random_variable,
          class QR_generator> double
RandomShiftQMCMapping<price_process,random_variable,QR_generator>
  ::mapping(random_variable x)
{
  size_t i;
  double result = 0.0;
  QRgen.reset();
  for (i=0;i<QRgen.number_of_points();i++)
    result += mc_mapping.mapping(QRgen.random(x));
  return result/QRgen.number_of_points();
}
```

Listing 7.26: Mapping a given random shift to a quasi–random Monte Carlo estimate

No new code is required to implement the outer loop — this is simply a generic Monte Carlo algorithm, which can be instantiated by

$$\text{MCGeneric<Array<double,2>,double,}$$
$$\text{RandomArray<ranlib::Uniform<double>,double>>} \quad (7.91)$$
$$\text{randomQMC(QRfunc,unigen);}$$

where unigen is a RandomArray<ranlib::Uniform<double>,double> with member function random() returning an Array<double,2>& of the appropriate size — as before in SobolArray, the convention is to have a $(j \times n)$ Array when each draw of the Monte Carlo simulation requires n independent points of dimension j, e.g. for n time steps of a j–dimensional stochastic process. Running

$$\text{randomQMC.simulate(mcgathererQRran,32);} \quad (7.92)$$

on the MCGatherer mcgathererQRran then results in a reasonably meaningful standard deviation for the Monte Carlo estimator mcgathererQRran.mean() being returned by mcgathererQRran.stddev().

The inner, quasi–random simulation loop is contained in the functor

$$\text{boost::function<double (Array<double,2>)> QRfunc} \quad (7.93)$$

To instantiate QRfunc by

```
QRfunc =
boost::bind(&RandomShiftQMCMapping<GeometricBrownianMotion,
            Array<double,2>,SobolArrayNormal>::mapping,
            &QR_mapping,_1);
```

we need an instance QR_mapping of the template class

$$\text{RandomShiftQMCMapping<GeometricBrownianMotion,}$$
$$\text{Array<double,2>,SobolArrayNormal>} \quad (7.94)$$

The public interface of this class template mirrors MCMapping (see Listing 7.5), and it maps a realisation of the random shift v to a quasi–random estimate of a discounted expected payoff. The constructor of RandomShiftQMCMapping

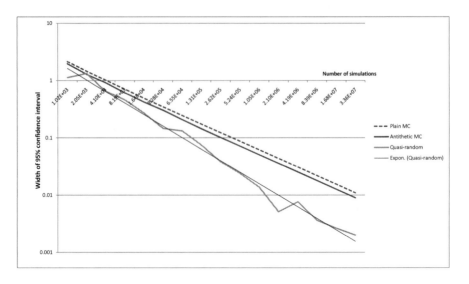

Figure 7.6 *Convergence of confidence intervals for Monte Carlo estimate of Margrabe option price, log/log scale, trendline added for quasi-random case.*

(given in Listing 7.25) creates the appropriate `MCMapping` from objects representing the payoff, the price process of the underlying(s), the term structure of interest rates and the index of the numeraire asset associated with the martingale measure under which the simulation is conducted. The member function `mapping()` (given in Listing 7.26) and analogously `mappingArray()` (for multiple payoffs) calculate the quasi–random estimate for a given shift of the quasi–random sequence generator, e.g. `SobolArrayNormal`, by overloading the member function `random()`, e.g.

$$\text{Array<double,2>\& SobolArrayNormal::} \atop \text{random(Array<double,2>\& shift)} \tag{7.95}$$

Listing 7.27 gives an example of pricing a Margrabe option in a two–factor Black/Scholes model using randomised quasi–random Monte Carlo — this is the "randomised" version of Listing 7.24. The full working code is given in `QRMargrabeExample.cpp` on the website. This example program compares pseudo-random Monte Carlo (with and without antithetic sampling) with randomised quasi–random Monte Carlo. Figure 7.6 plots the convergence of the confidence intervals in the three cases — note how quasi–random sampling improves the speed of convergence, while antithetic sampling reduces the level of the initial error, but converges at the same $\mathcal{O}(N^{-\frac{1}{2}})$ rate as "plain" Monte Carlo.

Randomisation allows us to use control variates in combination with quasi–random sampling — this is implemented in the file `QRCVTest.cpp` on the website, for the case of a call option on the arithmetic average. Note that using the template–based implementation of the generic Monte Carlo algo-

```
// Margrabe option price by randomised quasi-random Monte Carlo
// asset price process
GeometricBrownianMotion gbm(underlying);
// create MCPayoffList from MBinaries
boost::shared_ptr<MBinaryPayoff> M1payoff = M1.get_payoff();
boost::shared_ptr<MBinaryPayoff> M2payoff = M2.get_payoff();
MCPayoffList mcpayofflist;
mcpayofflist.push_back(M1payoff);
mcpayofflist.push_back(M2payoff,-K);
// For nested construction of randomised QMC using random shift
MCGatherer<double> mcgathererQRran;
ranlib::Uniform<double> ugen;
/* generator for array of uniform random numbers to shift each coordinate
   of the Sobol points */
RandomArray<ranlib::Uniform<double>,double> unigen(ugen,2,N);
unsigned long n = minpaths;
cout << "randomised QR MC value,95% CI lower bound,";
cout << "95% CI upper bound,Difference in standard errors,CI width\n";
boost::math::normal normal;
double d = boost::math::quantile(normal,0.95);
double CFprice = margrabe.price();
// loop to demostrate convergence
while (n<=maxpaths) {
  // Nested construction of randomised QMC using random shift
  SobolArrayNormal sobolr(2,N,n/32);
  /* MCMapping to map uniform random numbers to a shifted quasi-random
     simulation estimate of discounted payoffs */
  RandomShiftQMCMapping<GeometricBrownianMotion,Array<double,2>,
                       SobolArrayNormal>
    QR_mapping(sobolr,mcpayofflist,gbm,ts,numeraire_index);
  // mapping functor
  boost::function<double (Array<double,2>)> QRfunc =
    boost::bind(&RandomShiftQMCMapping<GeometricBrownianMotion,
                                       Array<double,2>,
                                       SobolArrayNormal>::mapping,
                &QR_mapping,_1);
  // generic Monte Carlo algorithm object
  MCGeneric<Array<double,2>,double,
            RandomArray<ranlib::Uniform<double>,double> >
    randomQMC(QRfunc,unigen);
  mcgathererQRran.reset();
  randomQMC.simulate(mcgathererQRran,32);
  cout << mcgathererQRran.mean() << ','
       << mcgathererQRran.mean()-d*mcgathererQRran.stddev() << ','
       << mcgathererQRran.mean()+d*mcgathererQRran.stddev() << ','
       << (mcgathererQRran.mean()-CFprice)/mcgathererQRran.stddev() << ','
       << 2.0*d*mcgathererQRran.stddev() << endl;
  n *= 2; }
```

Listing 7.27: Pricing a Margrabe option by randomised quasi–random Monte Carlo simulation

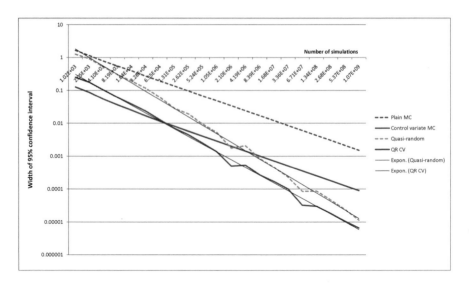

Figure 7.7 *Convergence of confidence intervals for Monte Carlo estimate of Margrabe option price, with and without control variates (CV), log/log scale, trendline added for quasi-random (QR) cases.*

rithm (i.e. `MCGeneric` and `MCGatherer`), no new code is required to combine control variates with quasi–random sampling. Figure 7.7 compares the variance reduction achieved using control variates, quasi–random sampling, and a combination of the two. The number of paths in each of the three cases is normalised to maintain the same number of evaluations of the payoff functions in each case. Initially, pseudo–random Monte Carlo with control variate is more accurate than quasi–random Monte Carlo, even if a control variate is used with the latter. This is because the control variate is only effective at the level of the randomisation, not at the level of the inner, quasi–random loop, and the randomisation only uses a limited number of draws (here 64). However, as quasi–random sampling provides better convergence, its accuracy eventually surpasses that of pseudo-random sampling with control variate. The contribution of the control variate to variance reduction, when combining quasi–random sampling and control variate, decreases as the number of quasi–random points is increased. One may reach the point where it is counterproductive, since the control variate requires an additional evaluation of the payoff function, so in our normalised comparison the number of quasi–random points without control variate is twice as large as in the case with control variate.

CHAPTER 8

The Heath/Jarrow/Morton model

Modelling a stochastic evolution of the term structure of interest rates in continuous time initially was based on an "economic equilibrium" approach, beginning with the seminal paper of Vasicek (1977). In that model, continuously compounded interest rates are normally distributed, and thus may become negative with positive probability. Subsequently, the model of Cox, Ingersoll and Ross (1985) guaranteed non-negative interest rates. In both cases, however, the initial term structure of interest rates was a model output, rather than input, making it difficult to fit the model to a term structure currently observed in the market and thus severely limiting the models' practical usefulness. Hull and White (1990) extended the Vasicek model to allow an exact fit to a current term structure by calibrating a time–dependent drift term. In all these papers, the quantity that was directly modelled was the continuously compounded short rate.[1] By postulating different dynamics for the continuously compounded short rate, various other so–called *short rate models* were proposed by a number of authors, including Brennan and Schwartz (1977), Black and Karasinski (1991) and Chan, Karolyi, Longstaff and Sanders (1992).

Going beyond modelling the short rate, Heath, Jarrow and Morton (1992) developed a *model framework* for the stochastic evolution of the entire term structure, specified in the form of instantaneous forward rates for all maturities.[2] Under mild technical conditions, it can be shown that all arbitrage–free diffusion–based term structure models are special cases of the general Heath/Jarrow/Morton (HJM) framework:[3] The models of Vasicek (1977) and Hull and White (1990) can be considered one–factor Gauss/Markov HJM models — "Gauss" because continuously compounded interest rates in these models follow a normal (Gaussian) distribution, and one–factor "Markov" because the term structure dynamics are Markovian in a single "factor." The model of Cox, Ingersoll and Ross (1985) is also a one–factor Markovian HJM model, with a non-central chi–square–distributed short rate. The less tractable short rate models of Brennan and Schwartz (1977), Black and Karasinski

---

[1] See Definition 2.4 in Chapter 2.

[2] I.e. a representation of the term structure of interest rates as per Definition 2.9 in Chapter 2.

[3] The HJM approach can be extended to jump–diffusion dynamics for default–free interest rates (see e.g. Chiarella and Nikitopoulos (2003)), as well as to the modelling of term structures of defaultable interest rates (see e.g. Schönbucher (2003) and Chiarella, Schlögl and Nikitopoulos (2007)).

(1991) and Chan, Karolyi, Longstaff and Sanders (1992) must also obey the
no–arbitrage conditions of the HJM framework — implicitly, as the forward
rate dynamics in these models are not known in closed form. In the present
chapter, after introducing the general HJM setup, we will consider closed–form
and Monte Carlo pricing of contingent claims in the most tractable subset,
the multifactor Gauss/Markov HJM term structure models (of which Vasicek
(1977) and Hull and White (1990) are single–factor special cases).

8.1 The model framework

Assumption 8.1 *For every fixed $T \leq T^*$, the dynamics of the instantaneous
forward rate $f(t, T)$ are given by*

$$f(t, T) = f(0, T) + \int_0^t \alpha(u, T)du + \int_0^t \sigma(u, T)dW(u) \quad \forall\, t \in [0, T] \quad (8.1)$$

Assumption 8.2 *For any maturity T, $\alpha(\cdot, T)$ and $\sigma(\cdot, T)$ follow adapted processes, such that*

$$\int_0^T |\alpha(u, T)|du + \int_0^T |\sigma(u, T)|^2 du < \infty \qquad \textbf{P}\text{--}a.s. \quad (8.2)$$

Note that $\sigma(u, T)$ and W are, in general, vector–valued and $\sigma(u, T)dW(u)$ is
to be interpreted as a sum product.

Since the zero coupon bond price $B(t, T)$ can be written as

$$
\begin{aligned}
B(t, T) &= \exp\left\{-\int_t^T f(t, u)du\right\} \\
&= \exp\left\{-\int_t^T f(0, s)ds - \int_t^T \int_0^t \alpha(u, s)duds \right. \\
&\qquad \left. + \int_t^T \int_0^t \sigma(u, s)dW(u)ds\right\} \\
&= \frac{B(0, T)}{B(0, t)} \exp\left\{-\int_t^T \int_0^t \alpha(u, s)duds + \int_t^T \int_0^t \sigma(u, s)dW(u)ds\right\},
\end{aligned}
\quad (8.3)
$$

the resulting *bond price dynamics* are

$$
\begin{aligned}
dB(t, T) = B(t, T) &\left(f(0, t)dt + \int_0^t \alpha(u, t)dudt - \int_0^t \sigma(u, t)dW(u)dt \right. \\
&\left. - \int_t^T \alpha(t, s)dsdt - \int_t^T \sigma(t, s)dsdW(t) + \frac{1}{2}\left(\int_t^T \sigma(t, s)ds\right)^2 dt\right)
\end{aligned}
$$

so we can write

$$dB(t, T) = B(t, T)(\mu(t, T)dt + b(t, T)dW(t)) \quad (8.4)$$

where μ and b are given by the following formulae

$$\mu(t,T) = f(t,t) - \alpha^*(t,T) + \frac{1}{2}|\sigma^*(t,T)|^2 , \quad b(t,T) = -\sigma^*(t,T) \qquad (8.5)$$

and for any $t \in [0,T]$ we have

$$\alpha^*(t,T) = \int_t^T \alpha(t,u)du , \quad \sigma^*(t,T) = \int_t^T \sigma(t,u)du$$

To derive the forward rate dynamics under the spot martingale measure, i.e. the equivalent martingale measure $\mathbf{P}_\beta$ associated with the numeraire

$$\beta(t) = \exp\left\{\int_0^t r(s)ds\right\} \qquad (8.6)$$

note that the bond price dynamics must be

$$dB(t,T) = B(t,T)(r(t)dt - \sigma^*(t,T)dW_\beta(t)) \qquad (8.7)$$

where W_β is a standard Brownian motion under $\mathbf{P}_\beta$, and so under this measure

$$\alpha^*(t,T) = \frac{1}{2}|\sigma^*(t,T)|^2 \qquad (8.8)$$

Differentiating with respect to T on either side of (8.8) yields $\alpha(t,T)$ in terms of $\sigma(t,\cdot)$, and so the forward rate satisfies

$$df(t,T) = \sigma(t,T)\sigma^*(t,T)dt + \sigma(t,T)dW_\beta(t) \qquad (8.9)$$

The continuously compounded short rate is given by

$$r(t) = f(0,t) + \int_0^t \sigma(u,t)\sigma^*(u,t)du + \int_0^t \sigma(u,t)dW_\beta(u) \qquad (8.10)$$

8.2 Gauss/Markov HJM

The most tractable — in the sense of ease of derivation of closed–form pricing formulas for interest rate derivatives — of the models within the HJM framework is one in which the volatility function $\sigma(t,T)$ is specified in such a way that $\sigma(t,T)$ is a deterministic function of process time t and maturity T, and the model is *Markovian*[4] with respect to a limited set of state variables, say just $r(t)$ in the one–factor case. To achieve the desired properties, set

$$b_i(t,T) = \frac{\nu_i(t)}{a_i}(e^{-a_i(T-t)} - 1) \qquad \forall\, t \le T \le T^* \qquad (8.11)$$

---

[4] A stochastic process X has the *Markov property* (i.e. is *Markovian*) with respect to a probability measure $\mathbf{Q}$ if for all $s \le t \le T^*$

$$E_Q[h(X(t))|\mathcal{F}_s] = E_Q[h(X(t))|X(s)]$$

for any (bounded Borel–measurable) function

$$h : \mathbf{R} \to \mathbf{R}$$

Note that, in general, this is a measure–dependent property.

for each component b_i of the bond volatility vector b, some constants a_i and deterministic functions $\nu_i(t)$.

This results in

$$
\begin{aligned}
\sigma_i(t,T) &= -\partial_2 b_i(t,T) \\
&= \nu_i(t)e^{-a_i(T-t)}
\end{aligned} \tag{8.12}
$$

for each component of the forward rate volatility function $\sigma(t,T)$.

In the one–factor case

$$
\begin{aligned}
f(t,T) =& f(0,T) + \int_0^t \nu(u)e^{-a(T-u)}\frac{\nu(u)}{a}(1-e^{-a(T-u)})du \\
& + \int_0^t \nu(u)e^{-a(T-u)}dW_\beta(u)
\end{aligned} \tag{8.13}
$$

$$
\begin{aligned}
r(t) = f(t,t) =& f(0,t) + \int_0^t \nu(u)e^{-a(t-u)}\frac{\nu(u)}{a}(1-e^{-a(t-u)})du \\
& + \int_0^t \nu(u)e^{-a(t-u)}dW_\beta(u)
\end{aligned} \tag{8.14}
$$

So we can represent the forward rates as functions of the short rate

$$
\begin{aligned}
f(t,T) =&\ e^{-a(T-t)}(r(t) - f(0,t)) + f(0,T) \\
& + \int_0^t \frac{\nu(u)^2}{a}\left(e^{-a(T-u)}(e^{-a(t-u)} - e^{-a(T-u)})\right)du \quad (8.15)
\end{aligned}
$$

In the multifactor case, W_β is a vector Brownian motion with d independent components $W_\beta^{(i)}$. Then

$$
\begin{aligned}
f(t,T) =&\ f(0,T) + \sum_{i=1}^d \int_0^t \frac{\nu_i(u)^2}{a_i}e^{-a_i(T-u)}(1-e^{-a_i(T-u)})du \\
& + \sum_{i=1}^d \int_0^t \nu_i(u)e^{-a_i(T-u)}dW_\beta^{(i)}(u)
\end{aligned} \tag{8.16}
$$

Thus we need d state variables, say $f(0,T_i)$ for $i = 1,\ldots,d$ and for any $t \le \min_i T_i$ and $T \ge t$ we can write $f(t,T)$ as a linear function of the $f(t,T_i)$. Alternatively, we can write $f(t,T)$ as a linear combination of d independent, normally distributed state variables $z_i(t)$:

$$
\begin{aligned}
f(t,T) =&\ f(0,T) + \sum_{i=1}^d \int_0^t \frac{\nu_i(u)^2}{a_i}e^{-a_i(T-u)}(1-e^{-a_i(T-u)})du \\
& + \sum_{i=1}^d e^{-a_i T}z_i(t)
\end{aligned} \tag{8.17}
$$

where

$$z_i(t) \quad = \quad \int_0^t \nu_i(u)e^{a_i u}dW_\beta^{(i)}(u) \qquad (8.18)$$

$$\Rightarrow \quad z_i(t) \quad \overset{\mathbf{P}_\beta}{\sim} \quad \mathcal{N}\left(0, \int_0^t \nu_i(u)^2 e^{2a_i u}du\right)$$

8.2.1 Zero coupon bond prices — one-factor case

By (8.15), we have in the one-factor case

$$f(t,u) = \mathcal{A}'(t,u) + \mathcal{B}'(t,u)r(t) \qquad (8.19)$$

with

$$\mathcal{A}'(t,u) \quad = \quad f(0,u) - e^{-a(u-t)}f(0,t) \qquad (8.20)$$

$$+ \int_0^t \frac{\nu(s)^2}{a}\left(e^{-a(u-s)}(e^{-a(t-s)} - e^{-a(u-s)})\right)ds$$

$$\mathcal{B}'(t,u) \quad = \quad e^{-a(u-t)} \qquad (8.21)$$

By (8.3) and (8.19), the zero coupon bond price is thus an *exponential–affine function* of the short rate:

$$B(t,T) = \mathcal{A}(t,T)e^{-\mathcal{B}(t,T)r(t)}$$

with

$$\mathcal{A}(t,T) \quad = \quad \exp\left\{-\int_t^T \mathcal{A}'(t,u)du\right\} \qquad (8.22)$$

$$\mathcal{B}(t,T) \quad = \quad \int_t^T \mathcal{B}'(t,u)du \qquad (8.23)$$

8.2.2 Zero coupon bond prices — multifactor case

Analogously we can define for the multifactor case

$$\mathcal{A}(t,T) \quad = \quad \exp\left\{-\int_t^T f(0,u)du\right\}$$

$$\cdot \exp\left\{-\int_t^T \sum_{i=1}^d \int_0^t \frac{\nu_i(s)^2}{a_i} e^{-a_i(u-s)}(1 - e^{-a_i(u-s)})dsdu\right\} \qquad (8.24)$$

$$= \quad \frac{B(0,T)}{B(0,t)}\exp\left\{-\int_t^T \sum_{i=1}^d \int_0^t \frac{\nu_i(s)^2}{a_i} e^{-a_i(u-s)}(1 - e^{-a_i(u-s)})dsdu\right\}$$

$$\mathcal{B}_i(t,T) \quad = \quad \int_t^T e^{-a_i u}du = \frac{1}{a_i}(e^{-a_i t} - e^{-a_i T}) \qquad (8.25)$$

I.e. $\mathcal{B}(t,T)$ is a d–dimensional vector with components $\mathcal{B}_i(t,T)$ and $\mathcal{A}(t,T)$ is a scalar. To make (8.24) more explicit, one would need to make assumptions about $\nu_i(s)$, e.g. stepwise constant.

We now have

$$B(t,T) = \mathcal{A}(t,T)e^{-\mathcal{B}(t,T)z(t)} \tag{8.26}$$

where $z(t)$ is a d–dimensional random vector with

$$z_i(t) \overset{\mathbf{P}_\beta}{\sim} \mathcal{N}\left(0, \int_0^t \nu_i(u)^2 e^{2a_i u}du\right) \tag{8.27}$$

8.3 Option pricing in the Gaussian HJM framework

Note that the only assumption made in this section (except where explicitly stated otherwise) is that the forward rate volatility function $\sigma(t,T)$ is deterministic. In particular, the dynamics need not be Markovian in any particular set of factors.

8.3.1 Pricing a European call option on a zero coupon bond

The option payoff is

$$\max(0, B(T_0, T_1) - K) \tag{8.28}$$

where T_0 is option expiry and T_1 is the maturity of the underlying bond. The option price at time 0 is given by

$$
\begin{aligned}
C \;=\; & E_\beta\left[\exp\left\{-\int_0^{T_0} r(s)ds\right\} B(T_0,T_1)\mathbb{I}_{\{B(T_0,T_1)>K\}}\right] \\
& -E_\beta\left[\exp\left\{-\int_0^{T_0} r(s)ds\right\} K\mathbb{I}_{\{B(T_0,T_1)>K\}}\right]
\end{aligned} \tag{8.29}
$$

We evaluate each of the expectations using a measure change. For the second expectation,

$$
\begin{aligned}
& E_\beta\left[\exp\left\{-\int_0^{T_0} r(s)ds\right\} K\mathbb{I}_{\{B(T_0,T_1)>K\}}\right] \\
=\; & E_{T_0}\left[\frac{1}{\beta(T_0)}K\mathbb{I}_{\{B(T_0,T_1)>K\}}\frac{\beta(T_0)B(0,T_0)}{\beta(0)B(T_0,T_0)}\right] \\
=\; & B(0,T_0)K\mathbf{P}_{T_0}\{B(T_0,T_1) > K\}
\end{aligned} \tag{8.30}
$$

For the first expectation,

$$E_\beta \left[\exp\left\{ -\int_0^{T_0} r(s)ds \right\} B(T_0, T_1)\mathbb{I}_{\{B(T_0,T_1)>K\}} \right]$$

$$= B(0, T_0)E_{T_0}\left[B(T_0, T_1)\mathbb{I}_{\{B(T_0,T_1)>K\}} \right]$$

$$= B(0, T_0)E_{T_1}\left[B(T_0, T_1)\mathbb{I}_{\{B(T_0,T_1)>K\}} \frac{B(0,T_1)B(T_0,T_0)}{B(T_0,T_1)B(0,T_0)} \right]$$

$$= B(0, T_1)\mathbf{P}_{T_1}\{B(T_0, T_1) > K\} \qquad (8.31)$$

Thus we need, in general, the dynamics of $B(t, T_i)$ under $\mathbf{P}_{T_j}$, where $\mathbf{P}_{T_j}$ denotes the time T_j forward measure, i.e. the equivalent martingale measure associated with taking $B(t, T_j)$ as the numeraire.

For the measure change from $\mathbf{P}_\beta$

$$\left. \frac{d\mathbf{P}_{T_j}}{d\mathbf{P}_\beta} \right|_{\mathcal{F}_t} = \frac{B(t, T_j)\beta(0)}{B(0, T_j)\beta(t)} \qquad \mathbf{P}_\beta\text{-a.s.} \qquad (8.32)$$

$\frac{B(t, T_j)}{\beta(t)}$ is a martingale under $\mathbf{P}_\beta$, so

$$d\frac{B(t, T_j)}{\beta(t)} = \frac{B(t, T_j)}{\beta(t)}(-\sigma^*(t, T_j)dW_\beta(t)) \qquad (8.33)$$

$$\Longleftrightarrow \quad \frac{B(t, T_j)}{\beta(t)} = \frac{B(0, T_j)}{\beta(0)} \exp\left\{ -\int_0^t \sigma^*(s, T_j)dW_\beta(s) \right. \qquad (8.34)$$

$$\left. - \frac{1}{2}\int_0^t |\sigma^*(s, T_j)|^2 ds \right\}$$

Thus by Girsanov's Theorem

$$dW_{T_j}(t) = dW_\beta(t) + \sigma^*(t, T_j)dt \qquad (8.35)$$

and this implies

$$dW_{T_k}(t) = dW_\beta(t) + \sigma^*(t, T_k)dt$$
$$= dW_{T_j}(t) + (\sigma^*(t, T_k) - \sigma^*(t, T_j))dt \qquad (8.36)$$

Note that $\frac{B(t,T_1)}{B(t,T_0)}$ must be a martingale under $\mathbf{P}_{T_0}$. Thus

$$\frac{B(T_0, T_1)}{B(T_0, T_0)} = \frac{B(0, T_1)}{B(0, T_0)} \exp\left\{ -\int_0^{T_0} (\sigma^*(u, T_1) - \sigma^*(u, T_0))dW_{T_0}(u) \right.$$

$$\left. - \frac{1}{2}\int_0^{T_0} (\sigma^*(u, T_1) - \sigma^*(u, T_0))^2 du \right\} \qquad (8.37)$$

Under $\mathbf{P}_{T_1}$ we have

$$\frac{B(T_0,T_1)}{B(T_0,T_0)} = \frac{B(0,T_1)}{B(0,T_0)} \exp\left\{-\int_0^{T_0}(\sigma^*(u,T_1)-\sigma^*(u,T_0))dW_{T_1}(u)\right.$$

$$\left. + \frac{1}{2}\int_0^{T_0}(\sigma^*(u,T_1)-\sigma^*(u,T_0))^2du\right\} \quad (8.38)$$

Thus

$$\ln B(T_0,T_1) \stackrel{\mathbf{P}_{T_0}}{\sim} \mathcal{N}\left(\ln\frac{B(0,T_1)}{B(0,T_0)} - \frac{1}{2}\int_0^{T_0}(\sigma^*(u,T_1)-\sigma^*(u,T_0))^2du,\right.$$

$$\left.\int_0^{T_0}(\sigma^*(u,T_1)-\sigma^*(u,T_0))^2du\right) \quad (8.39)$$

$$\Rightarrow \mathbf{P}_{T_0}\{B(T_0,T_1) > K\} = N\left(\frac{\ln\frac{B(0,T_1)}{B(0,T_0)K} - \frac{1}{2}\int_0^{T_0}(\sigma^*(u,T_1)-\sigma^*(u,T_0))^2du}{\sqrt{\int_0^{T_0}(\sigma^*(u,T_1)-\sigma^*(u,T_0))^2du}}\right)$$

and similarly

$$\mathbf{P}_{T_1}\{B(T_0,T_1) > K\} = N\left(\frac{\ln\frac{B(0,T_1)}{B(0,T_0)K} + \frac{1}{2}\int_0^{T_0}(\sigma^*(u,T_1)-\sigma^*(u,T_0))^2du}{\sqrt{\int_0^{T_0}(\sigma^*(u,T_1)-\sigma^*(u,T_0))^2du}}\right)$$

This yields the Black/Scholes–type formula

$$C = B(0,T_1)N(h_1) - B(0,T_0)KN(h_2) \quad (8.40)$$

$$h_{1,2} = \frac{\ln\frac{B(0,T_1)}{B(0,T_0)K} \pm \frac{1}{2}\int_0^{T_0}(\sigma^*(u,T_1)-\sigma^*(u,T_0))^2du}{\sqrt{\int_0^{T_0}(\sigma^*(u,T_1)-\sigma^*(u,T_0))^2du}}$$

8.3.2 Options on coupon bonds

Given a time line of tenor dates $\mathbb{T} = \{T_0,T_1,\ldots,T_N\}$, a coupon bond with coupon k pays $c_i = (T_i - T_{i-1})k$ at times T_i, $1 \le i < N$, and $c_N = 1 + (T_N - T_{N-1})k$ at T_N. Its value at time T_0 is given by

$$V(T_0,\mathbb{T},k) = \sum_{i=1}^{N} c_i B(T_0,T_i) \quad (8.41)$$

Consider a call option on a coupon bond, i.e. the payoff at expiry is

$$\max(0, V(T_0,\mathbb{T},k) - K) = \max\left(0, \sum_{i=1}^{N} c_i B(T_0,T_i) - K\right) \quad (8.42)$$

In the general multifactor case, this is difficult to price, since the distribution of $V(T_0,\mathbb{T},k)$ is not explicitly known under any measure. Restricting ourselves to the *Markovian one–factor case*, however, we can follow Jamshidian (1989) and solve

$$\sum_{i=1}^{N} c_i B(T_0, T_i) = K \tag{8.43}$$

$$\Longleftrightarrow \quad \sum_{i=1}^{N} c_i \mathcal{A}(T_0, T_i) e^{-\mathcal{B}(T_0, T_i) r(T_0)} = K \tag{8.44}$$

for $r(T_0)$. Call this solution $r^*(T_0)$ and write

$$B^*(T_0, T_i) = \mathcal{A}(T_0, T_i) e^{-\mathcal{B}(T_0, T_i) r^*(T_0)} \tag{8.45}$$

Then $\mathbb{I}_{\{V(T_0, T_N, k) > K\}} = \mathbb{I}_{\{r(T_0) < r^*(T_0)\}} = \mathbb{I}_{\{B(T_0, T_i) > B^*(T_0, T_i)\}}$ and

$$B(0, T_0) E_{T_0} \left[\left(\sum_{i=1}^{N} c_i B(T_0, T_i) - K \right) \mathbb{I}_{\{r(T_0) < r^*(T_0)\}} \right] \tag{8.46}$$

$$= \quad B(0, T_0) E_{T_0} \left[\left(\sum_{i=1}^{N} c_i B(T_0, T_i) - \left(\sum_{i=1}^{N} c_i B^*(T_0, T_i) \right) \right) \mathbb{I}_{\{r(T_0) < r^*(T_0)\}} \right]$$

$$= \quad \sum_{i=1}^{N} c_i B(0, T_0) E_{T_0} \left[(B(T_0, T_i) - B^*(T_0, T_i)) \mathbb{I}_{\{B(T_0, T_i) > B^*(T_0, T_i)\}} \right]$$

since zero coupon bond prices are monotonic in the short rate. Therefore we can price the option on a coupon bond as a portfolio of zero coupon bond options.

In the two–factor case, Brigo and Mercurio (2001) show how the problem can be reduced to a one–dimensional integral, which must be solved numerically (for example by Gauss/Hermite quadrature). The call option price is given by

$$B(0, T_0) E_{T_0} \left[\left[\sum_{i=1}^{N} c_i B(T_0, T_i) - K \right]^+ \right]$$

$$= \quad B(0, T_0) E_{T_0} \left[\left[\sum_{i=1}^{N} c_i \mathcal{A}(T_0, T_i) e^{-\mathcal{B}_1(T_0, T_i) z_1(T_0) - \mathcal{B}_2(T_0, T_i) z_2(T_0)} - K \right]^+ \right]$$

$$= \quad B(0, T_0) E_{T_0} \left[E_{T_0} \left[\left(\sum_{i=1}^{N} c_i \mathcal{A}(T_0, T_i) e^{-\mathcal{B}_1(T_0, T_i) z_1(T_0) - \mathcal{B}_2(T_0, T_i) z_2(T_0)} \right. \right. \right.$$

$$\left. \left. \left. - K \right) \mathbb{I}_{\{z_2(T_0) \le z_2^*(z_1(T_0))\}} \middle| z_1(T_0) \right] \right] \tag{8.47}$$

where $z_2^*(z_1(T_0))$ is the z_2^* which solves

$$\sum_{i=1}^{N} c_i \mathcal{A}(T_0, T_i) e^{-\mathcal{B}_1(T_0, T_i) z_1(T_0) - \mathcal{B}_2(T_0, T_i) z_2^*} = K \tag{8.48}$$

The inner expectation of (8.47) can be evaluated as

$$E_{T_0}\left[\sum_{i=1}^{N} c_i\mathcal{A}(T_0,T_i)e^{-\mathcal{B}_1(T_0,T_i)z_1(T_0)-\mathcal{B}_2(T_0,T_i)z_2(T_0)}\mathbb{I}_{\{z_2(T_0)\leq z_2^*(z_1(T_0))\}}\middle|z_1(T_0)\right]$$
$$-KP_{T_0}\{z_2(T_0)\leq z_2^*(z_1(T_0))|z_1(T_0)\}\quad(8.49)$$

Thus we need, in general, the distribution of $z_i(t)$ under some $\mathbf{P}_{T_j}$. Substituting (8.35), (8.11) and (8.5) into (8.18), we have

$$z_i(t)=\int_0^t \nu_i(u)e^{a_iu}dW_{T_j}^{(i)}(u)+\int_0^t \frac{\nu_i^2(u)}{a_i}e^{a_iu}\left(e^{-a_i(T_j-u)}-1\right)du,\quad(8.50)$$

therefore

$$P_{T_0}\{z_2(T_0)\leq z_2^*(z_1(T_0))|z_1(T_0)\}=$$
$$\Phi\left(\frac{z_2^*(z_1(T_0))-\int_0^{T_0}\frac{\nu_i^2(u)}{a_i}e^{a_iu}\left(e^{-a_i(T_0-u)}-1\right)du}{\sqrt{\int_0^{T_0}\nu_i^2(u)e^{2a_iu}du}}\right)\quad(8.51)$$

Applying the *Gaussian Shift Theorem*, i.e. Equation (4.123) of Chapter 4, to the first part of (8.49),

$$E_{T_0}\left[c_i\mathcal{A}(T_0,T_i)e^{-\mathcal{B}_1(T_0,T_i)z_1(T_0)-\mathcal{B}_2(T_0,T_i)z_2(T_0)}\mathbb{I}_{\{z_2(T_0)\leq z_2^*(z_1(T_0))\}}\middle|z_1(T_0)\right]$$
$$=\int_{-\infty}^{z_2^*(z_1(T_0))}c_i\mathcal{A}(T_0,T_i)e^{-\mathcal{B}_1(T_0,T_i)z_1(T_0)-\mathcal{B}_2(T_0,T_i)z_2(T_0)}$$
$$\frac{1}{\sqrt{2\pi\mathrm{Var}_{T_0}[z_2(T_0)]}}\exp\left\{-\frac{(z_2(T_0)-E_{T_0}[z_2(T_0)])^2}{2\mathrm{Var}_{T_0}[z_2(T_0)]}\right\}dz_2(T_0)$$

$$=c_i\mathcal{A}(T_0,T_i)e^{-\mathcal{B}_1(T_0,T_i)z_1(T_0)}$$
$$\exp\left\{-E_{T_0}[z_2(T_0)]\mathcal{B}_2(T_0,T_i)+\frac{1}{2}\mathcal{B}_2^2(T_0,T_i)\mathrm{Var}_{T_0}[z_2(T_0)]\right\}$$
$$\cdot\Phi\left(\frac{z_2^*(z_1(T_0))-E_{T_0}[z_2(T_0)]+\mathcal{B}_2(T_0,T_i)\mathrm{Var}_{T_0}[z_2(T_0)]}{\sqrt{\mathrm{Var}_{T_0}[z_2(T_0)]}}\right)\quad(8.52)$$

where

$$E_{T_0}[z_i(T_0)]=\int_0^{T_0}\frac{\nu_i^2(u)}{a_i}e^{a_iu}\left(e^{-a_i(T_0-u)}-1\right)du\quad(8.53)$$

$$\mathrm{Var}_{T_0}[z_i(T_0)]=\int_0^{T_0}\nu_i^2(u)e^{2a_iu}du\quad(8.54)$$

Combining (8.47), (8.51) and (8.52), the two–factor coupon bond option formula becomes

$$B(0,T_0)E_{T_0}\left[\left[\sum_{i=1}^{N}c_iB(T_0,T_i)-K\right]^+\right]$$

$$= B(0,T_0)E_{T_0}\left[\sum_{i=1}^{N}c_i\mathcal{A}(T_0,T_i)e^{-\mathcal{B}_1(T_0,T_i)z_1(T_0)}\right.$$

$$\exp\left\{-E_{T_0}[z_2(T_0)]\mathcal{B}_2(T_0,T_i)+\frac{1}{2}\mathcal{B}_2^2(T_0,T_i)\mathrm{Var}_{T_0}[z_2(T_0)]\right\}$$

$$\Phi\left(\frac{z_2^*(z_1(T_0))-E_{T_0}[z_2(T_0)]+\mathcal{B}_2(T_0,T_i)\mathrm{Var}_{T_0}[z_2(T_0)]}{\sqrt{\mathrm{Var}_{T_0}[z_2(T_0)]}}\right)$$

$$\left.-K\Phi\left(\frac{z_2^*(z_1(T_0))-E_{T_0}[z_2(T_0)]}{\sqrt{\mathrm{Var}_{T_0}[z_2(T_0)]}}\right)\right]$$

which can be written as the one–dimensional integral

$$= B(0,T_0)\int_{-\infty}^{+\infty}\left(\sum_{i=1}^{N}c_i\mathcal{A}(T_0,T_i)e^{-\mathcal{B}_1(T_0,T_i)z_1(T_0)}\right. \tag{8.55}$$

$$\exp\left\{-E_{T_0}[z_2(T_0)]\mathcal{B}_2(T_0,T_i)+\frac{1}{2}\mathcal{B}_2^2(T_0,T_i)\mathrm{Var}_{T_0}[z_2(T_0)]\right\}$$

$$\Phi\left(\frac{z_2^*(z_1(T_0))-E_{T_0}[z_2(T_0)]+\mathcal{B}_2(T_0,T_i)\mathrm{Var}_{T_0}[z_2(T_0)]}{\sqrt{\mathrm{Var}_{T_0}[z_2(T_0)]}}\right)$$

$$\left.-K\Phi\left(\frac{z_2^*(z_1(T_0))-E_{T_0}[z_2(T_0)]}{\sqrt{\mathrm{Var}_{T_0}[z_2(T_0)]}}\right)\right)$$

$$\frac{1}{\sqrt{2\pi\mathrm{Var}_{T_0}[z_1(T_0)]}}\exp\left\{-\frac{(z_1(T_0)-E_{T_0}[z_1(T_0)])^2}{2\mathrm{Var}_{T_0}[z_1(T_0)]}\right\}dz_1(T_0)$$

8.3.3 Caps, floors and swaptions

Let the notional amount be 1. Then a *cap* with payment dates T_i, $i \in \{2,\ldots,n\}$, and *cap level* κ pays

$$(T_i-T_{i-1})\cdot\max(0,L(T_{i-1},T_i)-\kappa) \quad \text{in each } T_i \tag{8.56}$$

and the corresponding floor pays

$$(T_i-T_{i-1})\cdot\max(0,\kappa-L(T_{i-1},T_i)) \tag{8.57}$$

where $L(T_{i-1}, T_i)$ is the simply compounded rate (e.g. LIBOR) given in terms of zero coupon bond prices by

$$L(t, T_i) = \frac{1}{T_i - T_{i-1}} \left(\frac{B(t, T_{i-1})}{B(t, T_i)} - 1 \right) \tag{8.58}$$

The component caplets of a cap can be represented as a portfolio of put options on zero coupon bonds. Note that the payoff in T_i is already known in T_{i-1}. Its value in T_{i-1} is

$$B(T_{i-1}, T_i) \cdot (T_i - T_{i-1}) \cdot \max(0, L(T_{i-1}, T_i) - \kappa) \tag{8.59}$$

$$= \frac{1}{1 + (T_i - T_{i-1})L(T_{i-1}, T_i)} (T_i - T_{i-1}) \cdot \max(0, L(T_{i-1}, T_i) - \kappa)$$

$$= \max \left(0, \frac{1 + (T_i - T_{i-1})L(T_{i-1}, T_i) - (1 + (T_i - T_{i-1})\kappa)}{1 + (T_i - T_{i-1})L(T_{i-1}, T_i)} \right)$$

$$= (1 + (T_i - T_{i-1})\kappa) \max \left(0, \frac{1}{1 + (T_i - T_{i-1})\kappa} - B(T_{i-1}, T_i) \right)$$

Thus a caplet equals $(1 + (T_i - T_{i-1})\kappa)$ (European) put options on the zero coupon bond $B(T_{i-1}, T_i)$ with an exercise price of $(1 + (T_i - T_{i-1})\kappa)^{-1}$. The option expires in T_{i-1}. Analogously, a floor can be decomposed into call options on zero coupon bonds.

A European *swaption* is the option to enter a swap contract at a predetermined point in time and a predetermined swap rate ℓ_k. Thus a receiver swaption is equivalent to a call and a payer swaption is equivalent to a put with exercise price 1 on a coupon bond with a coupon rate ℓ_k.

8.4 Adding a foreign currency

Let $X(t)$ denote the spot exchange rate at time t in units of domestic currency per unit of foreign currency and assume that $X(t)$ has a deterministic relative volatility $\sigma_X(t)$, i.e the dynamics of $X(t)$ are given by

$$dX(t) = X(t) \left(\mu(t)dt + \sigma_X(t)dW(t) \right)$$

As before, $W(t)$ is a d–dimensional Brownian motion, σ_X is a d–dimensional vector and $\sigma_X(t)dW(t)$ is to be interpreted as a sum product.

Under the domestic spot risk–neutral measure, the foreign savings account $\tilde{\beta}$, converted to domestic currency by multiplying by the exchange rate[5] and

---

[5] Note that the exchange rate $X(t)$ itself is not an asset, but rather a conversion factor to convert asset prices from one unit of measurement (foreign currency) into another (domestic currency).

discounted by the domestic savings account β, must be a martingale. Therefore

$$d\left(\frac{X(t)\tilde{\beta}(t)}{\beta(t)}\right) = \frac{X(t)\tilde{\beta}(t)}{\beta(t)}\sigma_X(t)dW_{beta}(t) \tag{8.60}$$

$$\Rightarrow \quad \mu(t) = r(t) - \tilde{r}(t) \tag{8.61}$$

where $\tilde{r}(t)$ is the foreign continuously compounded short rate of interest.

For each currency, the dynamics of the term structure of interest rates are assumed to be given by Gauss/Markov HJM models of the type presented in Section 8.2.

To relate the dynamics of interest rates in different currencies, we will need the following results from standard theory:

$$dW_{T_j}(t) = dW_\beta(t) + \sigma^*(t, T_j)dt \tag{8.62}$$

$$d\tilde{W}_\beta(t) = dW_\beta(t) - \sigma_X(t)dt \tag{8.63}$$

(8.62) mirrors (8.35), and (8.63) follows from Theorem 4.6 as in the case of deterministic interest rates considered in Chapter 4, p. 121.

8.4.1 Standard FX options

Applying Theorem 4.7 immediately yields formulas for European FX call and put options. Consider European options on one unit of foreign currency, expiring at time T_1, with exercise price K. The time 0 call option price is given by

$$C = X(0)\tilde{B}(0, T_1)N(h_1) - B(0, T_1)KN(h_2)$$

and the put option price by

$$P = B(0, T_1)KN(-h_2) - X(0)\tilde{B}(0, T_1)N(-h_1)$$

with

$$h_{1,2} = \frac{\ln\frac{X(0)\tilde{B}(0,T_1)}{B(0,T_1)K} \pm \frac{1}{2}\int_0^{T_1}(\sigma_X(u) + \tilde{\sigma}_B(u, T_1) - \sigma_B(u, T_1))^2 du}{\sqrt{\int_0^{T_1}(\sigma_X(u) + \tilde{\sigma}_B(u, T_1) - \sigma_B(u, T_1))^2 du}}$$

$\sigma_B(u, T_1) = -\sigma^*(u, T_1)$ denoting the volatility of the zero coupon bond maturing in T_1. Note that

$$\sigma_X(u) + \tilde{\sigma}_B(u, T_1) - \sigma_B(u, T_1)$$

is the instantaneous volatility of the forward exchange rate

$$\frac{X(t)\tilde{B}(t, T_1)}{B(t, T_1)}$$

8.4.2 A "Quintessential Formula" in the multicurrency Gauss/Markov HJM model

Noting that in the present context the drift adjustments (8.62) and (8.63) are deterministic when moving between any probability measure relevant for pricing (i.e. the spot and forward risk neutral measures), we see that the $X(t)$, $B(t,T)$ and $\tilde{B}(t,T)$ are lognormally distributed under all such measures. Furthermore, $X(t)$ and $\tilde{B}(t,T)$ can alternatively be interpreted, respectively, as an exchange rate and a foreign zero coupon bond, or as an asset and its corresponding "dividend discount factor."[6] In the latter case, $X(t)$ and $\tilde{B}(t,T)$ correspond to the $X_i(t)$ and $D_i(t,T)$ in Section 4.3 of Chapter 4. Since lognormality is maintained under all forward measures, one can adapt the "Quintessential Option Pricing Formula" to encompass the present case of stochastic (lognormal) $B(t,T)$ and $D_i(t,T)$.

Specifically, we assume a set of assets X_i, the stochastic dynamics of which are given by diffusion processes with deterministic relative volatility, and which pay continuous dividend yields d_{X_i}. Without loss of generality, we will call those X_i with a deterministic term structure of d_{X_i} *assets* and those with stochastic d_{X_i} *exchange rates*.[7] The interest rate dynamics and similarly the dynamics of the stochastic instantaneous dividend yields d_{X_i}, respectively, are each assumed to be given by a Gauss/Markov HJM model.

These assumptions imply that the market is complete and the dynamics under the spot risk neutral measure are given by (8.9) for the term structure of interest rates, and

$$dX_i(t) = X_i(t)((r(t) - d_{X_i}(t))dt + \sigma_{X_i}(t)dW_\beta(t)) \qquad (8.64)$$

for each asset/exchange rate X_i. When chosen to be stochastic, the "dividend discount factors" $D_i(t,T)$ (i.e. the "foreign zero coupon bonds"), now defined in terms of d_{X_i} as

$$D_i(t,T) = E_\beta\left[\exp\left\{-\int_t^T d_{X_i}(s)ds\right\}\middle|\mathcal{F}_t\right] \qquad (8.65)$$

follow the Gauss/Markov HJM zero coupon bond dynamics, i.e.

$$dD_i(t,T) = D_i(t,T)(d_{X_i}(t)dt - \sigma_{D_i}^*(t,T)dW_{X_i}(t)) \qquad (8.66)$$

Here, W_{X_i} is a standard Brownian motion under the risk neutral measure

---

[6] By "reinterpreting" $X(t)$ and $\tilde{B}(t,T)$, the model can (and has been) applied in a broad range of contexts: Identifying $X(t)$ as the Consumer Price Index and the $\tilde{B}(t,T)$ as discount factors associated with the real (as opposed to nominal) interest rates, one obtains a stochastic inflation model (see e.g. Jarrow and Yildirim (2003)); identifying $X(t)$ as a commodity price and the $\tilde{B}(t,T)$ as discount factors associated with convenience yields gives a model of stochastic convenience yields (see e.g. Miltersen and Schwartz (1998)). In each case the mathematics of the model remain the same.

[7] This is a modelling convenience, allowing us to capture assets with deterministic dividend yields and assets with stochastic d_{X_i} in the same model.

associated with the numeraire[8]

$$X_i(t) \exp\left\{ \int_0^t dx_i(s)ds \right\} \tag{8.67}$$

Expressing (8.66) under the domestic spot risk neutral measure, we have

$$dD_i(t,T) = D_i(t,T)(dx_i(t)dt - \sigma_{D_i}^*(t,T)(dW_{X_i}(t) - \sigma_{X_i}(t)dt)) \tag{8.68}$$

Note that $\sigma_{X_i}(t)$ is the volatility of X_i, while $-\sigma_{D_i}^*(t,T)$ is the volatility of $D_i(t,T)$, and relates to the volatilities $\sigma_{D_i}(t,T)$ of the instantaneous forward dividend yields via[9]

$$\sigma_{D_i}^*(t,T) = \int_t^T \sigma_{D_i}(t,u)du \tag{8.69}$$

In this modification of the context of Section 4.3 of Chapter 4, consider the valuation of the M–Binary payoff V_{T_M} as specified in (4.51). In the present context, it is convenient to modify the notation of (4.51) slightly, to

$$V_{T_M} = \underbrace{\left(\prod_{j=1}^n (Y_{\bar{\imath}(j)}(T_{\bar{k}(j)}))^{\alpha_j} \right)}_{\text{payoff amount}}$$

$$\cdot \underbrace{\prod_{l=1}^m \mathbb{I}\left\{ S_{l,l} \prod_{j=1}^n (Y_{\bar{\imath}(j)}(T_{\bar{k}(j)}))^{A_{l,j}} > S_{l,l}a_l \right\}}_{\text{payoff indicator}} \tag{8.70}$$

where the $Y_{\bar{\imath}(j)}$ are the basic "building block" financial variables, in terms of which all M–Binaries will be specified. Denoting by $\overline{T}$ the time horizon of the model, these building blocks are

Terminal forward prices (in currency $\bar{c}(i)$)	$\dfrac{X_i(t)D_{X_i}(t,\overline{T})}{B_{\bar{c}(i)}(t,\overline{T})}$
Zero coupon bond prices (in currency c)	$B_c(t,T_k)$

We can write

$$V_t = B(t,T_M)E_{T_M}[V_{T_M}] \tag{8.71}$$

where E_{T_M} denotes the expectation under the time T_M forward measure. Since the Y_i remain lognormal under all forward measures, V_t can still be written

---

[8] If X_i is an exchange rate, this is the foreign continuously compounded savings account (expressed in domestic currency).

[9] In the "exchange rate" interpretation of X_i, $D_i(t,T)$ is a foreign zero coupon bond $\tilde{B}(t,T)$, $-\sigma_{D_i}^*(t,T)$ its volatility, and $\sigma_{D_i}(t,T)$ is the volatility of the foreign instantaneous forward rate $\tilde{f}(t,T)$.

in the form (4.52), i.e.

$$V_t = \xi \beta \mathcal{N}_m(S\overline{h}; S\overline{C}S) \prod_{j=1}^{n} (Y_{\bar{\imath}(j)}(t))^{\alpha_j} \tag{8.72}$$

with modified coefficients

$$\xi = B(t, T) \exp\left\{\alpha^\top \overline{\mu} + \frac{1}{2}\alpha^\top \overline{\Gamma}\alpha\right\} \qquad \text{scalar}$$

$$\overline{h} = \overline{D}^{-1}(\ell + A(\overline{\mu} + \overline{\Gamma}\alpha)) \qquad \text{vector}$$

$$\overline{C} = \overline{D}^{-1}(A\overline{\Gamma}A^\top)\overline{D}^{-1} \qquad \text{matrix}$$

A and S are matrices and α is a vector as per (4.51), ℓ is an m-dimensional vector which remains unchanged as per (4.55), and $\overline{D}$ is an $m \times m$ diagonal matrix given by

$$\overline{D}^2 = \text{diag}(A\overline{\Gamma}A^\top) \tag{8.73}$$

Thus the modifications are restricted to $\overline{\Gamma}$ and $\overline{\mu}$. These are, respectively, the $n \times n$ covariance matrix of logarithmic asset prices for the n (asset, time) combinations, and the n-dimensional vector entering the expectations of the logarithmic asset prices under the time T_M forward measure. In order to derive $\overline{\Gamma}$ and $\overline{\mu}$ for the chosen building blocks, consider first $X_{\bar{\imath}(j)}(\overline{T})$. We have

$$X_{\bar{\imath}(j)}(\overline{T}) = X_{\bar{\imath}(j)}(\overline{T})D_{\bar{\imath}(j)}(\overline{T}, \overline{T}) \tag{8.74}$$

Divided by the numeraire $B_{\overline{c}(\bar{\imath}(j))}(\cdot, \overline{T})$, this becomes an exponential martingale under the time $\overline{T}$ forward measure, i.e.

$$\frac{X_{\bar{\imath}(j)}(t)D_{\bar{\imath}(j)}(t, \overline{T})}{B_{\overline{c}(\bar{\imath}(j))}(t, \overline{T})} = \frac{X_{\bar{\imath}(j)}(0)D_{\bar{\imath}(j)}(0, \overline{T})}{B_{\overline{c}(\bar{\imath}(j))}(0, \overline{T})}$$

$$\cdot \exp\left\{\int_0^t \overline{\sigma}_{\bar{\imath}(j)}(s)dW_{\overline{T}}(s) - \frac{1}{2}\int_0^t \overline{\sigma}_{\bar{\imath}(j)}^2(s)ds\right\} \tag{8.75}$$

Here $\overline{\sigma}_{\bar{\imath}(j)}$ is the volatility of the quotient on the left hand side of the equation, i.e.

$$\overline{\sigma}_i = \sigma_{X_i} - \sigma_{D_i}^*(s, \overline{T}) + \sigma^*(s, \overline{T}) \tag{8.76}$$

We have

$$\begin{aligned}
\overline{\mu}_j &= E_{T_M}[\ln Y_{\bar{\imath}(j)}(T_{\bar{k}(j)})] \\
&= \ln \frac{X_{\bar{\imath}(j)}(0)D_{\bar{\imath}(j)}(0, \overline{T})}{B(0, \overline{T})} - \frac{1}{2}\int_0^{T_{\bar{k}(j)}} \overline{\sigma}_{\bar{\imath}(j)}^2(s)ds \\
&\quad + E_{T_M}\left[\int_0^{T_{\bar{k}(j)}} \overline{\sigma}_{\bar{\imath}(j)}(s)dW_{\overline{T}}(s)\right]
\end{aligned} \tag{8.77}$$

for terminal forward prices, while for zero coupon bonds

$$
\begin{aligned}
\bar{\mu}_j &= E_{T_M}[\ln Y_{\bar{\imath}(j)}(T_{\bar{k}(j)})] \\
&= E_{T_M}[\ln B_{\bar{c}(\bar{\imath}(j))}(T_{\bar{k}(j)}, T_{\bar{m}(j)})] \\
&= \ln \frac{B_{\bar{c}(\bar{\imath}(j))}(0, T_{\bar{m}(j)})}{B_{\bar{c}(\bar{\imath}(j))}(0, T_{\bar{k}(j)})} \\
&\quad - \frac{1}{2} \int_0^{T_{\bar{k}(j)}} (\sigma^*_{\bar{c}(\bar{\imath}(j))}(s, T_{\bar{k}(j)}) - \sigma^*_{\bar{c}(\bar{\imath}(j))}(s, T_{\bar{m}(j)}))^2 ds \\
&\quad + E_{T_M}\left[\int_0^{T_{\bar{k}(j)}} (\sigma^*_{\bar{c}(\bar{\imath}(j))}(s, T_{\bar{k}(j)}) - \sigma^*_{\bar{c}(\bar{\imath}(j))}(s, T_{\bar{m}(j)})) dW_{T_{\bar{k}(j)}}^{(\bar{c}(\bar{\imath}(j)))}(s)\right]
\end{aligned}
\tag{8.78}
$$

where $-\sigma^*_{\bar{c}(\bar{\imath}(j))}(s, T)$ is the volatility of the zero coupon bond in currency $\bar{c}(\bar{\imath}(j))$, $B_{\bar{c}(\bar{\imath}(j))}(s, T)$. $W_{T_{\bar{k}(j)}}^{(\bar{c}(\bar{\imath}(j)))}$ is a standard Brownian motion under the currency $\bar{c}(\bar{\imath}(j))$, time $T_{\bar{k}(j)}$ forward measure. By (8.36), we have

$$
\begin{aligned}
&E_{T_M}\left[\int_0^{T_{\bar{k}(j)}} (\sigma^*_{\bar{c}(\bar{\imath}(j))}(s, T_{\bar{k}(j)}) - \sigma^*_{\bar{c}(\bar{\imath}(j))}(s, T_{\bar{m}(j)})) dW_{T_{\bar{k}(j)}}^{(\bar{c}(\bar{\imath}(j)))}(s)\right] \\
&= E_{T_M}\left[\int_0^{T_{\bar{k}(j)}} (\sigma^*_{\bar{c}(\bar{\imath}(j))}(s, T_{\bar{k}(j)}) - \sigma^*_{\bar{c}(\bar{\imath}(j))}(s, T_{\bar{m}(j)})) dW_{T_M}^{(\bar{c}(\bar{\imath}(j)))}(s)\right] \\
&\quad + \int_0^{T_{\bar{k}(j)}} (\sigma^*_{\bar{c}(\bar{\imath}(j))}(s, T_{\bar{k}(j)}) - \sigma^*_{\bar{c}(\bar{\imath}(j))}(s, T_{\bar{m}(j)})) \\
&\quad \quad (\sigma^*_{\bar{c}(\bar{\imath}(j))}(s, T_{\bar{k}(j)}) - \sigma^*_{\bar{c}(\bar{\imath}(j))}(s, T_M)) ds
\end{aligned}
\tag{8.79}
$$

If $\bar{c}(\bar{\imath}(j)) = 0$ (i.e. domestic currency), the expectation of the stochastic integral with respect to W_{T_M} is zero. Otherwise, we use (8.62) and (8.63) to change measures one more time:

$$
\begin{aligned}
&E_{T_M}\left[\int_0^{T_{\bar{k}(j)}} (\sigma^*_{\bar{c}(\bar{\imath}(j))}(s, T_{\bar{k}(j)}) - \sigma^*_{\bar{c}(\bar{\imath}(j))}(s, T_{\bar{m}(j)})) dW_{T_M}^{(\bar{c}(\bar{\imath}(j)))}(s)\right] \\
&= E_{T_M}\left[\int_0^{T_{\bar{k}(j)}} (\sigma^*_{\bar{c}(\bar{\imath}(j))}(s, T_{\bar{k}(j)}) - \sigma^*_{\bar{c}(\bar{\imath}(j))}(s, T_{\bar{m}(j)})) dW_{T_M}^{(0)}(s)\right] \\
&\quad + \int_0^{T_{\bar{k}(j)}} (\sigma^*_{\bar{c}(\bar{\imath}(j))}(s, T_{\bar{k}(j)}) - \sigma^*_{\bar{c}(\bar{\imath}(j))}(s, T_{\bar{m}(j)})) \\
&\quad \quad (\sigma^*_{\bar{c}(\bar{\imath}(j))}(s, T_M) - \sigma_{X_{\bar{c}(\bar{\imath}(j))}}(s) - \sigma_0^*(s, T_M)) ds
\end{aligned}
\tag{8.80}
$$

where $\sigma_{X_{\bar{c}(\bar{\imath}(j))}}$ is the volatility of the exchange rate between domestic currency and currency $\bar{c}(\bar{\imath}(j))$. Thus $\bar{\mu}_j$ is given by the logarithm of the initial forward zero coupon bond price in (8.78) and the volatility integrals in (8.78), (8.79) and (8.80). For the matrix $\overline{\Gamma}$, we have

$$
\begin{aligned}
\overline{\Gamma}_{jl} &= \text{Cov}[\ln Y_{\bar{\imath}(j)}(T_{\bar{k}(j)}), \ln Y_{\bar{\imath}(l)}(T_{\bar{k}(l)})] \\
&= \int_0^{\min(T_{\bar{k}(j)}, T_{\bar{k}(l)})} \sigma_{Y_{\bar{\imath}(j)}}(u) \sigma_{Y_{\bar{\imath}(l)}}(u) du
\end{aligned} \tag{8.81}
$$

If Y_i is a terminal forward price, $\sigma_{Y_i} = \overline{\sigma}_i$, given by (8.76). Note that when Y_i is the terminal forward price of an asset paying a deterministic dividend yield, $\sigma_{D_i}^*$ in (8.76) will be zero. If Y_i is a zero coupon bond, i.e. as in the above notation

$$
Y_{\bar{\imath}(j)}(T_{\bar{k}(j)}) = B_{\overline{c}(\bar{\imath}(j))}(T_{\bar{k}(j)}, T_{\bar{m}(j)}),
$$

then $\sigma_{Y_{\bar{\imath}(j)}}$ is the volatility of the $T_{\bar{k}(j)}$ forward zero coupon bond price, i.e.

$$
\sigma_{Y_{\bar{\imath}(j)}}(s) = \sigma_{\overline{c}(\bar{\imath}(j))}^*(s, T_{\bar{k}(j)}) - \sigma_{\overline{c}(\bar{\imath}(j))}^*(s, T_{\bar{m}(j)})
$$

Thus the M–Binary pricing formula in a multicurrency Gauss/Markov HJM setting is now complete, and we can price a large class of payoffs by simply casting them in M–Binary form using terminal forward assets, terminal forward exchange rates, and zero coupon bonds as building blocks. This is straightforward and best illustrated by way of example. Consider a model with two currencies, domestic and foreign, with a domestic equity asset and a foreign equity asset. In this context we wish to price each of the following options:

1. a European call option, expiring at time T_1, on a domestic equity asset

2. an FX European call option, expiring at time T_1

3. a European call option, expiring at time T_1, on a foreign equity asset

4. a European call option, expiring at time T_1, on a domestic zero coupon bond maturing at time T_2

5. a European call option, expiring at time T_1, on a foreign zero coupon bond maturing at time T_2 (strike paid in foreign currency)

6. a European option where the payoff is given by the greater of zero and the price of the foreign equity asset, converted to domestic currency by a fixed exchange rate X_0, minus a strike K in domestic currency (this is known as a *quanto* option)

7. a European option paying the difference (if positive) of the foreign interest rate with simple compounding over a given future accrual period, and a given strike level. The payoff occurs at the end of the accrual period and is converted to domestic currency at a fixed exchange rate X_0 (this is known as a *quanto caplet*).

Firstly, we denote

- $X_1(t)$ is the exchange rate in units of domestic currency per unit of foreign currency, $\sigma_{X_1}(t)$ is its volatility. Thus $Y_1(t)$ is the corresponding terminal forward exchange rate:

$$
Y_1(t) = \frac{X_1(t)\tilde{B}(t, \overline{T})}{B(t, \overline{T})}
$$

- $X_2(t)$ is the domestic equity asset price in units of domestic currency, $\sigma_{X_2}(t)$ is its volatility. Thus $Y_2(t)$ is the corresponding terminal forward price:

$$Y_2(t) = \frac{X_2(t)D_{X_2}(t,\overline{T})}{B(t,\overline{T})}$$

- $X_3(t)$ is the foreign equity asset price in units of foreign currency, $\sigma_{X_3}(t)$ is its volatility. Thus $Y_3(t)$ is the corresponding terminal forward price:

$$Y_3(t) = \frac{X_3(t)D_{X_3}(t,\overline{T})}{\tilde{B}(t,\overline{T})}$$

- $Y_4(t)$ is the price of the domestic zero coupon bond maturing in $\overline{T}$, i.e. $Y_4(t) = B(t,\overline{T})$

- $Y_5(t)$ is the price of the foreign zero coupon bond maturing in $\overline{T}$, i.e. $Y_5(t) = \tilde{B}(t,\overline{T})$

- $Y_6(t)$ is the price of the domestic zero coupon bond maturing in T_2, i.e. $Y_6(t) = B(t,T_2)$

- $Y_7(t)$ is the price of the foreign zero coupon bond maturing in T_2, i.e. $Y_7(t) = \tilde{B}(t,T_2)$

We can then proceed to express the payoffs in terms of the building blocks as follows.

For the call option on the domestic equity asset, we have (note that D_{X_2} is assumed to be deterministic)

$$[X_2(T_1) - K]^+$$
$$= \frac{1}{D_{X_2}(T_1,\overline{T})} \left[\frac{X_2(T_1)D_{X_2}(T_1,\overline{T})}{B(T_1,\overline{T})} B(T_1,\overline{T}) - D_{X_2}(T_1,\overline{T})K \right]^+$$
$$= \frac{1}{D_{X_2}(T_1,\overline{T})} \left[Y_2(T_1)Y_4(T_1) - D_{X_2}(T_1,\overline{T})K \right]^+ \qquad (8.82)$$

If $T_1 = \overline{T}$, this simplifies to

$$[X_2(T_1) - K]^+ = [Y_2(\overline{T}) - K]^+$$

For the FX option, we have

$$[X_1(T_1) - K]^+ = \left[\frac{X_1(T_1)\tilde{B}(T_1,\overline{T})}{B(T_1,\overline{T})} \frac{B(T_1,\overline{T})}{\tilde{B}(T_1,\overline{T})} - K \right]^+$$
$$= \left[Y_1(T_1)Y_4(T_1)Y_5^{-1}(T_1) - K \right]^+ \qquad (8.83)$$

If $T_1 = \overline{T}$, this simplifies to

$$[X_1(T_1) - K]^+ = [Y_1(\overline{T}) - K]^+$$

For the call option on the foreign equity asset, note that the M–Binary formula is set up to return a value in domestic currency, so the payoff must also be expressed in domestic currency:

$$X_1(T_1)[X_3(T_1) - \tilde{K}]^+$$

$$= \left[\frac{X_1(T_1)\tilde{B}(T_1,\overline{T})}{B(T_1,\overline{T})} \frac{X_3(T_1)D_{X_3}(T_1,\overline{T})}{\tilde{B}(T_1,\overline{T})} \frac{B(T_1,\overline{T})}{D_{X_3}(T_1,\overline{T})} \right.$$

$$\left. - \frac{X_1(T_1)\tilde{B}(T_1,\overline{T})}{B(T_1,\overline{T})} \frac{B(T_1,\overline{T})}{\tilde{B}(T_1,\overline{T})} \tilde{K} \right]^+ \qquad (8.84)$$

$$= \frac{\left[Y_1(T_1)Y_2(T_1)Y_4(T_1) - Y_1(T_1)Y_4(T_1)Y_5^{-1}(T_1)D_{X_3}(T_1,\overline{T})\tilde{K} \right]^+}{D_{X_3}(T_1,\overline{T})}$$

If $T_1 = \overline{T}$, this simplifies to

$$X_1(T_1)[X_3(T_1) - \tilde{K}]^+ = [Y_1(\overline{T})Y_3(\overline{T}) - Y_1(\overline{T})\tilde{K}]^+$$

For the call option on a domestic zero coupon bond, we have

$$[B(T_1,T_2) - K]^+ = [Y_6(T_1) - K]^+ \qquad (8.85)$$

and for the call option on a foreign zero coupon bond,

$$X_1(T_1)[\tilde{B}(T_1,T_2) - K]^+ =$$

$$[Y_1(T_1)Y_4(T_1)Y_5^{-1}(T_1)Y_7(T_1) - Y_1(T_1)Y_4(T_1)Y_5^{-1}(T_1)\tilde{K}]^+ \quad (8.86)$$

If $T_1 = \overline{T}$, the latter simplifies to

$$X_1(T_1)[\tilde{B}(T_1,T_2) - K]^+ = [Y_1(\overline{T})Y_7(\overline{T}) - Y_1(\overline{T})\tilde{K}]^+$$

For the quanto option, we have

$$X_0[X_3(T_1) - \tilde{K}]^+ = \frac{X_0}{D_{X_3}(T_1,\overline{T})}[Y_3(T_1)Y_5(T_1) - D_{X_3}(T_1,\overline{T})\tilde{K}]^+ \qquad (8.87)$$

If $T_1 = \overline{T}$, this simplifies to

$$X_0[X_3(T_1) - \tilde{K}]^+ = X_0[Y_3(\overline{T}) - \tilde{K}]^+$$

For the quanto caplet, consider the accrual period $[T_1, T_1 + \delta]$. The foreign interest rate with simple compounding over this period is given by

$$\frac{1}{\delta}\left(\frac{1}{\tilde{B}(T_1, T_1 + \delta)} - 1 \right)$$

The payoff at time $T_1 + \delta$, converted to domestic currency at the fixed exchange rate X_0, is

$$X_0\delta\left[\frac{1}{\delta}\left(\frac{1}{\tilde{B}(T_1, T_1 + \delta)} - 1 \right) - \tilde{\kappa} \right]^+$$

```
class GaussMarkovWorld {
public:
  /// Constructor
  GaussMarkovWorld(std::vector<boost::shared_ptr<GaussianEconomy> >&
                   xeconomies,
                   std::vector<boost::shared_ptr<DeterministicAssetVol> >&
                   xvols,
                   Array<double,1> xinitial_exchange_rates);
  /// Construct from a CSV file specification
  GaussMarkovWorld(const char* path);
  /// Destructor
  ~GaussMarkovWorld();
  /// Determine which asset values should be reported back by operator()
  int set_reporting(int currency_index,
                    int asset_index,
                    double maturity = -1.0);
  /// Set drifts, etc. to simulate under a particular choice of numeraire.
  void set_numeraire(int num);
  /// Query the number of random variables (per period) driving the process.
  inline int factors() const { return max_rank; };
  /// Set process timeline
  bool set_timeline(const Array<double,1>& timeline);
  /// Get number of steps in process time discretisation.
  int number_of_steps() const;
  /** Generate a realisation of the process under the martingale
      measure associated with a given numeraire asset.
      underlying_values is an asset x (time points) Array. */
  void operator()(Array<double,2>& underlying_values,
                  Array<double,1>& numeraire_values,
                  const Array<double,2>& x,
                  const TermStructure& ts,
                  int numeraire_index);
};
```

Listing 8.1: Selected public member functions of the class GaussMarkovWorld.

This payoff is already known at time T_1, and its time T_1 discounted value is

$$X_0\delta\left[\frac{B(T_1, T_1 + \delta)}{\tilde{B}(T_1, T_1 + \delta)} - B(T_1, T_1 + \delta)(1 + \delta\tilde{\kappa})\right]^+$$

Defining the building blocks $Y_8(t) = B(t, T_1 + \delta)$ and $Y_9(t) = \tilde{B}(t, T_1 + \delta)$, this becomes

$$X_0[Y_8(T_1)Y_9^{-1}(T_1) - Y_8(T_1)(1 + \delta\tilde{\kappa})]^+ \tag{8.88}$$

8.5 Implementing closed–form solutions

In order to implement the closed form solutions discussed in the preceding sections (and subsequently Monte Carlo simulations in the same model context), the multicurrency Gauss/Markov HJM model is represented by a collection of C++ classes. At the top level the entire multicurrency model is encapsulated in the class GaussMarkovWorld (see Listing 8.1). An instance

```
/** Class representing assets quoted in a particular "currency."
    These assets include a term structure of zero coupon bond
    prices and assets with deterministic continuous dividend
    yield (which may be zero). */
class GaussianEconomy {
public:
  GaussianEconomy(std::vector<boost::shared_ptr<BlackScholesAsset> >&
                  xunderlying,
                  boost::shared_ptr<DeterministicAssetVol> xv,
                  boost::shared_ptr<TermStructure> xinitialTS);
  GaussianEconomy(const char* path);
  /// Vector of pointers to underlying assets.
  std::vector<boost::shared_ptr<BlackScholesAsset> >        underlying;
  /** Pointer to deterministic volatility function object
      (for Gaussian term structure dynamics). */
  boost::shared_ptr<DeterministicAssetVol>                         v;
  /** Vector of pointers to DeterministicAssetVol objects where all
      components except one are zero - corresponds to volatilies of
      the state variables z. */
  std::vector<boost::shared_ptr<DeterministicAssetVol> > component_vol;
  /** Pointer to term structure object containing the initial term
      structure of interest rates. */
  boost::shared_ptr<TermStructure>                         initialTS;
  /// For closed-form solutions.
  boost::shared_ptr<GaussianHJM>                                 hjm;
};
```

Listing 8.2: The class `GaussianEconomy`.

of `GaussMarkovWorld` is constructed from its constituent (single–currency) economies, and from the exchange rates (of each foreign currency in units of domestic currency), characterised by their initial values and their volatilities. Specifically, the `GaussMarkovWorld` constructor takes as arguments a reference to a `std::vector` of Boost shared pointers to instances of `GaussianEconomy` (representing each constituent economy), a reference to a `std::vector` of Boost shared pointers to instances of `DeterministicAssetVol` (representing the volatility of each exchange rate) and an `Array<double,1>` of initial values of the exchange rates. Each instance of the class `GaussianEconomy` (see Listing 8.2) is constructed from the equity assets in that economy, the Gauss/Markov HJM volatility for the interest rate dynamics in that economy, and its initial term structure of interest rates. Specially, the `GaussianEconomy` constructor takes as arguments a reference to a `std::vector` of Boost shared pointers to instances of `BlackScholesAsset` (representing the equity assets), a Boost shared pointer to an instance of `DeterministicAssetVol` (representing the Gauss/Markov HJM volatility), and a Boost shared pointer to an instance of a class derived from `TermStructure` (representing the initial term structure of interest rates for this economy). The class `GaussianEconomy` serves to collect the data members making up an economy, and therefore all the data members are exposed as `public`.

Closed form calculations in the Gauss/Markov HJM model are encapsulated in the class `GaussianHJM`, an instance of which is constructed by each instance

```
class GaussianHJM {
public:
    /// Constructor.
    GaussianHJM(DeterministicAssetVol* xv,TermStructure* ini);
    /** Number of factors, i.e. dimension of the driving Brownian motion.
        Passed through to the deterministic volatility function object
        DeterministicVol. */
    int factors() const;
    /// Access the volatility function.
    const DeterministicAssetVol& volatility_function() const;
    /// Calculate a zero coupon bond price given the state variables.
    double bond(const Array<double,1> &W,double t,double ttm) const;
    /** Initial zero coupon bond price for a given maturity.
        Passed through to the initial TermStructure object. */
    double bond(double mat) const;
    /// Calculate the ("time zero") value of a zero coupon bond option.
    double ZCBoption(double T1,double T2,double K,int sign = 1) const;
    /// Calculate the ("time zero") value of a coupon bond option.
    double CBoption(const Array<double,1>& tenor,
                    const Array<double,1>& coupon,
                    double K,int sign = 1) const;
    /// Calculate the ("time zero") value of swaption.
    double swaption(const Array<double,1>& tenor,   ///< Swap tenor.
                    double K,     ///< Strike (fixed side) level.
                    int sign = 1 ///< 1 = receiver, -1 = payer.
                    ) const;
    /// Calculate the ("time zero") value of a caplet.
    double caplet(double mat,double delta,double K) const;
    /// Calculate the ("time zero") value of an equity option.
    double option(const BlackScholesAsset& S,
                  double mat,double K,int sign = 1) const;
    /// Calculate the ("time zero") value of an FX option.
    double FXoption(double X0,double T1,double K,const GaussianHJM& fmodel,
                    const DeterministicAssetVol& fxvol,int sign = 1) const;
    /** Swap netting the difference between the foreign and domestic floating
        rates (reverse if sign = -1), paid on a domestic notional. */
    double DiffSwap(const Array<double,1>& T,const GaussianHJM& fmodel,
                    const DeterministicAssetVol& fxvol,int sign = 1) const;
    /** Foreign caplet with payoff converted to domestic currency at a
        guaranteed exchange rate X0. */
    double QuantoCaplet(double X0,double T,double delta,double lvl,
                        const GaussianHJM& fmodel,
                        const DeterministicAssetVol& fxvol) const;
};
```

Listing 8.3: Selected public member functions of the class `GaussianHJM`.

of `GaussianEconomy`. A particular instance of such a model is characterised by a volatility function for the instantaneous forward rates and by an initial term structure of interest rates, passed to the constructor of `GaussianEconomy` as pointers to instances of classes derived from `DeterministicAssetVol` and `TermStructure`, respectively. Listing 8.3 shows member functions of `GaussianHJM` implementing a selection of derivatives pricing solutions.

The member function `ZCBoption()` implements the closed form solution (8.40); see Listing 8.4. The key calculation is the variance of the logarithmic

```
double GaussianHJM::ZCBoption(double T1,  ///< Option expiry
                              double T2,  ///< Maturity of the underlying bond
                              double K,   ///< Exercise price
                              int sign    ///< 1 if call (default), -1 if put
                              ) const
{
  double vol = v->FwdBondVol(0.0,T1,T2);
  double B1  = initialTS->operator()(T1);
  double B2  = initialTS->operator()(T2);
  double h1  = (log(B2/(K*B1)) + 0.5*vol*vol) / vol;
  return sign * (B2*boost::math::cdf(N,sign*h1)-
                 K*B1*boost::math::cdf(N,sign*(h1-vol)));
}
```

Listing 8.4: GaussianHJM::ZCBoption().

zero coupon bond price in (8.39), given by an integral over the forward bond volatility. This is implemented by classes derived from the abstract base class DeterministicAssetVol (representing the Gauss/Markov HJM volatility) in the virtual member function FwdBondVol(). That is to say, the concrete classes derived from the abstract base class DeterministicAssetVol, e.g. ConstVol, ExponentialVol or PiecewiseConstVol, represent the HJM instantaneous forward rate volatility $\sigma(t, T)$, and implement FwdBondVol() to calculate the integral over the respective forward bond volatility accordingly. By delegating the calculation of volatility integrals to the volatility function objects, the zero coupon bond option pricing formula and all other closed form calculations in the Gauss/Markov HJM model are implemented independently of the specific functional form of the instantaneous forward rate volatility. A similar approach was taken in Chapter 4 (see in particular Section 4.4.1), and now is extended to include zero coupon bond volatilities $-\sigma^*(t, T)$.

As discussed in Section 8.3, options on coupon bonds can be priced in (almost) closed form in the one– and two–factor versions of the Gauss/Markov HJM model. These two cases are implemented in the member function CBoption(), which throws a std::logic_error if the model has more than two factors. The bond notional is assumed to be \$1, and the bond is characterised by a tenor structure Array<double,1> tenor defining the coupon accrual periods, and an Array<double,1> coupon of the associated coupon payments (i.e. accrual periods go from tenor(i) to tenor(i+1), and coupon(i) is paid at time tenor(i+1)). In the one–factor case, the coupon bond option can be represented as a sum of zero coupon bond options as per (8.46), each of which is priced by a call to ZCBoption(). In the two–factor case, CBoption() implements the method of Brigo and Mercurio (2001) as reproduced in equations (8.47) to (8.55). As noted in Chapter 2,[10] a swap can be represented in terms of a coupon bond; consequently an option to enter a fixed–for–floating interest rate swap (a swaption) as a receiver (payer) of the fixed leg can be represented

---

[10] See Section 2.6.1.

```
double GaussianHJM::FXoption(double X0, ///< Initial exchange rate.
                             double T1, ///< Option expiry.
                             double K,  ///< Strike.
                             const GaussianHJM& fmodel,
                             const DeterministicAssetVol& fxvol,
                             int sign   ///< Call = 1, put = -1.
                             ) const
{
    double B    = initialTS->operator()(T1);
    double fB   = fmodel.initialTS->operator()(T1);
    double var  = fxvol.volproduct(0.0,T1,fxvol);
    var += v->bondbondvolproduct(0.0,T1,T1,T1,*v);
    var += fmodel.v->bondbondvolproduct(0.0,T1,T1,T1,*(fmodel.v));
    var -= 2.0 * v->bondvolproduct(0.0,T1,T1,fxvol);
    var += 2.0 * fmodel.v->bondvolproduct(0.0,T1,T1,fxvol);
    var -= 2.0 * v->bondbondvolproduct(0.0,T1,T1,T1,*(fmodel.v));
    double sd   = sqrt(var);
    double h1   = (log(X0*fB/(K*B)) + 0.5*var) / sd;
    return sign * (X0*fB*boost::math::cdf(N,sign*h1)
                    -K*B*boost::math::cdf(N,sign*(h1-sd)));
}
```

Listing 8.5: <code>GaussianHJM::FXoption()</code>.

as a call (put) on a coupon bond struck at par. This is implemented in the member function swaption(), which calls CBoption().

A *caplet* pays the difference (if positive) between the interest rate with simple compounding $r_a(T, T+\delta)$ over a given future accrual period $[T, T+\delta]$, and a given strike level κ. The payoff occurs at time $T + \delta$, but the amount is already known at time T. The time $T + \delta$ payoff is

$$\delta \max(0, r_a(T, T+\delta) - \kappa)$$

Via Definition 2.3 in Chapter 2, this can be expressed in terms of a zero coupon bond price as

$$\delta \max\left(0, \frac{1}{\delta}\left(\frac{1}{B(T, T+\delta)} - 1\right) - \kappa\right) \tag{8.89}$$

Discounted to time T, this becomes

$$\max(0, 1 - B(T, T+\delta) - \delta\kappa B(T, T+\delta)) = (1 + \delta\kappa)\max\left(0, \frac{1}{1+\delta\kappa} - B(T, T+\delta)\right)$$

Thus the caplet is equivalent to $(1 + \delta\kappa)$ put options on a zero coupon bond maturing at $T + \delta$, struck at $1/(1 + \delta\kappa)$, where the option expires at time T. The member function caplet() implements this through a call to the member function ZCBoption().

The FX option formula of Section 8.4.1 is implemented in the member function FXoption(). The instance *this of GaussianHJM, from which this member function is called, is assumed to represent the domestic economy. As can be seen in Listing 8.5, FXoption() takes as arguments a reference to an instance of GaussianHJM representing the foreign economy, a reference to an

instance of (a class derived from) `DeterministicAssetVol` representing the
volatility of the exchange rate, and the initial value $X(0)$ of the exchange
rate. The bulk of the code in `FXoption()` serves to calculate the integral
(collected in the variable `var`) over the volatility of the forward exchange rate.
In order to be able to abstract from the specific form of the three volatility
functions involved in the pricing formula in Section 8.4.1 ($\sigma_X(u)$, $\tilde{\sigma}_B(u, T)$,
$\sigma_B(u, T)$), integrals over the products between these volatility functions are
calculated by virtual member functions of `DeterministicAssetVol`. These
member functions are

    `DeterministicAssetVol::volproduct()`
    `DeterministicAssetVol::bondvolproduct()`   (8.90)
    `DeterministicAssetVol::bondbondvolproduct()`

These versions are necessary because an instance of `DeterministicAssetVol`
may enter the calculation of the volatility integral directly (e.g. $\sigma_X(u)$), or
indirectly via a zero coupon bond volatility, which is itself an integral, i.e.

$$\sigma_B(u, T) = -\sigma^*(u, T) = \int_u^T \sigma(u, s)ds$$

where an instance of `DeterministicAssetVol` represents $\sigma(u, s)$, the HJM
volatility of the instantaneous forward rates. Thus, in Listing 8.5, the statement

    `fxvol.volproduct(0.0,T1,fxvol);`   (8.91)

calculates

$$\int_0^{T_1} \sigma_X^2(u)du$$

The general calling convention for `DeterministicAssetVol::volproduct()`
is (if the `DeterministicAssetVol` `vol1` represents the volatility function
$\sigma^{(1)}(u)$ and `vol2` the volatility function $\sigma^{(2)}(u)$) that

    `vol1.volproduct(t,`Δt`,vol2);`   (8.92)

calculates

$$\int_t^{t+\Delta t} \sigma^{(1)}(u)\sigma^{(2)}(u)du$$

If `vol1` represents the HJM volatility of instantaneous forward rates, we have

$$\sigma_B^{(1)}(u, T) = -\int_u^T \sigma^{(1)}(u, s)ds$$

and the statement

    `vol1.bondvolproduct(t,`Δt`,T,vol2);`   (8.93)

calculates

$$\int_t^{t+\Delta t} \sigma_B^{(1)}(u, T)\sigma^{(2)}(u)du$$

If in addition `vol2` represents the HJM volatility of foreign instantaneous
forward rates, the statement

calculates

$$\int_t^{t+\Delta t} \sigma_B^{(1)}(u, T_1)\sigma_B^{(2)}(u, T_2)du$$

Like the implementation of (8.92), the implementation of (8.93) and (8.94) is a non-trivial task, as it depends on the concrete specification (e.g. ConstVol, ExponentialVol or PiecewiseVol) of not only vol1, but also vol2 (and these may be mixed and matched). Section 4.4.1 of Chapter 4 discussed how for the implementation of volproduct(), the interaction of volatility function objects of different concrete types is handled through the class Deterministic-VolMediator. DeterministicVolMediator also handles this interaction for bondvolproduct() and bondbondproduct(). Thus, when a new concrete class derived from DeterministicAssetVol is added to the library, the corresponding functionality for calculating volatility integrals has to be added to DeterministicVolMediator. This is admittedly tedious, but still far less tedious than modifying the volatility integral calculations wherever they are needed — the latter is avoided by encapsulating the volatility integral calculations in the three member functions of the abstract base class Deterministic-AssetVol.

Similarly, in order to price a European option on an asset with deterministic volatility and a deterministic term structure of dividend yields (i.e. a BlackScholesAsset) under stochastic interest rates given by a Gauss/Markov HJM model, GaussianHJM::option() calculates the integral over the volatility of the forward asset price using the three member functions of DeterministicAssetVol, (8.92), (8.93) and (8.94).

GaussianHJM also supplies the member function QuantoCaplet(), which prices the payoff of a foreign caplet, converted to domestic currency at a fixed exchange rate X_0. To price the appropriately modified caplet payoff (8.89) by calculating the expectation under the domestic forward measure to the payoff time $T + \delta$, consider

$$B(0, T + \delta)X_0\delta E_{T+\delta}\left[\left[\frac{1}{\delta}\left(\frac{1}{\tilde{B}(T, T+\delta)} - 1\right) - \tilde{\kappa}\right]^+\right]$$

$$= B(0, T + \delta)X_0\delta E_{T+\delta}\left[\left[\frac{\tilde{B}(0, T)}{\tilde{B}(0, T+\delta)}\right.\right. \tag{8.95}$$

$$\exp\left\{\int_0^T (\tilde{\sigma}_B(u, T) - \tilde{\sigma}_B(u, T+\delta))d\tilde{W}_{T+\delta}(u)\right.$$

$$\left.-\frac{1}{2}\int_0^T (\tilde{\sigma}_B(u, T) - \tilde{\sigma}_B(u, T+\delta))^2 du\right\} - (1 + \delta\tilde{\kappa})\Bigg]^+\Bigg]$$

Defining the quanto correction

$$q(T) = \exp\left\{ \int_0^T (\tilde{\sigma}_B(u,T) - \tilde{\sigma}_B(u,T+\delta)) \right.$$

$$\left. (-\tilde{\sigma}_B(u,T+\delta) + \sigma_B(u,T+\delta) - \sigma_X(u))du \right\} \quad (8.96)$$

and transforming $\tilde{W}_{T+\delta}$ into $W_{T+\delta}$ by the appropriate drift correction, (8.95) becomes

$$B(0,T+\delta)X_0 q(T)\delta E_{T+\delta}\left[\left[\frac{\tilde{B}(0,T)}{\tilde{B}(0,T+\delta)}\right.\right.$$

$$\exp\left\{ \int_0^T (\tilde{\sigma}_B(u,T) - \tilde{\sigma}_B(u,T+\delta))dW_{T+\delta}(u) \right.$$

$$\left. -\frac{1}{2}\int_0^T (\tilde{\sigma}_B(u,T) - \tilde{\sigma}_B(u,T+\delta))^2 du \right\} - \left.\frac{1+\delta\tilde{\kappa}}{q(T)}\right]^+\right]$$

$$= B(0,T+\delta)X_0 q(T)\left(\frac{\tilde{B}(0,T)}{\tilde{B}(0,T+\delta)}\mathcal{N}(h_1) - \frac{1+\delta\tilde{\kappa}}{q(T)}\mathcal{N}(h_2)\right) \quad (8.97)$$

with

$$h_{1,2} = \frac{\ln\left(\frac{\tilde{B}(0,T)}{\tilde{B}(0,T+\delta)}\frac{q(T)}{1+\delta\tilde{\kappa}}\right) \pm \frac{1}{2}\int_0^T (\tilde{\sigma}_B(u,T) - \tilde{\sigma}_B(u,T+\delta))^2 du}{\int_0^T (\tilde{\sigma}_B(u,T) - \tilde{\sigma}_B(u,T+\delta))^2 du}$$

Note that the expectation in (8.97) resolves to a standard Black/Scholes call option formula on an "asset" with initial value $\tilde{B}(0,T)/\tilde{B}(0,T+\delta)$, volatility $\tilde{\sigma}_B(u,T) - \tilde{\sigma}_B(u,T+\delta)$, and a "zero" interest rate — this is how (8.97) is implemented in the function `QuantoCaplet()`.[11]

Furthermore, `GaussianHJM` implements closed–form valuation of a swap netting the difference between foreign and domestic floating interest rates paid on a domestic unit notional (a *differential swap* or *diff swap*) in the member function `DiffSwap()`. The swap is characterised by a tenor structure T of accrual periods $[T_i, T_{i+1}]$, with time T_{i+1} cashflows of

$$(T_{i+1} - T_i)(\tilde{r}_a(T_i, T_{i+1}) - r_a(T_i, T_{i+1}))$$

The time T_0 value of this cashflow is given by

$$\sum_{i=0}^{n-1} B(T_0, T_{i+1})(T_{i+1} - T_i)E_{T_{i+1}}[\tilde{r}_a(T_i, T_{i+1}) - r_a(T_i, T_{i+1})]$$

---

[11] As noted above, this instrument, and many others, also can be priced using M-Binaries, the implementation of which is discussed below.

As $r_a(T_i, T_{i+1})$ is martingale under the time T_{i+1} forward measure, this equals

$$\sum_{i=0}^{n-1} B(T_0, T_{i+1})(T_{i+1} - T_i) \left(E_{T_{i+1}}[\tilde{r}_a(T_i, T_{i+1})] - f_a(T_0, T_i, T_{i+1}) \right)$$

Thus it remains to calculate the expectation of the foreign interest rate with simple compounding under the domestic time T_{i+1} forward measure:

$$E_{T_{i+1}}[\tilde{r}_a(T_i, T_{i+1})] = \frac{1}{T_{i+1} - T_i} \left(E_{T_{i+1}}[\tilde{B}(T_i, T_{i+1})^{-1}] - 1 \right)$$

and we have

$$E_{T_{i+1}}[\tilde{B}(T_i, T_{i+1})^{-1}]$$

$$= E_{T_{i+1}} \left[\frac{\tilde{B}(T_0, T_i)}{\tilde{B}(T_0, T_{i+1})} \exp \left\{ \int_{T_0}^{T_i} (\tilde{\sigma}_B(u, T_i) - \tilde{\sigma}_B(u, T_{i+1})) d\tilde{W}_{T_{i+1}}(u) \right. \right.$$

$$\left. \left. - \frac{1}{2} \int_{T_0}^{T_i} (\tilde{\sigma}_B(u, T_i) - \tilde{\sigma}_B(u, T_{i+1}))^2 du \right\} \right]$$

$$= \frac{\tilde{B}(T_0, T_i)}{\tilde{B}(T_0, T_{i+1})} E_{T_{i+1}} \left[\exp \left\{ \int_{T_0}^{T_i} (\tilde{\sigma}_B(u, T_i) - \tilde{\sigma}_B(u, T_{i+1})) dW_{T_{i+1}}(u) \right. \right.$$

$$\left. \left. - \frac{1}{2} \int_{T_0}^{T_i} (\tilde{\sigma}_B(u, T_i) - \tilde{\sigma}_B(u, T_{i+1}))^2 du \right\} \right]$$

$$\exp \left\{ \int_{T_0}^{T_i} (\tilde{\sigma}_B(u, T_i) - \tilde{\sigma}_B(u, T_{i+1})) \right.$$

$$\left. (-\tilde{\sigma}_B(u, T_{i+1}) + \sigma_B(u, T_{i+1}) - \sigma_X(u)) du \right\}$$

$$= \frac{\tilde{B}(T_0, T_i)}{\tilde{B}(T_0, T_{i+1})} \exp \left\{ \int_{T_0}^{T_i} (\tilde{\sigma}_B(u, T_i) - \tilde{\sigma}_B(u, T_{i+1})) \right.$$

$$\left. (-\tilde{\sigma}_B(u, T_{i+1}) + \sigma_B(u, T_{i+1}) - \sigma_X(u)) du \right\}$$

which again can be calculated via the virtual member functions bondvol-product() and bondbondvolproduct() of DeterministicAssetVol.

For a more general approach to closed form solutions, the M–Binary pricing formula discussed in Section 8.4.2 is also implemented. With the move to the multicurrency Gauss/Markov HJM setting, the only changes in the M–Binary pricing formula are in the covariance matrix $\overline{\Gamma}$ of logarithmic asset prices and the vector $\overline{\mu}$ of the expectations of the logarithmic asset prices under the domestic time T_M forward measure. Thus all that is necessary is to overload the constructor of the class MBinary already implemented in the context of Chapter 4. The constructor

```
GaussMarkovWorld world("WorldBSF.csv");
Array<double,1> T(2);
T = 0.0, mat;
world.set_timeline(T);
int FXidx        = world.set_reporting(1,-2);
int iZCBindex    = world.set_reporting(0,-1,mat);
int iFZCBindex   = world.set_reporting(1,-1,mat);
// Price option using MBinary
boost::shared_ptr<MBinaryPayoff>
  A_IFXpayoff(new
    MBinaryPayoff(*world.get_economies()[0]->initialTS,2,3,1));
boost::shared_ptr<MBinaryPayoff>
  B_IFXpayoff(new
    MBinaryPayoff(*world.get_economies()[0]->initialTS,2,3,1));
// imat is time of option expiry
A_IFXpayoff->timeline = 0.0, imat;
A_IFXpayoff->index    = FXidx, iZCBindex, iFZCBindex,
                        1, 1, 1;
A_IFXpayoff->alpha    = 1.0, 1.0, -1.0;
A_IFXpayoff->S        = 1.0;
A_IFXpayoff->A        = 1.0, 1.0, -1.0;
A_IFXpayoff->a        = strikeIFX;
B_IFXpayoff->timeline = 0.0, imat;
B_IFXpayoff->index    = FX_index, iZCBindex, iFZCBindex,
                        1, 1, 1;
B_IFXpayoff->alpha    = 0.0;
B_IFXpayoff->S        = 1.0;
B_IFXpayoff->A        = 1.0, 1.0, -1.0;
B_IFXpayoff->a        = strikeIFX;
MBinary AIFX(world,*A_IFXpayoff);
MBinary BIFX(world,*B_IFXpayoff);
double AIFXvalue = AIFX.price();
double BIFXvalue = BIFX.price();
double MBCFcallIFX = AIFXvalue - strikeIFX*BIFXvalue;
cout << "FX intermediate maturity option closed form value using MBinary: ";
cout << MBCFcallIFX << endl;
```

Listing 8.6: Pricing an FX call option using M–Binaries.

$$MBinary(const\ GaussMarkovWorld\&\ world, \\ MBinaryPayoff\&\ xpayoff); \qquad (8.98)$$

calculates $\overline{\mu}$ and $\overline{\Gamma}$ according to (8.76) to (8.80), using the virtual member functions volproduct(), bondvolproduct() and bondbondvolproduct() of DeterministicAssetVol.

Suppose we wish to price an FX call option using M–Binaries. First, we construct an instance of GaussMarkovWorld. For simplicity, we use the constructor which takes the name of a CSV file as its argument:[12]

$$GaussMarkovWorld\ world("WorldBSF.csv"); \qquad (8.99)$$

This particular input file on the website for this book constructs an instance of GaussMarkovWorld with two currencies, and one equity asset in each currency.

---

[12] GaussMarkovWorld and its associated classes implement such a constructor. For an example of the CSV (common separated values) input files, see the folder "CSV inputs" on the website for this book.

For an initial ("now") time 0 and a time horizon `mat`, we initialise the time line of our `GaussMarkovWorld`:

```
Array<double,1>
T(2);
T = 0.0, mat;
world.set_timeline(T);
```
(8.100)

The member function `set_reporting()` of `GaussMarkovWorld` adds an asset in our world to the list of "reportable assets." `set_reporting()` returns an integer, which is the unique identifying index of this asset in the list of reportable assets. As per the payoff specification (8.83), for the M–Binary representation we need the terminal forward exchange rate, so we call

```
int FXidx = world.set_reporting(1,-2);
```
(8.101)

The first argument of `set_reporting()` is the integer index of the foreign currency; the second argument is the integer "asset index." This is defined such that if the asset index is non-negative, it is the integer index of an equity asset in the currency given by the first argument of `set_reporting()`; if the asset index is -2, it refers to the terminal forward exchange rate. If the asset index is -1, it refers to a zero coupon bond in the currency given by the first argument, and the maturity of this zero coupon bond is given by a third argument of `set_reporting()`. In the specification (8.83), we need the domestic and the foreign zero coupon bonds maturing at time $\mathtt{mat} = \overline{T}$, so we call

```
int iZCBindex = world.set_reporting(0,-1,mat);
int iFZCBindex = world.set_reporting(1,-1,mat);
```
(8.102)

The two parts of (8.83), i.e.

$$[Y_1(T_1)Y_4(T_1)Y_5^{-1}(T_1) - K]^+ =$$

$$Y_1(T_1)Y_4(T_1)Y_5^{-1}(T_1)\mathbb{I}_{\left\{\frac{Y_1(T_1)Y_4(T_1)}{Y_5(T_1)} > K\right\}} - K\mathbb{I}_{\left\{\frac{Y_1(T_1)Y_4(T_1)}{Y_5(T_1)} > K\right\}}$$

define two M–Binaries with payoff dimension 3 and exercise dimension 1, so we can proceed as in Listing 8.6, initialising these M–Binaries and calculating the value of the option by calling the member function `price()` and combining the results.

Tables 8.1 and 8.2 summarise the M–Binary inputs for the specifications (8.82) to (8.88). The example program `MBinaryGaussianTest.cpp` on the website calculates option prices for these payoff specifications using M–Binaries and alternatively by the member functions of `GaussianHJM` (where available). This program also prices the options by Monte Carlo simulation, which is discussed in the next section.

8.6 Monte Carlo simulation in the HJM framework

In the chosen model (i.e. Gauss/Markov HJM for the term structure(s) of interest rates, combined with lognormal exchange rates and equity prices), the distribution of the state variables (which we wish to simulate) is available in closed form. Given the state variables, the values for all other relevant

Equation	(8.82)		(8.83)		(8.84)	
# of assets $N =$	2		3		4	
Payoff dimension $n =$	2		3		4	
# of observation times $M =$	1		1		1	
Exercise dimension $m =$	1		1		1	
Asset index mapping $\bar{i}(i) =$	$(2\ 4)$		$(1\ 4\ 5)$		$(1\ 2\ 4\ 5)$	
Time index mapping $\bar{k}(k) =$	1		1		1	
	$V^{(1)}$	$V^{(2)}$	$V^{(1)}$	$V^{(2)}$	$V^{(1)}$	$V^{(2)}$
Payoff powers α	$(1\ 1)$	$(0\ 0)$	$(1\ 1\ -1)$	$(0\ 0\ 0)$	$(1\ 1\ 1\ 0)$	$(1\ 0\ 1\ -1)$
Sign matrix S	1	1	1	1	1	1
Indicator powers A	$(1\ 1)$	$(1\ 1)$	$(1\ 1\ -1)$	$(1\ 1\ -1)$	$(0\ 1\ 0\ 1)$	$(0\ 1\ 0\ 1)$
Thresholds a	$D_{X_2}(T_1,\overline{T})K$	$D_{X_2}(T_1,\overline{T})KV^{(2)}$	K	K	$D_{X_3}(T_1,\overline{T})\check{K}$	$D_{X_3}(T_1,\overline{T})\check{K}K$
Price	$(V^{(1)} - D_{X_2}(T_1,\overline{T})KV^{(2)})/D_{X_2}(T_1,\overline{T})$		$V^{(1)} - KV^{(2)}$		$(V^{(1)} - D_{X_3}(T_1,\overline{T})\check{K}V^{(2)})/D_{X_3}(T_1,\overline{T})$	

Table 8.1 *M–Binary inputs to calculate the prices for the specifications (8.82) to (8.84).*

Equation	(8.85)		(8.86)		(8.87)		(8.88)	
# of assets $N =$	1		4		2		2	
Payoff dimension $n =$	1		4		2		2	
# of observation times $M =$	1		1		1		1	
Exercise dimension $m =$	1		1		1		1	
Asset index mapping $\bar{\imath}(i) =$	6		$(1\ 4\ 5\ 7)$		$(3\ 5)$		$(8\ 9)$	
Time index mapping $\bar{k}(k) =$	1		1		1		1	
	$V^{(1)}$	$V^{(2)}$	$V^{(1)}$	$V^{(2)}$	$V^{(1)}$	$V^{(2)}$	$V^{(1)}$	$V^{(2)}$
Payoff powers α	1	0	$(1\,1 - 1\,1)$	$(1\,1 - 1\,0)$	$(1\,1)$	$(0\,0)$	$(1 - 1)$	$(1\,0)$
Sign matrix S	1	1	1	1	1	1	1	1
Indicator powers A	1	1	$(0\,0\,0\,1)$	$(0\,0\,0\,1)$	$(1\,1)$	$(1\,1)$	$(0 - 1)$	$(0 - 1)$
Thresholds a	K	K	$\tilde{K}$	$\tilde{K}$	$D_{X_3}(T_1,\overline{T})\tilde{K}$	$D_{X_3}(T_1,\overline{T})\tilde{K}$	$1+\delta\tilde{\kappa}$	$1+\delta\tilde{\kappa}$
Price	$V^{(1)} - KV^{(2)}$		$V^{(1)} - \tilde{K}V^{(2)}$		$\dfrac{X_0(V^{(1)} - D_{X_3}(T_1,\overline{T})\tilde{K}V^{(2)})}{D_{X_3}(T_1,\overline{T})}$		$X_0(V^{(1)} - (1+\delta\tilde{\kappa})V^{(2)})$	

Table 8.2 M–Binary inputs to calculate the prices for the specifications (8.85) to (8.88).

variables (e.g. bonds, interest rates, exchange rates, equity prices) can be calculated explicitly. However, the distribution of the continuously compounded
savings account (which serves as the numeraire of what is commonly taken
to be the risk-neutral measure under which the simulation is conducted) is
not Markovian in the state variables. Therefore, the simulation is instead
conducted under the *discrete rolling spot measure*, the numeraire for which
consists of a strategy of rolling over an investment into a zero coupon bond
maturing at the next date in the simulation time line.[13] In this case, the use
of explicit solutions eliminates all discretisation bias, i.e. the distribution of
rates for a time, say, 10 years from now, is correct whether we have a single 10
year time step in the simulation, or 1000 time steps of 0.01 years each. Note
that as the time step length goes to zero, the discrete rolling spot measure
approaches the continuous rolling spot measure.

8.6.1 Simulation of interest rate dynamics in a single currency

Denote by t_i, $0 \leq i \leq M$, the points along the time line for which we want to
simulate the interest rate dynamics. The simulation proceeds as follows: For
each time step,

1. Generate the state variable increments under the discrete rolling spot measure.

2. Propagate the state variables.

3. Propagate the numeraire process.

Step 1: State variable increments

By definition of the discrete rolling spot measure, the conditional distribution
of $z_j(t_{i+1})$ given $z_j(t_i)$ under this measure corresponds to the conditional distribution under the time t_{i+1} forward measure. From (8.50), the state variable
increments

$$\Delta z_j(t_i) = z_j(t_{i+1}) - z_j(t_i) \tag{8.103}$$

have the distribution

$$\Delta z_j(t_i) \overset{\mathbf{P}_{t_{i+1}}}{\underset{\sim}{}}$$

$$\mathcal{N}\left(\int_{t_i}^{t_{i+1}} \frac{\nu_j^2(u)}{a_j} e^{a_j u} \left(e^{-a_j(t_{i+1}-u)} - 1 \right) du, \int_{t_i}^{t_{i+1}} \nu_j^2(u) e^{2a_j u} du \right) \tag{8.104}$$

i.e. we can simulate the $\Delta z_j(t_i)$ as independent, normally distributed random
variables with mean and variance as given in (8.104).

---

[13] By adding a further state variable, the continuously compounded savings account could
also be simulated explicitly. However, this increases the computational overhead for no
practical gain.

Step 2: Propagate the state variables

Given $\Delta z_j(t_i)$ and $z_j(t_i)$, we can calculate $z_j(t_{i+1})$ using (8.103).

Step 3: Propagate the numeraire process

The value at time t_i of the discrete rollover account is given by

$$A(t_i) = \prod_{k=1}^{i} B(t_{k-1}, t_k)^{-1} \qquad (8.105)$$

Thus, to propagate the numeraire process from time t_i to t_{i+1}, we have

$$A(t_{i+1}) = \frac{A(t_i)}{B(t_i, t_{i+1})} = \frac{A(t_i)}{\mathcal{A}(t_i, t_{i+1})} e^{\mathcal{B}(t_i, t_{i+1})z(t_i)} \qquad (8.106)$$

with the deterministic functions $\mathcal{A}$ and $\mathcal{B}$ given by (8.24) and (8.25). Note that $\mathcal{B}(t_i, t_{i+1})z(t_i)$ is to be interpreted as the sum product between two vectors.

8.6.2 Joint simulation in several currencies

In a joint simulation in several currencies, all dynamics must be simulated under *one* chosen probability measure. We choose this measure to be the domestic discrete rolling spot measure. Thus, in general, we need the dynamics of the foreign state variables $\tilde{z}_j(t)$ under domestic forward measures. Analogously to the domestic case, we have

$$d\tilde{z}_j(t) = \tilde{\nu}_j(t)e^{\tilde{a}_j t}d\tilde{W}_\beta^{(j)}(t) \qquad (8.107)$$

Applying (8.62) and (8.63), this becomes

$$
\begin{aligned}
d\tilde{z}_j(t) &= \tilde{\nu}_j(t)e^{\tilde{a}_j t}dW_\beta^{(j)}(t) - \tilde{\nu}_j(t)e^{\tilde{a}_j t}\sigma_X^{(j)}(t)dt \\
&= \tilde{\nu}_j(t)e^{\tilde{a}_j t}dW_{T_i}^{(j)}(t) + \\
&\quad \frac{\tilde{\nu}_j(t)\nu_j(t)}{a_j}e^{\tilde{a}_j t}\left(e^{-a_j(T_i-t)} - 1\right)dt - \tilde{\nu}_j(t)e^{\tilde{a}_j t}\sigma_X^{(j)}(t)dt
\end{aligned} \qquad (8.108)
$$

so for the foreign state variable increments we have the distribution

$$
\Delta\tilde{z}_j(t_i) \overset{\mathbf{P}_{t_{i+1}}}{\sim}
$$
$$
\mathcal{N}\left(\int_{t_i}^{t_{i+1}} \tilde{\nu}_j(u)e^{\tilde{a}_j u}\left(\frac{\nu_j(u)}{a_j}\left(e^{-a_j(t_{i+1}-u)} - 1\right) - \sigma_X^{(j)}(u)\right)du,\right.
$$
$$
\left. \int_{t_i}^{t_{i+1}} \tilde{\nu}_j^2(u)e^{2\tilde{a}_j u}du\right) \qquad (8.109)
$$

However, since the same j–th component of the driving Brownian motion determines the evolution of $\Delta\tilde{z}_j(t_i)$ and $\Delta z_j(t_i)$, these two random variables are correlated. To determine the correlation coefficient, note first that

$$\text{Var}[\Delta z_j(t_i) + \Delta \tilde{z}_j(t_i)] = \text{Var}[\Delta z_j(t_i)] + \text{Var}[\Delta \tilde{z}_j(t_i)]$$
$$+ 2\text{Corr}[\Delta z_j(t_i), \Delta \tilde{z}_j(t_i)]\sqrt{\text{Var}[\Delta z_j(t_i)]\text{Var}[\Delta \tilde{z}_j(t_i)]} \quad (8.110)$$

By Itô's Lemma and the properties of stochastic integrals, it further holds that

$$\text{Var}[\Delta z_j(t_i) + \Delta \tilde{z}_j(t_i)] = \int_{t_i}^{t_{i+1}} \left(\nu_j(u)e^{a_j u} + \tilde{\nu}_j(u)e^{\tilde{a}_j u}\right)^2 du \quad (8.111)$$

allowing us to solve (8.110) for $\text{Corr}[\Delta z_j(t_i), \Delta \tilde{z}_j(t_i)]$.

The terminal forward exchange rate $X(t)\tilde{B}(t,\overline{T})/B(t,\overline{T})$ is a martingale under the terminal forward measure, i.e.

$$\frac{X(t)\tilde{B}(t,\overline{T})}{B(t,\overline{T})} = \frac{X(0)\tilde{B}(0,\overline{T})}{B(0,\overline{T})}$$
$$\exp\left\{ \int_0^t \left(\sigma_X(u) + \tilde{\sigma}_B(u,\overline{T}) - \sigma_B(u,\overline{T})\right) dW_{\overline{T}}(u) \right.$$
$$\left. - \frac{1}{2} \int_0^t \left(\sigma_X(u) + \tilde{\sigma}_B(u,\overline{T}) - \sigma_B(u,\overline{T})\right)^2 du \right\} \quad (8.112)$$

Define the state variable x for the terminal forward exchange rate as

$$x(t) := \int_0^t \left(\sigma_X(u) + \tilde{\sigma}_B(u,\overline{T}) - \sigma_B(u,\overline{T})\right) dW_{\overline{T}}(u)$$
$$- \frac{1}{2} \int_0^t \left(\sigma_X(u) + \tilde{\sigma}_B(u,\overline{T}) - \sigma_B(u,\overline{T})\right)^2 du \quad (8.113)$$

Then the state variable increment from time t_i to time t_{i+1} is

$$\Delta x(t_i) = \int_{t_i}^{t_{i+1}} \left(\sigma_X(u) + \tilde{\sigma}_B(u,\overline{T}) - \sigma_B(u,\overline{T})\right) dW_{\overline{T}}(u)$$
$$- \frac{1}{2} \int_{t_i}^{t_{i+1}} \left(\sigma_X(u) + \tilde{\sigma}_B(u,\overline{T}) - \sigma_B(u,\overline{T})\right)^2 du \quad (8.114)$$

These increments are normally distributed with variance

$$\int_{t_i}^{t_{i+1}} \left(\sigma_X(u) + \tilde{\sigma}_B(u,\overline{T}) - \sigma_B(u,\overline{T})\right)^2 du \quad (8.115)$$

The expectation of $\Delta x(t_i)$ under $\mathbf{P}_{t_{i+1}}$ (i.e., the discrete rolling spot measure under which the Monte Carlo simulation is being conducted) is given by

$$\int_{t_i}^{t_{i+1}} \left(\sigma_X(u) + \tilde{\sigma}_B(u,\overline{T}) - \sigma_B(u,\overline{T})\right) \left(\sigma_B(u,T_{i+1}) - \sigma_B(u,\overline{T})\right) du$$
$$- \frac{1}{2} \int_{t_i}^{t_{i+1}} \left(\sigma_X(u) + \tilde{\sigma}_B(u,\overline{T}) - \sigma_B(u,\overline{T})\right)^2 du \quad (8.116)$$

Similarly, considering the terminal forward price of a domestic equity asset X_j,

$$\frac{X_j(t)D_{X_j}(t,\overline{T})}{B(t,\overline{T})} = \frac{X_j(0)D_{X_j}(0,\overline{T})}{B(0,\overline{T})} \exp\left\{\int_0^t \left(\sigma_{X_j}(u) - \sigma_B(u,\overline{T})\right) dW_{\overline{T}}(u)\right.$$

$$\left. - \frac{1}{2}\int_0^t \left(\sigma_{X_j}(u) - \sigma_B(u,\overline{T})\right)^2 du\right\} \quad (8.117)$$

The increments of the corresponding state variable are

$$\Delta x_j(t_i) = \int_{t_i}^{t_{i+1}} \left(\sigma_{X_j}(u) - \sigma_B(u,\overline{T})\right) dW_{\overline{T}}(u)$$

$$- \frac{1}{2}\int_{t_i}^{t_{i+1}} \left(\sigma_{X_j}(u) - \sigma_B(u,\overline{T})\right)^2 du \quad (8.118)$$

which are normally distributed with variance

$$\int_{t_i}^{t_{i+1}} \left(\sigma_{X_j}(u) - \sigma_B(u,\overline{T})\right)^2 du \quad (8.119)$$

The expectation of $\Delta x_j(t_i)$ under $\mathbf{P}_{t_{i+1}}$ is given by

$$\int_{t_i}^{t_{i+1}} \left(\sigma_{X_j}(u) - \sigma_B(u,\overline{T})\right)\left(\sigma_B(u,T_{i+1}) - \sigma_B(u,\overline{T})\right) du$$

$$- \frac{1}{2}\int_{t_i}^{t_{i+1}} \left(\sigma_{X_j}(u) - \sigma_B(u,\overline{T})\right)^2 du \quad (8.120)$$

For the foreign equity asset $\tilde{X}_j$, we have

$$\frac{\tilde{X}_j(t)D_{\tilde{X}_j}(t,\overline{T})}{\tilde{B}(t,\overline{T})} = \frac{\tilde{X}_j(0)D_{\tilde{X}_j}(0,\overline{T})}{\tilde{B}(0,\overline{T})} \exp\left\{\int_0^t \left(\sigma_{\tilde{X}_j}(u) - \tilde{\sigma}_B(u,\overline{T})\right) d\tilde{W}_{\overline{T}}(u)\right.$$

$$\left. - \frac{1}{2}\int_0^t \left(\sigma_{\tilde{X}_j}(u) - \tilde{\sigma}_B(u,\overline{T})\right)^2 du\right\} \quad (8.121)$$

Thus the associated state variable increment $\Delta\tilde{x}_j(t_i)$ has variance

$$\int_{t_i}^{t_{i+1}} \left(\sigma_{\tilde{X}_j}(u) - \tilde{\sigma}_B(u,\overline{T})\right)^2 du \quad (8.122)$$

and its expectation under $\mathbf{P}_{t_{i+1}}$ (after an additional measure change from the foreign to the domestic discrete rolling spot measure) is given by (note that $\sigma_X(u)$ is the spot exchange rate volatility)

$$\int_{t_i}^{t_{i+1}} \left(\sigma_{\tilde{X}_j}(u) - \tilde{\sigma}_B(u,\overline{T})\right)\left(\sigma_B(u,T_{i+1}) - \tilde{\sigma}_B(u,\overline{T} - \sigma_X(u))\right) du$$

$$- \frac{1}{2}\int_{t_i}^{t_{i+1}} \left(\sigma_{\tilde{X}_j}(u) - \tilde{\sigma}_B(u,\overline{T})\right)^2 du \quad (8.123)$$

Lastly, in order to simulate the system of jointly normally distributed state variables z_j, $\tilde{z}_k$, x, x_ℓ, $\tilde{x}_m$, note that the relevant covariances are given by integrals over the products of volatility functions. Thus we have, for example,

$$\text{Cov}[\Delta z_j(t_i), \Delta \tilde{z}_k(t_i)] = \int_{t_i}^{t_{i+1}} \nu_j(u) e^{a_j u} \tilde{\nu}_k(u) e^{\tilde{a}_k u} du \qquad (8.124)$$

$$\text{Cov}[\Delta z_j(t_i), \Delta x(t_i)] = \int_{t_i}^{t_{i+1}} \nu_j(u) e^{a_j u}$$
$$\left(\sigma_X(u) + \tilde{\sigma}_B(u, \overline{T}) - \sigma_B(u, \overline{T}) \right) du \quad (8.125)$$

$$\text{Cov}[\Delta x(t_i), \Delta \tilde{x}_m(t_i)] = \int_{t_i}^{t_{i+1}} \left(\sigma_X(u) + \tilde{\sigma}_B(u, \overline{T}) - \sigma_B(u, \overline{T}) \right)$$
$$\left(\sigma_{\tilde{X}_m}(u) - \tilde{\sigma}_B(u, \overline{T}) \right) du \qquad (8.126)$$

with the remaining combinations constructed analogously.

The multicurrency simulation thus proceeds as follows: For each time step,

1. Generate the (correlated) state variable increments under the discrete rolling spot measure.

2. Propagate the state variables.

3. Propagate the numeraire process using (8.106).

8.7 Implementing Monte Carlo simulation

In keeping with the object–oriented programming paradigm, the classes and templates for Monte Carlo simulation introduced in Chapter 7 remain unchanged, and only the class representing the stochastic model is replaced, i.e. the class GeometricBrownianMotion (representing a Black/Scholes model of multiple assets driven by geometric Brownian motions) is replaced by the class GaussMarkovWorld. Thus the model–specific mapping of an Array<double,2> x of independent standard normal random variables to realisations of random paths of underlying financial variables is implemented in the public operator() of GaussMarkovWorld, i.e.

```
    void GaussMarkovWorld::
        operator()(Array <double,2>& underlying_values,
            Array<double,1>& numeraire_values,
            const Array<double,2>& x,
            const TermStructure& ts,
            int numeraire_index);
```
(8.127)

This generates a realisation of the paths of the chosen underlying financial variables under the martingale measure associated with a given numeraire asset.[14] The operator argument numeraire_index determines under which measure

---

[14] Note that the other, overloaded version of operator() provided by GeometricBrownian-Motion, which generates a realisation of financial variable paths under the martingale measure associated with deterministic bond prices as the numeraire, is not applicable in the context of stochastic interest rates.

the paths are generated. A `numeraire_index` of zero means simulation under the discrete domestic rolling spot measure. A negative `numeraire_index` means simulation under a foreign discrete rolling spot measure, i.e. $-i$ corresponds to the i-th foreign currency. In this case, (as in all other cases), the value of the numeraire nevertheless is expressed in domestic currency. A positive `numeraire_index` means simulation under the martingale measure associated with taking the i-th risky asset in domestic currency as the numeraire. Note that simulation under the continuous rolling spot martingale measure $\mathbf{P}_\beta$ is not implemented, because the value of the continuously compounded savings account is not Markovian in the state variables chosen for the Gauss/Markov HJM model — an additional state variable would be required, adding to the computational effort for no obvious gain.

The paths of the financial variables are placed by the `operator()` in the (assets × time points) array `underlying_values`, where the assets are in the order in which they were placed in the list of reportable assets using the public member function

```
    int GaussMarkovWorld::
```
$$
\begin{array}{ll}
\texttt{set\_reporting(int currency\_index,} & \\
\qquad\qquad\quad \texttt{int asset\_index,} & \text{(8.128)} \\
\qquad\qquad\quad \texttt{double maturity = -1.0);} &
\end{array}
$$

which was introduced in Section 8.5 above.[15] The integer returned by the function gives the index in the first dimension of `underlying_values`, by which the generated path of the reportable asset can be accessed.

The path of the value of the numeraire asset over all time points is placed by the `operator()` into the array `numeraire_values`. The argument `const TermStructure& ts` of the `operator()` is retained for compatibility purposes (to `GeometricBrownianMotion`) and has no effect on the behaviour of the operator, as all the required initial term structures are contained in the instances of `GaussianEconomy` making up a `GaussMarkovWorld`.

Lastly, the realisations of the driving random variables are passed to the `operator()` in the argument `const Array<double,2>& x`, which references a (number of factors × number of time steps) array of independent standard normal random variates. As discussed in Chapter 7, these could be "independent" pseudo–random variates, or a single realisation from a (number factors × number of time steps)–dimensional quasirandom sequence. The number of factors is the maximum rank of the variance/covariance matrices of the state variable increments for each time step, set in the member function `set_timeline()` of `GaussMarkovWorld`, as discussed below. This number can be accessed by the member function `int factors()`.

Suppose now that we wish to price the payoffs (8.83) and (8.88) by Monte Carlo simulation. Suppose further that the constituent M-Binaries `AIFX` and `BIFX` of (8.83) have been set up as in Listing 8.6, and the constituent M-Binaries `AIquanto` and `BIquanto` (with the `MBinaryPayoffs` `A_IQpayoff` and `B_IQpayoff`) of (8.88) have been set up analogously. As in Listing 8.7, we first

---

[15] See p. 295.

```
MCPayoffList mclist;
boost::shared_ptr<MCPayoffList> calloptionIFX(new MCPayoffList);
calloptionIFX->push_back(A_IFXpayoff);
calloptionIFX->push_back(B_IFXpayoff,-strikeIFX);
mclist.push_back(calloptionIFX);
boost::shared_ptr<MCPayoffList> calloptionIQ(new MCPayoffList);
calloptionIQ->push_back(A_IQpayoff);
calloptionIQ->push_back(B_IQpayoff,-strikeIQ);
mclist.push_back(calloptionIQ,fxspot);
MCMapping<GaussMarkovWorld,Array<double,2> >
  mc_mapping(mclist,world,*(world.get_economies()[0]->initialTS),
             numeraire_index);
boost::function<Array<double,1> (Array<double,2>)> func =
  boost::bind(&MCMapping<GaussMarkovWorld,Array<double,2> >::
              mappingArray,
              &mc_mapping,_1);
RandomArray<ranlib::NormalUnit<double>,double>
random_container(normalRNG,world.factors(),world.number_of_steps());
MCGeneric<Array<double,2>,
          Array<double,1>,
          RandomArray<ranlib::NormalUnit<double>,double> >
  mc(func,random_container);
MCGatherer<Array<double,1> > mcgatherer(number_of_options);
size_t n = minpaths;
while (mcgatherer.number_of_simulations()<maxpaths) {
  mc.simulate(mcgatherer,n);
  n = mcgatherer.number_of_simulations(); }
```

Listing 8.7: Pricing M–Binaries by Monte Carlo in a Gauss/Markov HJM model.

combine the constituent MBinaryPayoffs of, respectively, (8.83) and (8.88) in a separate MCPayoffList, each of which are then added to the MCPayoffList mclist used to create an MCMapping — MCPayoffList and MCMapping being a class and a template as introduced in Chapter 7. We then proceed as in Chapter 7 to construct an instance mc of MCGeneric and running the Monte Carlo simulation by calling mc.simulate(). The full working example program containing the steps in Listing 8.7 can be found in MBinaryGaussianTest.cpp on the website for this book.

8.7.1 Propagating the state variables and calculating the realisations of financial variables

The functionality to propagate the state variables and calculate the realisations of financial variables is implemented in four member functions/operators of GaussMarkovWorld. As discussed above, the operator() generates a realisation of the paths of the chosen (through set_reporting()) financial

```
void GaussMarkovWorld::
propagate_state_variables(const Array<double,2>& x)
{
  int j;
  Array<double,1> current_x(number_of_state_variables);
  // propagate state variables, loop over time line
  for (j=1;j<state_variables->extent(secondDim);j++) {
  /* ordering: interest rate & asset state variables for each economy
     first, then exchange rate state variables term structure state
     variables are defined under the native currency spot measure all
     asset and exchange rate state variables are defined under the
     domestic terminal measure */
  current_x = x(Range(0,max_rank-1),j-1);
  mvn[j-1]->transform_random_variables(current_x);
  (*state_variables)(Range::all(),j) =
    (*state_variables)(Range::all(),j-1)
      + (*state_variable_drifts)(Range::all(),j-1) + current_x; }
}
```

Listing 8.8: Propagating state variables in Monte Carlo simulation in a Gauss/Markov HJM model.

variables and the numeraire asset. These financial variables are either zero coupon bonds, terminal forward exchange rates or terminal forward prices of equity assets, calculated from the state variables via (8.26), (8.112)–(8.113) and (8.117)–(8.118), respectively. The state variables are generated by the private member function

$$\text{void propagate\_state\_variables(const Array<double,2>\& x);} \qquad (8.129)$$

of GaussMarkovWorld. This function is called by the operator(), which passes it the array x of random variates driving the generation of the path realisation. As can be seen in Listing 8.8, the state variables are propagated along a path by adding the state variable increment, e.g.

$$x(t_{i+1}) = x(t_i) + \Delta x(t_i) \qquad (8.130)$$

where the state variables are jointly normally distributed under the simulation measure. The expectations of the state variable increments vary according to the chosen simulation measure/numeraire, and are contained in the (number of state variables × number of process time steps) array state_variable_drifts, which is filled with the appropriate values when the simulation measure/numeraire is chosen using the member function set_numeraire(). The variance/covariance matrix for the state variable increments does not depend on the numeraire; it is therefore initialised in the member function set_timeline().

8.7.2 Initialising the variance/covariance matrices and state variable drifts

The public member function of GaussMarkovWorld,

$$\texttt{bool set\_timeline(const Array<double,1>\& timeline);} \qquad (8.131)$$

sets the simulation time line, i.e. the time points for which the state variables, and thus the chosen financial variables, are going to be simulated. For each time step, it creates an instance of the class `MultivariateNormal`, representing a joint normal distribution with zero means and the variance/covariance matrix for the state variable increments, which is filled with values calculated according to (8.124)–(8.126), etc. By calling the appropriate member function of `MultivariateNormal`, i.e.

$$\texttt{mvn[j-1]->transform\_random\_variables(current\_x);} \qquad (8.132)$$

in Listing 8.8, the input independent random variates are transformed into variates with the correct covariances.[16]

In addition, `set_timeline()` appropriately resizes the `Arrays state_variables` (to hold the state variable realisations) and `state_variable_drifts`, initialises the (time zero) terminal forward exchange rates and equity prices, and calls `set_numeraire(current_numeraire)` to refresh the expectations of the state variable increments contained in `state_variable_drifts`. If the simulation is conducted under the domestic discrete rolling spot martingale measure, these expectations are calculated in `set_numeraire()` according to equations (8.104), (8.109), (8.116), (8.120) and (8.123). If a different numeraire is chosen, the appropriate measure changes are applied to the `state_variable_drifts`. Note that the calculations of the expectations and covariances of the state variable increments are expressed in terms of integrals over sum products of vector–valued volatility functions, just as in Section 8.4.2, and therefore — as in Section 8.5 — the implementation of these calculations also abstracts from the concrete specification of the volatility function (e.g. `ConstVol`, `ExponentialVol` or `PiecewiseVol`)[17] by encapsulating the volatility integral calculations in three virtual member functions of the abstract base class `DeterministicAssetVol`. For the calculations involving the state variables z_j or $\tilde{z}_j$, the pure virtual member function of `DeterministicAssetVol`,

$$\texttt{boost::shared\_ptr<DeterministicAssetVol>} \\ \texttt{component\_vol(int i) const;} \qquad (8.133)$$

when called from a concrete instance of a class derived from `Deterministic-AssetVol` representing the interest rate term structure HJM volatility function, returns a Boost `shared_ptr` to an instance representing the volatility of z_i.[18]

```
// world is an instance of GaussMarkovWorld
world.set_reporting(0,0);
world.set_reporting(0,-1,world.time_horizon());
// random_container is an instance of the appropriate RandomArray
// generate training paths for LS regression
MCTrainingPaths<GaussMarkovWorld,RandomArray<...> >
   training_paths(world,T,train,random_container,
                    *(domestic_economy.initialTS),numeraire_index);
// create basis functions
std::vector<boost::function<double (const Array<double,1>&,
                                    const Array<double,2>&)> >
   basisfunctions;
Array<int,1> p(2);
// basis functions as polynomials in terminal forward equity price
p(1) = 0.0;
for (i=0;i<=degree;i++) {
   p(0) = i;
   add_polynomial_basis_function(basisfunctions,p); }
/* similarly, create basis functions as polynomials in terminal ZCB &
   spot equity price */
...
/* Create payoff. BlackScholesAsset inputs are not used by Monte Carlo
   payoff from Mopt */
exotics::ProductOption Mopt(*(domestic_economy.underlying[0]),
                            *(domestic_economy.underlying[0]),
                            T(0),world.time_horizon(),strike,
                            *(domestic_economy.initialTS),-1);
boost::function<double (const Array<double,1>&,const Array<double,2>&)>
   payoff = boost::bind(&exotics::ProductOption::early_exercise_payoff,
                        &Mopt,_1,_2);
// Create exercise strategy based on LS regression
RegressionExerciseBoundary boundary(T,training_paths.state_variables(),
                                    training_paths.numeraires(),
                                    payoff,basisfunctions);
LSExerciseStrategy<RegressionExerciseBoundary> exercise_strategy(boundary);
```

Listing 8.9: Estimating the exercise boundary for an American put option in the Gauss/Markov HJM model.

8.7.3 Options with early exercise features

Conceptually, the regression–based algorithm for pricing instruments with early exercise features, as described in Section 7.7.1, is not specific to the context of the (multi-asset) Black/Scholes model around which the exposition of Chapter 7 was focused. Thus, if generic programming principles were properly applied in the C++ templates implementing the regression–based algorithm in Chapter 7, then these templates should work unchanged in the present context

---

[16] See the discussion of the class MultivariateNormal on p. 57.

[17] Note that these volatility functions imply that the model is Markovian in the state variables. If one chooses a different volatility function, one must ensure that this property is maintained.

[18] The name of the member function, component_vol(), refers to the fact that this is the i-th component of the vector–valued HJM volatility function, with all other components set to zero.

```
// instantiate objects for MC simulation
MCMapping<GaussMarkovWorld,Array<double,2> >
  mc_mapping(exercise_strategy,
             world,*(domestic_economy.initialTS),numeraire_index);
boost::function<double (Array<double,2>)>
  func = boost::bind(&MCMapping<GaussMarkovWorld,Array<double,2> >::mapping,
                     &mc_mapping,_1);
MCGeneric<Array<double,2>,double,RandomArray<...> >
  mc(func,random_container);
MCGatherer<double> mcgatherer;
size_t n = minpaths;
cout << "Simulations,MC value,95% CI lower bound,95% CI upper bound"
     << endl;
while (mcgatherer.number_of_simulations()<maxpaths) {
  mc.simulate(mcgatherer,n);
  cout << mcgatherer.number_of_simulations() << ',' << mcgatherer.mean()
       << ',' << mcgatherer.mean()-d*mcgatherer.stddev() << ','
       << mcgatherer.mean()+d*mcgatherer.stddev();
  cout << endl;
  n = mcgatherer.number_of_simulations();
  std::cerr << n << endl; }
```

Listing 8.10: Pricing an American put option by Monte Carlo in the Gauss/Markov HJM model.

of Monte Carlo simulation in the Gauss/Markov HJM model. This is in fact the case, as illustrated in Listings 8.9 and 8.10.[19] As in Section 7.7.2, the template MCTrainingPaths serves to create the "training paths" of the underlying financial variables and the numeraire, as required by the constructors of the continuation value regression estimators LongstaffSchwartzExerciseBoundary or RegressionExerciseBoundary (Listing 8.9 uses the latter). The difference to the instantiation of MCTrainingPaths in Section 7.7.2 is that the class representing the underlying stochastic process is now GaussMarkovWorld, rather than GeometricBrownianMotion. Since the option to be valued in the present example is an American put on a domestic equity asset, the financial variables "reported" by GaussMarkovWorld were set (by calls to the member function set_reporting()) to be the terminal forward price of the underlying equity asset and the corresponding terminal maturity zero coupon bond, allowing us to recover the spot price of the equity asset when necessary.

Next, a set of basis functions for the continuation value regression needs to be specified. Following the article of Longstaff and Schwartz (2001), in this example we use polynomials in the simulated financial variables (the terminal forward price and the terminal maturity zero coupon bond), allowing for polynomials of varying degree of each of the two variables individually, and of the product of the two variables (which equals the spot price of the equity asset). As the results reported in Table 8.3 show, the quality with which the regression approximates the optimal exercise boundary, for a given number

---

[19] The full working example program, used to generate the numbers in Table 8.3, is to be found in the file LSMCHJM.cpp on the website.

Max. polynomial deg.			Training	MC	Max. polynomial deg.			Training	MC
Fwd	ZCB	Spot	paths	value	Fwd	ZCB	Spot	paths	value
0	0	2	1000	49.60	0	2	2	1000	52.02
0	0	2	10000	49.65	0	2	2	10000	52.04
0	0	2	200000	49.66	0	2	2	200000	52.08
0	2	0	1000	48.95	2	0	2	1000	50.56
0	2	0	10000	48.70	2	0	2	10000	50.33
0	2	0	200000	48.61	2	0	2	200000	50.62
2	0	0	1000	42.01	2	2	0	1000	47.76
2	0	0	10000	47.04	2	2	0	10000	50.91
2	0	0	200000	50.06	2	2	0	200000	51.35
2	2	2	1000	51.99	1	2	2	1000	52.04
2	2	2	10000	52.07	1	2	2	10000	52.07
2	2	2	200000	52.11	1	2	2	200000	52.11
3	3	3	1000	50.46	1	2	3	1000	51.34
3	3	3	10000	38.65	1	2	3	10000	49.70
3	3	3	200000	45.07	1	2	3	200000	50.78

Table 8.3 *Suboptimal values for an American put option based on Long-staff/Schwartz exercise strategies with different polynomial basis functions and different numbers of training paths. The MC values are based on 3276800 paths with 60 time steps on each path; the 95% confidence intervals are ±0.07 around the given value. The price of the corresponding European option is 46.66.*

of training paths, depends on a good and parsimonious choice of basis functions.[20] If a small, but not very suitable, number of basis functions is chosen, the resulting Monte Carlo estimate for the American option converges to a lower value. If a large, but thus not parsimonious, number of basis functions is chosen, a much larger number of training paths is needed to get a good regression estimate of the continuation value, and therefore the Monte Carlo estimate converges to a lower value if the number of training paths is too low.

Given the paths of the terminal forward price of the equity asset and the terminal maturity zero coupon bond price generated by MCTrainingPaths instantiated on GaussMarkovWorld, the American put option on the spot equity price is an option on the product of these underlying financial variables, the early exercise payoff of which is implemented in the class exotics::Product-Option. Since we are using the class RegressionExerciseBoundary in this example, the member function

---

[20] Recall that the early exercise rule resulting from the regression–based algorithm is suboptimal, and thus the resulting Monte Carlo estimate for the American option value converges to a lower bound of the "true" value. Therefore, a higher estimated value (modulo confidence bounds) for the American option means a better quality of approximation.

```
exotics::ProductOption::
early_exercise_payoff(const Array<double, 1>& T,
                      const Array<double,2>& history)
    const;
```
(8.134)

takes as arguments an array T containing the time line up to the point where early exercise is to be evaluated, and an array history of the path of the (two) underlying financial variables up to this point, even though only the last entries in the path history are needed to evaluate the early exercise payoff.[21] Given an instance of RegressionExerciseBoundary, an instance of the template LSExerciseStrategy is constructed in the same manner an in Section 7.7.2, which then allows us to instantiate an MCMapping and subsequently an MCGeneric, with the member function simulate() to run Monte Carlo simulations, the results of which are collected in an instance of MCGatherer.

---

[21] For comparison, the same option is priced using the simpler LongstaffSchwartz-ExerciseBoundary in LSMCHJMsimple.cpp on the website.

Interfacing between C++ and Microsoft Excel

A.1 The XLW project

When developing code for applications in Quantitative Finance, it is helpful to be able to focus on the core functionality, rather than on graphical user interfaces, data input and output. This was the approach taken throughout this book. However, for this code to be used in practice, some reasonably convenient interface has to be provided. One way to go about this is to use Microsoft Excel as the interface, and provide the user with access to the functionality developed using C++ by "wrapping" it in Excel add–in functions in an XLL dynamic link library.

Working directly with the Excel C API (application programming interface) is not straightforward, but fortunately there is an open source application to generate that API code automatically from user-supplied C++ functions satisfying a set of simple conventions. This is the XLW project, currently in release version 5.0, available at `xlw.sourceforge.net`.

For our purposes, it is sufficient to start with modifying the simplest example provided with the XLW project; once XLW is installed, it is found under `xlw/examples/Start Here - Example` and the subfolder `vc10` contains a Visual Studio Solution file, which works with Microsoft Visual C++ 2010 Express,[1] called `FirstExample.sln`. This Solution contains two Projects, XLL and `RunInterfaceGenerator`. The Project XLL has the XLL dynamic link library as its output. Its input consists of two C++ source files, `Test.cpp` and `xlwTest.cpp`. The former contains the actual C++ functions, which become dynamic link library functions which can be called from Excel. These functions are declared in the header file `Test.h`. `xlwTest.cpp` is automatically generated from `Test.h` by the Project `RunInterfaceGenerator`, which uses a MAKE file to run the XLW utility `InterfaceGenerator.exe` on `Test.h`. Thus `xlwTest.cpp` contains automatically generated C++ code for using the Excel API to convert the C++ functions in `Test.cpp` into dynamic link library functions callable from Excel.

The C++ functions in `Test.cpp` are global (i.e. non-member) functions. The C++ built-in data types used as function arguments are `short`, `int`,

---

[1] As of the time of writing, Visual Studio Express 2012 was not yet supported by XLW. However, the examples discussed in this section will build and run using Visual Studio Express 2012, if the paths in the Visual Studio project files are appropriately modified.

```
short // echoes a short
EchoShort(short x // number to be echoed
            )
{
    return x;
}


MyMatrix EchoMat(const MyMatrix& EchoEe)
{
    return EchoEe;
}


double // computes the circumference of a circle
Circ(double Diameter //the circle's diameter
            )
{
      return Diameter* 3.14159;
}
```

Listing A.1: Excerpt from the XLW example source file `Test.cpp`

unsigned long and double. The C++ built-in data types used as function return values are all of the above, and `bool`. Furthermore, Standard Library strings are supported as arguments and return values, in the form of `std::string` and `std::wstring`. For arrays as function arguments and return values, XLW (being an open source development effort which has evolved over time) offers several alternatives. `Test.cpp` includes functions which use `MyArray`, `MyMatrix`, `NEMatrix` and `CellMatrix`. `MyArray` is a synonym for a `std::vector<double>`.[2] `MyMatrix` and `NEMatrix` are synonyms for the XLW class `NCMatrix`, which is a matrix of `doubles`. `CellMatrix` is a matrix of `CellValues`, where `CellValue` is an abstract base class defining an interface (specified in the XLW file `CellValue.h`) for a C++ class wrapping Excel cell contents of arbitrary type. It includes member functions allowing to determine the type of the cell contents,

$$
\begin{aligned}
&\texttt{virtual bool IsANumber () const = 0;}\\
&\texttt{virtual bool IsBoolean () const = 0;}\\
&\texttt{virtual bool IsEmpty () const = 0;}\\
&\texttt{virtual bool IsError () const = 0;}
\end{aligned}
\tag{A.1}
$$

and overloads the type cast operators to access the value of the cell contents,

$$
\begin{aligned}
&\texttt{virtual operator std::string () const = 0;}\\
&\texttt{virtual operator std::wstring () const = 0;}\\
&\texttt{virtual operator bool () const = 0;}\\
&\texttt{virtual operator double () const = 0;}\\
&\texttt{virtual operator unsigned long () const = 0;}
\end{aligned}
\tag{A.2}
$$

The assignment `operator=` is overloaded to allow `CellValue` to be assigned

---

[2] `MyArray` and `MyMatrix` can be set via `typedef` to user–defined arrays, if desired.

```
short // echoes a short
EchoShort(short x // number to be echoed
      );


MyMatrix // echoes a matrix
EchoMat(const MyMatrix& Echoee // argument to be echoed
       );


double // computes the circumference of a circle
Circ(double Diameter //the circle's diameter
           );
```

Listing A.2: Excerpt from the XLW example source file `Test.h`

a `CellValue`, `std::string`, `std::wstring`, `double`, `unsigned long`, `bool` or `int`, as well as a C-style null–terminated string. `CellValue` is currently implemented by a single derived class, `MJCellValue`.

Listing A.1 gives three examples of functions from `Test.cpp`. `EchoShort()` simply echoes the value of its argument (a `short`) back to Excel, `Circ()` calculates the circumference of a circle given its diameter, and `EchoMat()` echoes a matrix (given by a range in Excel) back to Excel. For the latter, note that since `EchoMat()` is an array–valued function, it must be called as such from Excel, i.e. with an output range of appropriate size selected and by pressing `<Ctrl><Shift><Enter>` rather than just `<Enter>`. Information about the function, which appears in Excel when the function is called via the function wizard, can be specified by appropriately located comments in the function declarations in `Test.h`, as in Listing A.2. A comment after the return type declaration is used to describe the function, e.g. "computes the circumference of a circle." A comment after each function argument declaration is used to describe that function argument, e.g. "the circle's diameter."

Let us now proceed to build an XLL containing some simple examples of accessing some of the code developed in this book from Excel. This is done in a modified version of the `FirstExample.sln`, called `ESXLWproject.sln`. This solution and all the associated files are available for download from the website for this book. Listing A.3 shows a function to calculate a Black/Scholes European option price, implemented by calling the library code developed in Chapter 4. Similarly, the functions `ESCoxRossRubinstein()` and `ESCoxRoss-RubinsteinAmerican()`, implemented in `ESXLWProject.cpp`, calculate, respectively, European and American option prices in the Cox/Ross/Rubinstein model presented in Chapter 3.

As an example of an array–valued function, consider Listing A.4. The function `ESLogLinear()` takes as its arguments a timeline of maturities of supplied discount factors, the supplied discount factors, and a vector of maturities for which interpolated discount factors are desired. These discount factors are calculated by the loglinear interpolation implemented in the class `TSLoglinear` introduced in Chapter 2, and returned as `MyArray` vector of `double`s. The

```
double // Black/Scholes European option price
ESBlackScholes(double S // current price of underlying asset
    ,double K // exercise price
    ,double sigma // volatility of underlying asset
    ,double r // risk-free interest rate (continuously compounded)
    ,double T // time to maturity (in years)
    ,int callput // 1 for call, -1 for put
        )
{
    quantfin::ConstVol vol(sigma);
    quantfin::BlackScholesAsset stock(&vol,S);
    return stock.option(T,K,r,callput);
}
```

Listing A.3: Function to calculate a Black/Scholes European option price, implemented by calling the library code developed in Chapter 4

function `ESLoglinearMatrix()` in `ESXLWProject.cpp` gives an alternative implementation where the supplied discount factors and their associated maturities are passed to the function as a `MyMatrix` table. Note that if exceptions are thrown, either as a raw C-Style string by the function itself (as in `ESLogLinear()`) or via the Standard Library exceptions thrown by the library code called by the function, then the associated error message is displayed in the calling cell in Excel. For example, the exception thrown by the code implementing `TSLogLinear`, if the maturity for which an interpolated discount factor is requested is greater than the longest supplied maturity, displays the error message "cannot extrapolate."

A.2 The QuantLib ObjectHandler

While XLW makes it easy to wrap calculations in functions callable from Excel, it has the disadvantage that objects, i.e. instances of C++ classes, created by such calculations, do not persist outside of each function call. However, such persistence is necessary if one wishes to assemble a collection of objects in the spreadsheet, and then pass those objects to other functions for further calculations. For example, one might want to construct a multicurrency Gauss/Markov HJM model as discussed in Chapter 8, by first constructing the objects representing the initial interest rate term structures in each currency, objects representing the requisite volatility functions, the exchange rates and their volatilities, equity assets in each currency and so on. The open source project QuantLib provides an ObjectHandler, which can also be used in conjunction with the code presented in this book to make instances of C++ classes persist between calls from Excel to XLL functions.

The ObjectHandler package includes several example project files for Microsoft Visual Studio. The Project `ExampleXllStatic_vc9`, found in the ObjectHandler folder `ObjectHandler/Examples/xl`, creates a single client XLL

```
MyArray // log-linear interpolation given set of discount factors
ESLogLinear(
  const MyArray&
    given_times // timeline for supplied discount factors
  ,const MyArray&
    given_discount_factors  // supplied discount factors
  ,const MyArray&
    target // times for which discount factors are desired
        )
{
  int i;
  int n = given_times.size();
  if (n!=given_discount_factors.size())
    throw("Timeline and discount arrays must have same dimension");
  blitz::Array<double,1> T(n),B(n);
  MyArray result(target.size());
  for (i=0;i<n;i++) {
    T(i) = given_times[i];
    B(i) = given_discount_factors[i]; }
  quantfin::TSLogLinear ts(T,B);
  for (i=0;i<target.size();i++) result[i] = ts(target[i]);
  return result;
}
```

Listing A.4: Function to perform loglinear interpolation of discount factors, implemented by calling the library code developed in Chapter 2

bundling the core ObjectHandler, its Excel binding and code for specific objects, using Microsoft Visual C++ 2010 Express. We will now proceed to modify this example such that the specific objects made available via the XLL are instances of a C++ class presented in this book.[3] Note that the original ObjectHandler example project ExampleXllStatic_vc9 is embedded in the Visual C++ solution ObjectHandler_vc9, which also creates the requisite static link libraries (i.e. *.lib files). In order to build the modified project in its own stand–alone solution, we must include the requisite *.lib files in the project, i.e. xlsdk-vc90-mt.lib, aprutil-v90-mt-0_10_0.lib and apr-vc90-mt-0_10_0.lib. For the C++ classes presented in this book, we add Quantfin.lib. The main C++ source file for the project is addinstatic.cpp, modified from the original file supplied with ObjectHandler package. In this file resides the code defining the XLL functions which will be callable from Excel, and the code to register these functions with XLL. The functions serve to create, query, modify, delete and otherwise operate on objects in the object repository of the XLL.

The functions are registered by calls to the function Excel() inside the

---

[3] The modified Visual C++ solution, project and associated source files are available on the website for this book.

```
Excel(xlfRegister, 0, 7, &xDll,
  // function code name
  TempStrNoSize("\x0E""ohESTestObject"),
  // parameter codes: char* (char*,long*,double*)
  TempStrNoSize("\x05""CCNE#"),
  // function display name
  TempStrNoSize("\x0E""ohESTestObject"),
  // comma-delimited list of parameters
  TempStrNoSize("\x17""ObjectID,integer,double"),
  // function type
  // 0 = hidden function,
  // 1 = worksheet function,
  // 2 = command macro
  TempStrNoSize("\x01""1"),
  // function category
  TempStrNoSize("\x07""Example"));
```

Listing A.5: Registering a function with the XLL

function xlAutoOpen(), which also performs additional housekeeping and is called when the XLL is opened.

Listing A.5 shows the call to Excel() to register the function ohESTest-Object(). The requisite information about ohESTestObject() must be given in the fifth to tenth arguments of Excel(), each of which is wrapped as TempStrNoSize(). The hexadecimal code in TempStrNoSize() specifies the length of the string (i.e. the number of characters) in the argument. Thus the argument "CCNE#" contains 5 characters, so it is led with \x05. The argument "ohESTestObject" contains 14 characters, and 14 in hexadecimal is \x0E. The fifth argument is the function code name, i.e. the name (in the C++ source code) of the function to be registered with XLL. The sixth argument lists the parameter codes for the return value and the arguments of the function to be registered (in that order), terminated by the hash sign(#). The parameter codes are given in Table A.1 . Note that the value returned and all arguments of the function to be registered must be pointers. The seventh argument of Excel() is the function display name, i.e. the name by which the function can be called from Excel. This can, but does not have to, be the same as the function code name. The eighth argument contains a comma–delimited list of parameter names, which are displayed if the function is called from Excel via the Function Wizard. The ninth argument is the function type, with "1" indicating that it is a worksheet function. The tenth argument is the category under which the function is to be listed in the Function Wizard.

Consider now the C++ code for the function ohESTestObject(), given in Listing A.6, the task of which is to instantiate an ESTestObject in the ObjectHandler repository. The function takes three arguments. char* objectID points to a string of characters containing an identifier for the object. long* input_int points to a long integer and double* input_double points to a

Code	Parameter type
C	char*
N	long*
E	double*
L	bool*
P	OPER*

Table A.1 *Parameter codes for XLL function registration*

```
// function to instantiate ESTestObject in the repository
DLLEXPORT char *ohESTestObject(
        char *objectID,
        long *input_int,
        double *input_double)
{
boost::shared_ptr<ObjectHandler::FunctionCall> functionCall;
try {
  functionCall = boost::shared_ptr<ObjectHandler::FunctionCall>
    (new ObjectHandler::FunctionCall("ohESTestObject"));
  boost::shared_ptr<ObjectHandler::ValueObject> valueObject(
    new ESTestValueObject(objectID,*input_int,*input_double));
  boost::shared_ptr<ObjectHandler::Object> object(
    new ESTestObject(valueObject,*input_int,*input_double));
  std::string returnValue =
    ObjectHandler::RepositoryXL::instance().storeObject(objectID,
                                                        object);
  static char ret[XL_MAX_STR_LEN];
  ObjectHandler::stringToChar(returnValue, ret);
  return ret;
} catch (const std::exception &e) {
  ObjectHandler::RepositoryXL::instance().logError(e.what(),
                                                   functionCall);
  return 0;
}}
```

Listing A.6: Creating an object in the ObjectHandler repository

double. The function returns a char*, which will point to the identifier of the object in the repository.

First, an instance of ObjectHandler::FunctionCall is created for the function. Then, an ESTestObject is created and stored in the repository. This is done in two steps, creating what the ObjectHandler package calls a "value object" holding the names of object properties and their values, and then passing the "value object" to the constructor of the final object itself (in our case, an ESTestObject).

Thus the class ESTestValueObject is derived from the class ObjectHand-

```
DLLEXPORT double *ohESTestObjectQueryDouble(char *objectID,
                                            OPER *trigger)
{
boost::shared_ptr<ObjectHandler::FunctionCall> functionCall;
try {
  functionCall = boost::shared_ptr<ObjectHandler::FunctionCall>
    (new ObjectHandler::FunctionCall("ohESTestObjectQueryDouble"));
  OH_GET_OBJECT(testObject, // name of variable to hold shared_ptr
               objectID,     // object ID
               ESTestObject  // C++ class of object
         )
  static double ret;
  ret = testObject->double_value();
  return &ret;
} catch (const std::exception &e) {
ObjectHandler::RepositoryXL::instance().logError(e.what(),
                                               functionCall);

  return 0;     }
}
```

Listing A.7: Interrogating an object in the ObjectHandler repository

ler::ValueObject, see the class declaration in TestObject.hpp, and ES-
TestObject is derived from the class ObjectHandler::Library Object<ES-
TestVal>. The class ESTestVal serves to hold the actual object data, in this
case a double and a long integer. Once an ESTestObject is instantiated, it is
stored in the ObjectHandler repository by calling (in Listing A.6) the function

$$\text{ObjectHandler::RepositoryXL::} \atop \text{instance().storeObject(objectID,object);}} \qquad (A.3)$$

The return value of this function is the identifier of the stored object in
the repository, and this is passed back to Excel. This identifier can then
be used to access the object in subsequent XLL function calls, as for example via the function ohESTestObjectQueryDouble() in Listing A.7.[4] This
function, like the entire code associated with ESTestObject, serves only as
simple (but working) illustration, since object properties can be listed and
queried using generic functions supplied by ObjectHandler.[5] The function
takes the object identifier as its first argument – this should normally be
supplied by referencing the worksheet cell containing the XLL function call
creating the object, in order for worksheet recalculation dependencies to be
properly maintained by Excel. The second argument is an optional dummy,

---

[4] Note that in all of the C++ functions to be registered in the XLL, the code implementing
the function is enclosed in a try block. When an exception is caught, an error message is written to a log file. This log file was specified in the function xlAutoOpen() in
addinstatic.cpp, by a call to ObjectHandler::logSetFile().

[5] See the XLL function calls to ohObjectPropertyNames() and ohObjectPropertyValues()
under   the   heading   "Interrogate   the   object"   in   the   Excel   workbook
OHExampleWorkbook.xlsx supplied on the website for this book.

which has no effect on how the function is evaluated (i.e. the variable `trigger` is not used inside `ohESTestObjectQueryDouble()`); rather, it serves as a way to force recalculation dependencies which Excel might not recognise otherwise. In Listing A.7, first an instance of `ObjectHandler::FunctionCall` is created for the function. Then, a pointer to the object is retrieved from the repository using the macro `OH_GET_OBJECT()`. Once the pointer is obtained, one can access any public member function of its class. Here, we access the function `double_value()` of `ESTestObject`. Thus in this example `ohESTestObjectQueryDouble()` serves as an XLL function callable from Excel wrapping a member function of `ESTestObject` – only member functions, which are wrapped in this way, will be accessible from Excel.

Consider now the somewhat less trivial example of creating an object representing an interest rate term structure in the repository, and then querying that object for a (possibly interpolated) discount factor. The XLL function to create an instance of `TSLogLinear` (see Chapter 2) is called `TSLogLinear`, found in `addinstatic.cpp`.[6] The code is given in Listing A.8. The function takes three arguments, the first of which is the object identifier. The other two arguments are pointers to `OPER`. `xT` holds a one-dimensional array (given as a column of cells in the calling Excel worksheet) containing a time line of maturities, for which `xB` holds the corresponding discount factors (also given as a column in the worksheet). Once again, an instance of `ObjectHandler::FunctionCall` is created first. Then, `xT` and `xB` are converted to one-dimensional Blitz `Arrays`, from which an instance of `TSLogLinear` is then constructed. This is followed by the instantiation of an `ESTSValueObject` and an `ESTSObject`,[7] and the instance of `ESTSObject` is then stored in the ObjectHandler repository. Note that for simplicity, the instance `ts` of `TSLogLinear` is not stored as an ObjectHandler `ValueObject` property; rather, the pointer `ts` is stored as a data member of the `ESTSObject`. Thus the `ESTSValueObject` is only present for compatibility with the ObjectHandler generic functions. To retrieve a discount factor (interpolated if necessary) from an existing `TSLogLinear` object, we have the XLL function `TSLogLinearDF()` in Listing A.9.[8] Its arguments are the object identifier and the time point (i.e. the maturity) `*t` for which the discount factor is required. The object identifier permits retrieval of the corresponding instance `ts_object` of `ESTSObject` from the repository, and the desired discount factor is obtained by

$$ret = ts\_object\text{->}discount\_factor(*t);\qquad\qquad (A.4)$$

which passes `*t` through to the `operator()` of the underlying instance of `TSLogLinear`.

As noted above, error messages from the C++ code in the form of caught exceptions derived from `std::exception` are written to a log file. Thus calling

---

[6] This is not a name clash, since the C++ class `TSLogLinear` is in the `quantfin` namespace (as all library code created for this book).

[7] `ESTSValueObject` and `ESTSObject` are declared in `TSObject.hpp`.

[8] This function is also part of the code in `addinstatic.cpp`.

```
DLLEXPORT char *TSLogLinear(char *ObjectId,OPER *xT,OPER *xB) {
int i;
// declare a shared pointer to the Function Call object
boost::shared_ptr<ObjectHandler::FunctionCall> functionCall;
try {
  functionCall = boost::shared_ptr<ObjectHandler::FunctionCall>(
    new ObjectHandler::FunctionCall("TSLogLinear"));
  // convert input datatypes to C++ datatypes - second argument
  // is only relevant for outputting error msg if conversion fails
  std::vector<double> vecT =
    ObjectHandler::operToVector<double>(*xT,"timeline");
  std::vector<double> vecB =
    ObjectHandler::operToVector<double>(*xB,"discount factors");
  if (vecT.size()!=vecB.size())
    throw std::logic_error("Timeline and discount factors must
                          have the same number of entries");
  blitz::Array<double,1> T(vecT.size()),B(vecB.size());
  for (i=0;i<vecT.size();i++) T(i) = vecT[i];
  for (i=0;i<vecB.size();i++) B(i) = vecB[i];
  boost::shared_ptr<quantfin::TSLogLinear>
    ts(new quantfin::TSLogLinear(T,B));
  // Strip the Excel cell update counter suffix from Object IDs
  std::string ObjectIdStrip =
    ObjectHandler::CallingRange::getStub(ObjectId);
  boost::shared_ptr<ObjectHandler::ValueObject>
    valueObject(new ESTSValueObject(ObjectId));
  boost::shared_ptr<ObjectHandler::Object>
    object(new ESTSObject(valueObject,ts));
  // Store the Object in the Repository
  std::string returnValue = ObjectHandler::RepositoryXL::
    instance().storeObject(ObjectIdStrip,object);
  // Convert and return the return value
  static char ret[XL_MAX_STR_LEN];
  ObjectHandler::stringToChar(returnValue, ret);
  return ret;
} catch (const std::exception &e) {
ObjectHandler::RepositoryXL::instance().logError(e.what(),
                                    functionCall);
  return 0;     }}
```

Listing A.8: Create a `TSLogLinear` object in the ObjectHandler repository

`TSLogLinearDF` from the worksheet with a wrong object identifier, say with an identifier of an `ESTestObject` instead of an `ESTSObject`, causes one of the ObjectHandler functions to throw an exception and logs the error

```
DLLEXPORT double *TSLogLinearDF(char *objectID,double* t)
{
boost::shared_ptr<ObjectHandler::FunctionCall> functionCall;
try {
  functionCall = boost::shared_ptr<ObjectHandler::FunctionCall>
    (new ObjectHandler::FunctionCall("TSLogLinearDF"));
  OH_GET_OBJECT(ts_object, // name of variable to hold shared_ptr
               objectID, // object ID
               ESTSObject // C++ class of object
           )
  static double ret;
  ret = ts_object->discount_factor(*t);
  return &ret;
} catch (const std::exception &e) {
  ObjectHandler::RepositoryXL::instance().logError(e.what(),
                                                   functionCall);
  return 0;    }
}
```

Listing A.9: Function to query a discount factor for maturity *t from an existing TSLogLinear object

```
ERROR - [OHExampleWorkbook.xlsx]Sheet1!$H$10 - TSLogLinearDF
- Error retrieving object with id 'TestObject#0002' -
unable to convert reference to type 'class ESTSObject' found
instead 'class ESTestObject'
```

Similarly, calling TSLogLinearDF() from the worksheet with a maturity argument greater than the longest maturity for which an instance of ESTSObject was created, causes the operator() of TSLogLinear to throw an exception and logs the error

```
ERROR - [OHExampleWorkbook.xlsx]Sheet1!$H$10 - TSLogLinearDF
- cannot extrapolate
```

Automatic generation of documentation using Doxygen

Doxygen is a freely available,[1] widely used tool for generating documentation from annotated C++ sources.[2] It generates hyperlinked output with the most common output formats being HTML (which can be converted into the CHM format used by the Microsoft Windows Help browser) and PDF. This appendix briefly reviews some of the main functionality of Doxygen used to generate the documentation of the C++ code accompanying this book.

All C++ source files for the code accompanying this book are annotated to some extent with C++ comments in the Doxygen format.[3] Each file[4] starts with a Doxygen comment block of the type

```
/** \file Hello.cpp
    \Hello world!" program
*/
```

This is a multiline C++ comment, where the additional star in `/**` flags it as a Doxygen annotation. Doxygen keywords are prefaced by a backslash: The `\file` keyword indicates that this comment block provides documentation for the source file `Hello.cpp`; the `\brief` keyword means that Doxygen will treat all that follows as a brief (as opposed to detailed) description, until it encounters an empty line or the end of the comment block. The initial comment block in Listing B.1[5] gives the brief description of the class `GaussianHJM`, followed by its detailed description. Doxygen recognises this comment block as pertaining to the class `GaussianHJM`, because it immediately precedes the class declaration. Brief descriptions can also be specified as one-line C++ comments using three forward slashes instead of the usual two, as done for example for the member function `ZCBoption()` of `Gaussian HJM`. Documentation for a member function can be placed in the header file declaring the class, or alternatively in the C++ source file where the function is implemented, in this case `GaussianHJM.cpp`. The implementation and annotating comments of `GaussianHJM::ZCBoption()` are reproduced in Listing B.2. The one-line comments starting with `///<` give brief descriptions for each preceding function argument. The detailed description of the function includes mathematical

---

[1] Available at `http://www.doxygen.org/download.html`

[2] Doxygen also supports a number of other programming languages, including C, Objective–C, C#, Java and Python.

[3] The resulting output documentation can be found on the website for this book.

[4] See e.g. Listing 1.1 in Chapter 1.

[5] Found in GaussianHJM.hpp.

```
/** Class for a Gauss/Markov Heath/Jarrow/Morton (HJM) model.

    A particular instance of such a model is characterised by
    a volatility function for the instantaneous forward rates
    and by an initial term structure of interest rates.
 */
class GaussianHJM {
```

Listing B.1: Initial comment block for the class `GaussianHJM`

formulas, which are typeset in LaTeX. The Doxygen keywords \f[and \f]
mimic the delimiters for displayed formulas (i.e.$$) in LaTeX, and the Doxy-
gen keyword \f$ mimics the LaTeX delimiter $ for inline formulas.[6] Lastly, the
Doxygen keyword \sa adds a "See Also" section to the detailed description.
Note that the mention of `DeterministicAssetVol::FwdBondVol()`, both in
the detailed description and in the "See also" section, is recognised by Doxygen
as referring to a documented function, so a hyperlink to the documentation
of said function is generated automatically.

---

[6] For details on further features of LaTeX supported by Doxygen, see the Doxygen docu-
mentation.

```
/** Calculate the ("time zero") value of a zero coupon bond option.

    In the Gaussian HJM model, the "time zero" prices of a call and
    a put option on a zero coupon bond are given by the
    Black/Scholes-type formula
    \f[ C=B(0,T_2)N(h_1)-B(0,T_1)KN(h_2) \f] and
    \f[ P=B(0,T_1)KN(-h_2)-B(0,T_2)N(-h_1) \f]
    with
    \f[ h_{1,2}=\frac{\ln\frac{B(0,T_2)}{B(0,T_1)K}\
    pm\frac12\int_0^{T_1}
    (\sigma^*(u,T_2)-\sigma^*(u,T_1))^2du}{\sqrt{\int_0^{T_1}
    (\sigma^*(u,T_2)-\sigma^*(u,T_1))^2du}}\f]
    The value of the integral
    \f[ \int_0^{T_1}(\sigma^*(u,T_2)-\sigma^*(u,T_1))^2du \f]
    depends on the choice of \f$ \sigma(u,s) \f$ and is therefore
    passed through to DeterministicVol::FwdBondVol().
 */
double GaussianHJM::ZCBoption(
  double T1,    ///< Option expiry
  double T2,    ///< Maturity of the underlying bond
  double K,     ///< Exercise price
  int sign      ///< 1 if call option (default), -1 if put option
  ) const
{
  double vol = v->FwdBondVol(0.0,T1,T2);
  double B1  = initialTS->operator()(T1);
  double B2  = initialTS->operator()(T2);
  double h1  = (log(B2/(K*B1)) + 0.5*vol*vol) / vol;
  return sign * (B2*boost::math::cdf(N,sign*h1)
                 -K*B1*boost::math::cdf(N,sign*(h1-vol)));
}
```

Listing B.2: The annotated function `GaussianHJM::ZCBoption()`

References

Abramowitz, M. and Stegun, I. A. (eds) (1964), *Handbook of Mathematical Functions*, National Bureau of Standards.

Adams, K. J. and van Deventer, D. R. (1994), Fitting Yield Curves and Forward Rate Curves with Maximum Smoothness, *The Journal of Fixed Income* 4(1), 52–62.

Arrow, K. (1964), The Role of Securities in the Optimal Allocation of Risk–Bearing, *Review of Economic Studies* **31**, 91–96.

Ash, R. B. and Doléans-Dade, C. A. (2000), *Probability and Measure Theory*, 2nd edn, Harcourt Academic Press.

Bahra, B. (1997), Implied Risk–Neutral Probability Density Functions from Option Prices: Theory and Application, Bank of England, working paper.

Bajeux-Besnainou, I. and Portait, R. (1997), The Numeraire Portfolio: A New Perspective on Financial Theory, *The European Journal of Finance* **3**(4), 291–309.

Barone-Adesi, G. and Whaley, R. E. (1987), Efficient Analytic Approximation of American Option Values, *Journal of Finance* **42**, 301–320.

Beaglehole, D. R. and Tenney, M. S. (1991), General Solutions of Some Interest Rate–Contingent Claim Pricing Equations, *Journal of Fixed Income* pp. 69–83.

Belomestny, D., Bender, C. and Schoenmakers, J. (2009), True Upper Bounds for Bermudan Products Via Non-Nested Monte Carlo, *Mathematical Finance* **19**(1), 53–71.

Berntsen, J., Espelid, T. O. and Genz, A. C. (1991), An Adaptive Algorithm for the Approximate Calculation of Multiple Integrals, *ACM Transactions on Mathematical Software* **17**(4), 437–451.

Black, F. (1976), The Pricing of Commodity Contracts, *Journal of Financial Economics* **3**, 167–179.

Black, F. and Karasinski, P. (1991), Bond and Option Pricing when Short Rates are Lognormal, *Financial Analysts Journal* pp. 52–59.

Black, F. and Scholes, M. (1973), The Pricing of Options and Corporate Liabilities, *Journal of Political Economy* **81**(3), 637–654.

Black, F., Derman, E. and Toy, W. (1990), A One–Factor Model of Interest Rates and Its Application to Treasury Bond Options, *Financial Analysts Journal* pp. 33–39.

Breeden, D. T. and Litzenberger, R. H. (1978), Prices of State-contingent Claims Implicit in Option Prices, *Journal of Business* **51**(4), 621–651.

Brennan, M. J. and Schwartz, E. S. (1977), Savings Bonds, Retractable Bonds and Callable Bonds, *Journal of Financial Economics* pp. 67–88.

Brigo, D. and Mercurio, F. (2001), *Interest Rate Models: Theory and Practice*, Springer-Verlag.

Brown, C. A. and Robinson, D. M. (2002), Skewness and Kurtosis Implied by Option Prices: A Correction, *Journal of Financial Research* **25**(2), 279–282.

Buchen, P. (2001), Image Options and the Road to Barriers, *Risk* pp. 127–130.

Buchen, P. (2012), *An Introduction to Exotic Option Pricing*, Chapman & Hall/CRC Financial Mathematics Series, CRC Press.

Buchen, P. and Konstandatos, O. (2005), A New Method of Pricing Lookback Options, *Mathematical Finance* **15**(2), 245–259.

Carr, P. and Schoutens, W. (2007), Hedging under the Heston Model with Jump-to-Default, Courant Institute of Mathematical Sciences, working paper.

Carriére, J. (1996), Valuation of Early–Exercise Price of Options Using Simulations and Nonparametric Regression, *Insurance: Mathematics and Economics* **19**, 19–30.

Chan, K., Karolyi, G., Longstaff, F. and Sanders, A. (1992), An Empirical Comparison of Alternative Models of the Short–Term Interest Rate, *Journal of Finance* **47**, 1209–1227.

Chiarella, C. and Nikitopoulos, C. S. (2003), A Class of Jump–Diffusion Bond Pricing Models within the HJM Framework with State Dependent Volatilities, *Asia–Pacific Financial Markets* **10**(2-3), 87–127.

Chiarella, C., Schlögl, E. and Nikitopoulos, C. (2007), A Markovian Defaultable Term Structure Model with State Dependent Volatilities, *International Journal of Theoretical and Applied Finance* **10**(1), 155–202.

Clewlow, L. and Strickland, C. (1998), *Implementing Derivatives Models*, John Wiley & Sons.

Conze, A. and Viswanathan, R. (1991), Path Dependent Options: The Case of Lookback Options, *The Journal of Finance* **46**(5), 1893–1907.

Corrado, C. (2007), The Hidden Martingale Restriction in Gram–Charlier Option Prices, *Journal of Futures Markets* **27**(6), 517–534.

Corrado, C. and Su, T. (1996), Skewness and Kurtosis in S&P 500 Index Returns Implied by Option Prices, *Journal of Financial Research* **19**(2), 175–92.

Cox, J. C., Ingersoll, J. E. and Ross, S. A. (1985), A Theory of the Term Structure of Interest Rates, *Econometrica* **53**(2), 385–407.

Cox, J. C., Ross, S. A. and Rubinstein, M. (1979), Option Pricing: A Simplified Approach, *Journal of Financial Economics* **7**, 229–263.

Crank, J. and Nicolson, P. (1947), A Practical Method for Numerical Evaluation of Solutions of Partial Differential Equations of the Heat-Conduction Type, *Proceedings of the Cambridge Philosophical Society* **43**, 50–67.

Debreu, G. (1959), *Theory of Value*, John Wiley & Sons.

Derman, E. and Kani, I. (1994), Riding on a Smile, *Risk* pp. 32–39.

Fama, E. (1970), Efficient Capital Markets: A Review of Theory and Empirical Work, *Journal of Finance* **25**(2), 383–417.

Faure, H. (1982), Discrépence de suites associées à un système de numération (en dimesion s), *Acta Arithmetica* **41**, 337–351.

Fengler, M. R. (2005), Arbitrage–Free Smoothing of the Implied Volatility Surface, Humboldt–Universität zu Berlin, SFB 649, working paper 2005-019.

Frachot, A. (1994), *Les modèles de la structure des taux*, PhD thesis, Ecole Nationale de la Statistique et de l'Administration Economique (ENSAE).

Frey, R. and Sommer, D. (1996), A Systematic Approach to Pricing and Hedging International Derivatives with Interest Rate Risk: Analysis of International Derivatives Under Stochastic Interest Rates, *Applied Mathematical Finance* **3**(4), 295–317.

Gamma, E., Helm, R., Johnson, R. and Vlissides, J. (1995), *Design Patterns: Elements of Reusable Object–Oriented Software*, Addison–Wesley.

Garman, M. B. and Kohlhagen, S. W. (1983), Foreign Currency Option Values, *Journal of International Money and Finance* **2**, 231–237.

Gatheral, J. (2006), *The Volatility Surface: A Practitioner's Guide*, John Wiley & Sons.

Geman, H., El Karoui, N. and Rochet, J.-C. (1995), Changes of Numeraire, Changes of Probability Measure and Option Pricing, *Journal of Applied Probability* **32**(2), 443–458.

Genz, A. (2004), Numerical Computation of Rectangular Bivariate and Trivariate Normal and *t* Probabilities, *Statistics and Computing* **14**, 151–160.

Genz, A. C. and Malik, A. A. (1980), An Adaptive Algorithm for Numeric Integration Over an N-dimensional Rectangular Region, *Journal of Computational and Applied Mathematics* **6**(4), 295–302.

Geske, R. (1979), The Valuation of Compound Options, *Journal of Financial Economics* pp. 1511–1524.

Geske, R. and Johnson, H. E. (1984), The American Put Option Valued Analytically, *The Journal of Finance* **39**(5), 1511–1524.

Glasserman, P. (2004), *Monte Carlo Methods in Financial Engineering*, Vol. 53 of *Applications of Mathematics*, Springer Verlag.

Harrison, J. M. and Pliska, S. R. (1981), Martingales and Stochastic Integrals in the Theory of Continuous Trading, *Stochastic Processes and their Applications* **11**, 215–260.

Harrison, J. M. and Pliska, S. R. (1983), A Stochastic Calculus Model of Continuous Trading: Complete Markets, *Stochastic Processes and their Applications* **15**, 313–316.

Haugh, M. B. and Kogan, L. (2004), Pricing American Options: A Duality Approach, *Operations Research* **52**(2), 258–270.

Heath, D., Jarrow, R. and Morton, A. (1992), Bond Pricing and the Term Structure of Interest Rates: A New Methodology for Contingent Claims Valuation, *Econometrica* **60**(1), 77–105.

Heston, S. (1993), A Closed–Form Solution for Options With Stochastic Volatility, With Application to Bond and Currency Options, *The Review of Financial Studies* **6**, 327–343.

Ho, T. S. and Lee, S.-B. (1986), Term Structure Movements and Pricing Interest Rate Contingent Claims, *Journal of Finance* **XLI**(5), 1011–1029.

Hull, J. and White, A. (1990), Pricing Interest–Rate Derivative Securities, *The Review of Financial Studies* **3**(4), 573–592.

International Standards Organization (2005), *Proposed Draft Technical Report on C++ Library Extensions*, ISO/IEC PDTR 19768.

Jäckel, P. (2002), *Monte Carlo Methods in Finance*, John Wiley & Sons.

Jamshidian, F. (1989), An Exact Bond Option Formula, *Journal of Finance* **44**, 205–209.

Jamshidian, F. (1991), Forward Induction and Construction of Yield Curve Diffusion Models, *The Journal of Fixed Income* **1**, 62–74.

Jarrow, R. and Rudd, A. (1983), *Option Pricing*, Dow–Jones Irwin, Homewood, Illinois.

Jarrow, R. and Yildirim, Y. (2003), Pricing Treasury Inflation Protected Securities and Related Derivatives using an HJM Model, *Journal of Financial and*

Quantitative Analysis **38**(2), 337–358.

Joe, S. and Kuo, F. Y. (2008), Constructing Sobol' Sequences with Better Two–Dimensional Projections, *SIAM Journal of Scientific Computing* **30**(5), 2635–2654.

Johnson, H. (1987), Options on the Maximum or Minimum of Several Assets, *Journal of Financial and Quantitative Analysis* **22**, 277–283.

Jondeau, E. and Rockinger, M. (2001), Gram–Charlier Densities, *Journal of Economic Dynamics and Control* **25**, 1457–1483.

Joshi, M. S. (2004), *C++ Design Patterns and Derivatives Pricing*, Cambridge University Press.

Jurczenko, E., Maillet, B. and Negrea, B. (2002), Multi-moment Approximate Option Pricing Models: A General Comparison (Part 1), CNRS — University of Paris I Panthéon–Sorbonne, working paper.

Kahalé, N. (2004), An Arbitrage–free Interpolation of Volatilities, *Risk* pp. 102–106.

Karatzas, I. and Shreve, S. E. (1991), *Brownian Motion and Stochastic Calculus*, Vol. 113 of *Graduate Texts in Mathematics*, 2nd edn, Springer-Verlag.

Kendall, M. G. and Stuart, A. (1969), *The Advanced Theory of Statistics*, Vol. 1, 3 edn, Charles Griffin & Company, London.

Kloeden, P. E. and Platen, E. (1992), *Numerical Solution of Stochastic Differential Equations*, Vol. 23 of *Applications of Mathematics*, Springer–Verlag, New York, New York, USA.

Konstandatos, O. (2004), *A New Framework for Pricing Barrier and Lookback Options*, PhD thesis, School of Mathematics and Statistics, University of Sydney.

Konstandatos, O. (2008), *Pricing Path Dependent Options: A Comprehensive Framework*, Applications of Mathematics, VDM-Verlag.

Kotz, S., Balakrishnan, N. and Johnson, N. L. (2000), *Continuous Multivariate Distributions: Models and Applications*, John Wiley and Sons, Inc.

Leisen, D. P. J. and Reimer, M. (1996), Binomial Models for Option Valuation — Examining and Improving Convergence, *Applied Mathematical Finance* **3**(4), 319–346.

Lischner, R. (2003), *C++ in a Nutshell*, O'Reilly, Sebastopol, California.

Longstaff, F. and Schwartz, E. (2001), American Options by Simulation: A Simple Least-squares Approach, *Review of Financial Studies* **14**, 113–147.

Margrabe, W. (1978), The Value of an Option to Exchange one Asset for Another, *Journal of Finance* **XXXIII**(1), 177–186.

McCulloch, J. H. (1971), Measuring the Term Structure of Interest Rates, *Journal of Business* **44**, 19–31.

McCulloch, J. H. (1975), The Tax–Adjusted Yield Curve, *The Journal of Finance* **XXX**(3), 811–830.

Merton, R. C. (1973a), An Intertemporal Capital Asset Pricing Model, *Econometrica* **41**(5), 867–886.

Merton, R. C. (1973b), Theory of Rational Option Pricing, *Bell Journal of Economics and Management Science* **4**, 141–183.

Miltersen, K. and Schwartz, E. (1998), Pricing of Options on Commodity Futures with Stochastic Term Structures of Convenience Yield and Interest Rates, *Journal of Financial and Quantitative Analysis* **33**, 33–59.

Musiela, M. and Rutkowski, M. (1997), *Martingale Methods in Financial Modelling*, Vol. 36 of *Applications of Mathematics*, Springer–Verlag, New York, New

York, USA.

Niederreiter, H. (1987), Point Sets and Sequences with Low Discrepancy, *Monatshefte für Mathematik* **104**, 273–337.

Niederreiter, H. (1992), *Random Number Generation and Quasi–Monte Carlo Methods*, Society for Industrial and Applied Mathematics.

Peizer, D. B. and Pratt, J. W. (1968), A Normal Approximation for Binomial, F, Beta, and Other Common Related Tail Probabilities, I, *Journal of the American Statistical Association* **63**, 1416–1456.

Press, W. H., Teukolsky, S. A., Vetterling, W. T. and Flannery, B. P. (2007), *Numerical Recipes: The Art of Scientific Computing*, 3rd edn, Cambridge University Press.

Protter, P. (1990), *Stochastic Integration and Differential Equations: A New Approach*, Vol. 21 of *Applications of Mathematics*, Springer–Verlag.

Rady, S. (1997), Option Pricing in the Presence of Natural Boundaries and a Quadratic Diffusion Term, *Finance and Stochastics* **1**(4), 331–344.

Reiner, E. and Rubinstein, M. (1991), Breaking Down the Barrier, *Risk* **4**(8), 28–35.

Revuz, D. and Yor, M. (1994), *Continuous Martingales and Brownian Motion*, Vol. 293 of *A Series of Comprehensive Studies in Mathematics*, 2nd edn, Springer–Verlag.

Rogers, L. C. G. (2002), Monte Carlo Valuation of American Options, *Mathematical Finance* **12**, 271–286.

Rubinstein, M. (1991), Options for the Undecided, *Risk* **4**(4), 43.

Sandmann, K. and Schlögl, E. (1996), Zustandspreise und die Modellierung des Zinsänderungsrisikos (State Prices and the Modelling of Interest Rate Risk), *Zeitschrift für Betriebswirtschaft* **66**(7), 813–836.

Sandmann, K. and Sondermann, D. (1989), A Term Structure Model and the Pricing of Interest Rate Options, Rheinische Friedrich–Wilhelms–Universität Bonn, working paper B–129.

Sandmann, K. and Sondermann, D. (1991), A Term Structure Model and the Pricing of Interest Rate Derivatives, Rheinische Friedrich–Wilhelms–Universität Bonn, working paper B–180.

Sandmann, K. and Sondermann, D. (1997), A Note on the Stability of Lognormal Interest Rate Models and the Pricing of Eurodollar Futures, *Mathematical Finance* **7**(2), 119–122.

Schlögl, E. (2013), Option Pricing Where the Underlying Assets Follow a Gram/Charlier Density of Arbitrary Order, *Journal of Economic Dynamics and Control* **37**(3), 611–632.

Schönbucher, P. J. (2003), *Credit Derivatives Pricing Models: Models, Pricing and Implementation*, John Wiley & Sons.

Seydel, R. (2006), *Tools for Computational Finance*, Universitext, 3 edn, Springer–Verlag.

Shea, G. S. (1984), Pitfalls in Smoothing Interest Rate Term Structure Data: Equilibrium Models and Spline Approximations, *Journal of Financial and Quantitative Analysis* **19**(3), 253–269.

Shea, G. S. (1985), Interest Rate Term Structure Estimation with Exponential Splines: A Note, *The Journal of Finance* **XL**(1), 319–325.

Shtern, V. (2000), *Core C++: A Software Engineering Approach*, Prentice Hall PTR, Upper Saddle River, NJ.

Skipper, M. and Buchen, P. (2003), The Quintessential Option Pricing Formula, University of Sydney, working paper.

Sobol', I. M. (1967), On the Distribution of Points in a Cube and the Approximate Evaluation of Integrals, *USSR Journal of Computational Mathematics and Mathematical Physics* **7**, 784–802. English translation.

Sommer, D. (1994), Continuous–time Limits in the Generalized Ho–Lee Framework under the Forward Measure, SFB 303, University of Bonn, working paper B–276.

Stroustrup, B. (1997), *The C++ Programming Language*, 3rd edn, Addison–Wesley, Reading, Massachusetts.

Tian, Y. (1993), A Modified Lattice Approach to Option Pricing, *Journal of Futures Markets* **13**, 563–577.

Tsitsiklis, J. and van Roy, B. (1999), Optimal Stopping of Markov Processes: Hilbert Space Theory, Approximation Algorithms, and an Application to Pricing High–Dimensional Financial Derivatives, *IEEE Transactions on Automatic Control* **44**, 1840–1851.

van Perlo-ten Kleij, F. (2004), *Contributions to Multivariate Analysis with Applications in Marketing*, PhD thesis, Rijksuniversiteit Groningen.

Vasicek, O. (1977), An Equilibrium Characterization of the Term Structure, *Journal of Financial Economics* **5**, 177–188.

Vasicek, O. A. and Fong, H. G. (1982), Term Structure Modeling Using Exponential Splines, *The Journal of Finance* **XXXVII**(2), 339–348.

Veldhuizen, T. L. (1995), Expression templates, *C++ Report* **7**(5), 26–31. Reprinted in C++ Gems, ed. Stanley Lippman.

Veldhuizen, T. L. (1998), Arrays in Blitz++, *Proceedings of the 2nd International Scientific Computing in Object-Oriented Parallel Environments (ISCOPE'98)*, Lecture Notes in Computer Science, Springer-Verlag.

Veldhuizen, T. L. and Gannon, D. (1998), Active Libraries: Rethinking the roles of compilers and libraries, *Proceedings of the SIAM Workshop on Object Oriented Methods for Inter-operable Scientific and Engineering Computing (OO'98)*, SIAM Press.

West, G. (2005), Better Approximations to Cumulative Normal Functions, *WILMOTT Magazine* pp. 70–76.

Wilmott, P. (2000), *Paul Wilmott on Quantitative Finance*, John Wiley & Sons.

Wilmott, P., Howison, S. and Dewynne, J. (1995), *The Mathematics of Financial Derivatives: A Student Introduction*, Cambridge University Press.

Zhang, P. G. (1998), *Exotic Options: A Guide to Second Generation Options*, 2nd edn, World Scientific.

Index